NEWTON
INNOVATION AND CONTROVERSY

NEWTON
INNOVATION AND CONTROVERSY

Peter Rowlands

University of Liverpool, UK

World Scientific

NEW JERSEY · LONDON · SINGAPORE · BEIJING · SHANGHAI · HONG KONG · TAIPEI · CHENNAI · TOKYO

Published by

World Scientific Publishing Europe Ltd.
57 Shelton Street, Covent Garden, London WC2H 9HE
Head office: 5 Toh Tuck Link, Singapore 596224
USA office: 27 Warren Street, Suite 401-402, Hackensack, NJ 07601

Library of Congress Cataloging-in-Publication Data
Names: Rowlands, Peter, author.
Title: Newton — innovation and controversy / by Peter Rowlands (University of Liverpool, UK).
Other titles: Newton — innovation and controversy
Description: Singapore ; Hackensack, NJ : World Scientific, [2017] |
    Includes bibliographical references and index.
Identifiers: LCCN 2017029435| ISBN 9781786344014 (hc ; alk. paper) |
    ISBN 1786344017 (hc ; alk. paper)
Subjects: LCSH: Newton, Isaac, 1642-1727.
Classification: LCC QC16.N7 R688 2017 | DDC 530.092 [B] --dc23
LC record available at https://lccn.loc.gov/2017029435

British Library Cataloguing-in-Publication Data
A catalogue record for this book is available from the British Library.

First published 2018 (Hardcover)
Reprinted 2018 (in paperback edition)
ISBN 9781786344021 (pbk)

Copyright © 2018 by World Scientific Publishing Europe Ltd.

All rights reserved. This book, or parts thereof, may not be reproduced in any form or by any means, electronic or mechanical, including photocopying, recording or any information storage and retrieval system now known or to be invented, without written permission from the Publisher.

For photocopying of material in this volume, please pay a copying fee through the Copyright Clearance Center, Inc., 222 Rosewood Drive, Danvers, MA 01923, USA. In this case permission to photocopy is not required from the publisher.

Desk Editors: Suraj Kumar/Jennifer Brough/Shi Ying Koe

Typeset by Stallion Press
Email: enquiries@stallionpress.com

Preface

Newton — Innovation and Controversy completes the account of Newton's scientific work begun in the two earlier books in the series: *Newton and Modern Physics* and *Newton and the Great World System*. The first of these emphasized the analytical approach to problems, largely developing from his optical work, where Newton had no simple way of applying mathematics to generate a series of results based on a generic underlying theory. The second was concerned with showing how the study of the planetary motions in terms of his new concept of universal gravity gave him just such an opportunity. The result was the greatest revolution in scientific history, with mathematical physics established as a universal subject, potentially capable of explaining any aspect of Nature from first principles.

To a large extent, it was the *synthetic* aspect of this development that most interested Newton's contemporaries and immediate successors. That is, taking Newton's analytically-derived fundamental principles as a given, the main object of physics became the explanation of as many physical phenomena, both celestial and terrestrial, as possible. Eventually, however, further breakthroughs required new fundamental principles, derived by the same kind of analytical reasoning as had been used originally by Newton.

The remarkable thing is that this seems to have most often occurred by a kind of attrition after exhausting the alternative possibilities rather than as a choice consciously made, while a study of Newton's works shows that his own mental processes seem to have been pushing him in the direction of these developments. This suggests that a study of Newton's work, in the dual context of both contemporary and subsequent developments, might provide insights into his extraordinary analytical thinking. In this capacity Newton seems to me to be absolutely singular. There have been other great mathematicians, other great mathematical physicists in the synthetic sense and other great experimentalists, but surely no analytical thinker who approaches anywhere near his level.

The aim of this three-volume study is to fully express the analytical power that made possible not only the great breakthroughs that were fully acknowledged by posterity, but also the ones that were suppressed or abandoned as a result of the controversies that such radical innovation necessarily generated. The value of this is found not only in the fuller picture it gives of Newton himself, but also in the possible leads it suggests for developing physics in the future. Physics is always in need of analytical inputs, never more so than at the present when we have models which have led to an incredible number of accurate experimental predictions, but no analytical insight as to why they are there. If we could understand the sources of such thinking in the practice of one of its greatest proponents, then we might be inspired to finding ways in which it could be emulated.

This is a study primarily of Newton's physics, seen especially from an analytic perspective, but also in relation to how the combination of analytical and synthetic approaches could lead to an unstoppable combination given favourable circumstances. It naturally builds upon the work of many great Newton scholars, especially during the last fifty years, but I would like to think that the reader will find a considerable number of new insights and interpretations in the three books. Newton is so prolific and multifaceted that there is always a new twist or new angle even to a familiar story. I would also like to think that my experience as a physicist with a special interest in the foundations of the subject has been as valuable in this enterprise as my experience as a historian.

I don't claim as much expertise in the non-scientific aspects of Newton's career, treated in the seventh chapter of the present book, where I have clearly drawn on the research of a number of outstanding historians and scholars, but I certainly think that any picture of Newton would be incomplete without them. Even in these areas, however, there are manifestations of Newton's analytical powers, sometimes leading to results as 'valid' as those in the more recognizably 'scientific' fields.

Like the other volumes in this series, this book has benefited from the cooperation and support of many people, in particular Niccolò Guicciardini for his profound comments and suggestions on the mathematical sections, Stephen Snobelen for providing funding for a conference in Halifax, Nova Scotia, and also Mike Houlden, Colin Pask, Mervyn Hobden, John Spencer and my wife Sydney.

Peter Rowlands
Oliver Lodge Laboratory
University of Liverpool

About the Author

Dr. Peter Rowlands obtained his BSc and PhD degrees from the University of Manchester, UK. He then spent some time working in industry and further education. He became a Research Fellow at the Department of Physics, University of Liverpool in 1987, and still works there. He has also been elected as Honorary Governor of Harris Manchester College, University of Oxford, a post he has held since 1993. Dr. Rowlands has published around 200 research papers and 12 books. Some of his recent works, published by World Scientific, include *Zero to Infinity The Foundations of Physics* (2007), *The Foundations of Physical Law* (2014), and *How Schrödinger's Cat Escaped the Box* (2015). As a theoretical physicist, his research interests include, but are not limited to, foundations of physics, quantum mechanics, particle physics, and gravity. He has also done extensive work on the subjects of history and philosophy of science, and has published several books on these topics.

Contents

Preface   v

About the Author   vii

Chapter 1   Innovation and Controversy   1
1.1. Newton and Controversy 1
1.2. Newton and his Modern Critics 3
1.3. An Early Controversy 6
1.4. Gravity Established as a Universal Theory 11

Chapter 2   Analysis and Synthesis   17
2.1. Newton and Leibniz 17
2.2. Mathematicians at War 25
2.3. Analytic and Synthetic 35
2.4. The Dual Methods 42
2.5. Newton's Mathematical Legacy 45

Chapter 3   Thermodynamics   51
3.1. Heat as Motion 51
3.2. Transfer of Heat 55
3.3. Temperature and the Law of Cooling 59
3.4. The Dynamics of Cooling 65
3.5. The Impossibility of Perpetual Motion 69
3.6. The Decay of Motion 73
3.7. The Laws of Thermodynamics 79

Chapter 4   The Microstructure of Matter   87
4.1. Atoms and Fundamental Particles 87
4.2. Internal Forces and Matter 92
4.3. A Hierarchical Structure 98

4.4. Capillarity and the Force of Attraction	110
4.5. Newton's Influence on the Explanation of Material Structure	115

Chapter 5   The States of Matter — 121

5.1. Solids	121
5.2. Liquids	125
5.3. Gases	128
5.4. Resistance in Fluids	140
5.5. Vortices	159

Chapter 6   Microforces: Cohesion and Chemistry — 169

6.1. Cohesion	169
6.2. Attraction and Repulsion in the Different States of Matter	172
6.3. Double Forces	179
6.4. Alchemy and Chemistry	182
6.5. Chemistry and Active Principles	195
6.6. Basic Chemical Principles	204
6.7. Newton and the Chemical Revolution	208

Chapter 7   Nonscience — 213

7.1. Alchemy and Ancient Wisdom	213
7.2. Theology	218
7.3. Chronology and Prophecy	229
7.4. Psychology	240
7.5. Economics	243

Chapter 8   Securing the Legacy — 249

8.1. Science Becomes Politics	249
8.2. Completing the Picture	256
8.3. Newton and the Enlightenment	260
8.4. The Power of Abstraction	263

Appendix: Some Results of Recent Scholarship — 271

Bibliography — 281

Index — 299

Chapter 1

Innovation and Controversy

1.1. Newton and Controversy

Newton was one of the most significant innovators in scientific history, particularly famous for his work in optics, gravity, mathematics and dynamics, though he also created significant ideas in many other areas. Much of this still resonates today, but there was a price to pay. Controversy is an inevitable consequence of setting out, as Newton did, to change the entire intellectual map of the universe, perhaps even of the human mind itself. Everything that he published was controversial; some of it still is. Newton's optics was attacked from the instant his first published paper appeared in 1672 and he was still defending it 50 years later when his assistant John T. Desaguliers demonstrated the relevant experiments to a party of French visitors. Newton's theory of universal gravity was attacked from 1687 by some of Europe's leading intellectuals, including Huygens and Leibniz, because it rigorously excluded any hypothetical mechanism, and implied that forces could be transmitted between material particles across empty space. The same work also caused feuds with Hooke and Flamsteed in England over distribution of credit and access to data.[1]

The disputes with Hooke and Flamsteed have been discussed at length by commentators and used as examples of Newton's quarrelsome nature, but they are not the most significant controversies of his career, as they were essentially only personal. Much more significant was another personal quarrel, again with Leibniz, over the invention of the calculus, as this also led to profound discussions concerning the nature of mathematics and the relative significance of the processes of analysis and synthesis. The personal attack on Newton in this case and Newton's response were also particularly significant, as depriving Newton of credit for his innovatory mathematics

---

[1] See *Newton and the Great World System*, Chapters 5 and 6.

would have taken from him the only section of his work that had remained intrinsically noncontroversial, and consequently damaged his credibility with respect to the rest. At the same time Leibniz was contriving to subvert aspects of Newtonian dynamics and gravitational theory in an attempt at preserving the Cartesian world-view which Newton had set out to replace. Theology was another driving factor in this second quarrel, though based only on a few hints that Newton had included in his *Opticks* and *Principia*. Newton, however, avoided any personal involvement in the argument, leaving Samuel Clarke, his friend and the rector of his local parish, to handle the dynamical subtleties. Together with experiments on thermometry, published anonymously in a low-key manner, this ultimately led in the direction of the nineteenth century theory of thermodynamics.

Despite being involved in so many controversies, no one could have been more averse to them than Newton or gone out of his way to prevent them. Those he anticipated he set out to avoid either by not publishing potentially offending material or by publishing it only in Query form and toning down his language. He knew that his views on theology and chronology were necessarily going to become discordant in matters where there was a huge diversity of opinions and no experimental method to support the ideas. He wisely held back from publishing them in his lifetime but immediate controversy erupted in both areas on posthumous publication of key works. In fact, the chronology was attacked ferociously even in his own lifetime with a surreptitious edition of a French translation of an abstract, which was published only so that it could be emphatically refuted.[2] This set the tone for comments on Newton's chronology almost down to the present day. The theological work was published only in the context of the *Observations on the Prophecy of Daniel and the Apocalypse of St. John*, of 1733, and the *Two Notable Corruptions*, of 1754, but it didn't take the eighteenth century divines long to suspect Newton's secret heterodoxy. The condemnation of this, however, was offset by approval of his powerful support for the argument from design.[3]

Even when he set out to secure his scientific legacy, Newton deliberately sacrificed large sections of his work, including the theory of matter and the electrified universe, in order that the more securely established parts would not be challenged. He knew that his views on the nature of matter and the composition of material substances would provoke opposition, because, like

---

[2] *Abregé de la chronologie de M. le Chevalier Newton*, Paris, 1725.
[3] Mandelbrote (2016).

his gravitational theory, they emptied space and assumed forces acting at a distance, so he drew back on numerous occasions from publishing them except in a few hints in the *Opticks*. He published some material on fluids in Book II of the *Principia* because he wanted to understand the aethereal fluids that he modified but never abandoned. Ultimately, this became the most criticized part of his accepted work because it was completely misinterpreted — essentially as erroneous Cartesian-style hypothesizing. He published only hints of his ideas on chemistry and cohesion because the former was based on procedures that could not be defended as totally rational, and the latter depended again on forces acting at a distance which had given him endless trouble with his theory of gravity. Unlike Robert Boyle, he was careful about never publishing anything connected with alchemy, and it was not until the nineteenth and twentieth centuries that this became one of the most controversial aspects of his career. In all these areas Newton was a major innovator, and the degree of controversy his work generated can be taken as a measure of its originality and significance.

1.2. Newton and his Modern Critics

Much of the criticism of Newton — both personality and work — is modern in origin. Newton, we have said, is seen by many as the greatest scientist who ever lived.[4] A colossal figure, a towering intellect — the superlatives do not do justice to this extraordinary individual, who remains massively influential nearly three hundred years after his death. Yet there has always been a critical undercurrent. Of course, like everyone else, Newton had flaws, both as a man and as a scientist, but, for reasons that have little to do with establishing historical truth, he seems to have been singled out for some fairly intensive negative criticism over the last fifty years. We hear that he was a magus, a believer in alchemy and millennial prophecies, not at all the rational scientist of popular legend. He was unsociable, never married, quarrelled with lots of people, ruthlessly persecuted enemies of the state, was bipolar, autistic, obsessive, abnormal, had a disturbed childhood, was a puritan and dogmatic, and represented the worst aspects of monolithic science. He refused to publish his mathematical work, was unfair to Hooke, Leibniz, Flamsteed and others, got involved in bitter priority disputes,

---

[4]Iliffe and Smith (2016), for example, rank him 'among the top two or three greatest theoretical scientists ever' (with Maxwell and Einstein), among the top two mathematicians (with Gauss), and 'in the first rank of experimentalists', with no one else figuring in more than one of these categories (15). He was, in particular, 'an extraordinary problem-solver, as good, it would appear, as humanity has ever produced' (30).

repressed British mathematics, and had a negative influence in supporting the corpuscular theory of light. His celestial mechanics left a lot to be worked out by his successors, his mechanics had to be translated into the proper algebraic form.

In addition the term 'Newtonian' is regularly used as the expression for theories which have necessarily been superseded by more modern thinking, for example, by relativity theory, quantum theory, or the theory of chaotic systems. Conventional philosophy has never rated his work in that field, while some authorities claim that he was not particularly successful as an economist at the Mint. In science, he pushed his data beyond the limits to which it could go, and left a lot of his work in a rather sketchy form, as well as being wrong on numerous occasions. There was also the desire for a more modern hero, so Einsteinolatry appeared in the early twentieth century to replace the Newtonolatry of the previous two hundred years, amid claims that Newton's most important work had been superseded by special relativity, general relativity and quantum mechanics.

Clearly some of the iconoclasm is simply an overreaction to earlier hyperbole. But many of the comments made about both the man and his work are massively wide of the mark. Assumptions have been made on the flimsiest of evidence or lack of it, not to mention the prejudices of the writer and the desire to produce a sensationalist slant. Speculations by one author, that Newton destroyed Hooke's portrait or that he made up the story of being inspired by the fall of an apple, become 'facts' in another. Most of the judgements of the man stem from a single work by Frank E. Manual, which, though brilliantly written, is nevertheless based on the pseudoscientific premises of Freudian psychoanalysis.[5] Also, though much brilliant work has been done on the science by serious students of the sources, others have allowed superficial investigations to colour their assessments. In fact, much of Newton's work is very difficult and even the best-informed scholars can be wrong on crucial points. In addition, mathematical and physical researchers have been slow in using their special knowledge to add their perspective.

While making critical comments where they are appropriate, I will not be following this line. A study of the sources shows repeatedly that Newton is a difficult author who demands close attention, and cannot be

---

[5] Manuel (1968). To give an idea of the level of absurdity to which this can lead, we may quote, as Gleick does (2003, 240), Manuel's question: 'In the act of fornication between his friend Halifax and his niece was Newton vicariously having carnal intercourse with his mother?' (1968, 262).

understood on a superficial reading. We now know, just to take a few random examples, that his alchemical researches were a work of science and not an occult study, merging into chemistry (and even high energy physics); that he was an important philosopher, a capable economist, that even his more bizarre interests yielded some valid results; that a number of significant mathematical results previously considered faulty are in fact correct; and that his abstract mathematical models (for example, that of gases in the second book of the *Principia*) should not be confused with physical ones and are completely correct in their own terms. In addition, the problem with the 'replacement' of Newton as all-conquering hero with Einstein, is that it doesn't quite fit the facts of twentieth century science. Though relativity received most public attention, partly for extra-scientific reasons, quantum mechanics was really the great revolution of the period, though it arrived without any of the media circus attending the earlier theory. General relativity, a much more specialised and restricted theory, doesn't have the same status, for all its cultural cachet, and talk of the two theories as the 'twin pillars' of modern physical thought. Einstein came too early to be a really significant contributor to quantum mechanics. The hero cannot be totally aligned with the heroic story. Einstein's work is of great significance, but this significance does not lie in being opposed to the work of his predecessor.

The special and general theories of relativity led to a vast number of superficial comments about Newtonian theory. The word 'Newtonian', as we have seen, is often used to mean something other than what was being done in the revolutionary period of the twentieth century. 'Non-Newtonian' is used for relativistic versus nonrelativistic, quantum versus classical, chaotic versus standard deterministic, but none of the modern developments is actually non-Newtonian in the sense implied. If 'Newtonian' means, as it should, pertaining to Newton's own ideas, we find that none of the modern theories is outside his interests and none really contradicts his vision. We need to recall that a lot of the 'Newtonian ideas overthrown' aspect of early twentieth century philosophy was political rather than scientific. People felt that there ought to be a spirit of revolution about in a period that defined itself by the word 'modern', and the modernist revolution would not be just another stage in human history but a truly defining moment which could not, by definition, be superseded. We still haven't eradicated that mode of thinking despite its many disastrous consequences. Despite a mass of truly great scholarship, a lot of what was said about Newton and Newtonian science in the twentieth century was based on fallacy. Thomas Kuhn saw science as being created by revolutions separated by long periods of stasis or 'normal' science; Popper saw science as defined by potential falsification.[6] Both attempted to fit Newton in this picture, but his work does not fit well with them. The idea that a single experimental result can decide on philosophical issues, or change an entire philosophy, is a fantasy. It is also dangerous to assume that current ways of thinking are intrinsically superior to earlier ones before a fundamental basis is established for physics.

1.3. An Early Controversy

The first and in many ways the most significant contemporary controversy came, in 1672, with Newton's first paper on optics, containing his new theory of colour.[7] For a relatively unknown individual to develop an extended scientific theory based on work done over a period of time using a new methodology is dangerous, because it requires going through the initial stages to reach the results of real importance. In the case of Newton, it meant also having his experimental expertise. The much-lauded system of 'peer review' is often useless in these cases, especially where the new methodology requires inductive or recursive thought, because the sheer novelty of the concepts means that no reference can be found within existing knowledge for comparison. In fact, the history of physics during the last century shows that many of the most important discoveries made during the period had a difficult time in making it through the system or being accepted after they had, and we can only imagine what the blind faith in such an arbitrary process as reviewing by a few individuals, with all the problems posed by their prejudices, ignorance, self-interest, and sheer lack of imagination, has cost us in lost opportunities. Science, like every other field of human endeavour, is dominated by power structures. The only difference is that science still seems to operate under the delusion that the power structures are benign.

Newton's career shows that it is not a completely new problem, for a disastrous peer review of this first published paper nearly stopped his scientific career in its tracks. The very serious quarrel which followed his disclosure of the composite nature of white light saw Newton ranged against Hooke, Huygens, Pardies and a whole host of English Jesuits, and only ended when Newton terminated the correspondence in a fit of rage. Essentially, the critiques of Hooke and the others involved, not only the rejection of his supposed 'hypotheses', but also his embryonic universal method. Only his

---

[6] Kuhn (1962); Popper (1963).
[7] See *Newton and Modern Physics* and *Newton and the Great World System* for details.

'impact' invention of the reflecting telescope survived the onslaught.[8] This episode has to be taken as the source for many of his subsequent actions, and the later priority battles should also be seen in this light.

The origin of the problem was a difference in methodology. Newton believed that certain concepts, such as refrangibility, could be derived by abstraction from empirical phenomena and then used without further reference to mechanistic models. In this he was correct, and such thinking became the route to making physics the key to universal truth, rather than an empirical study aimed at the kind of immediate practical utility that seems to have been the main objective of the early fellowship of the Royal Society. Newton's method allowed him to assume properties like periodicity and corpuscularity, and later polarisability, to exist as abstract concepts, without requiring a model. He *did* devise models on occasions, but they were never the primary purpose of his scientific investigations, and they often represented his least successful efforts at explanation. However, thinking in the abstract style appears to be very different from the default mode normally assumed by humankind, and Newton's contemporaries simply did not see the point. Many later writers have not seen the point either, even to the present day. Newton's abstract methodology has never been fully appreciated either before or since. But, simply by a process of attrition and natural selection, it ultimately became the classic method for creating general solutions of wide applicability rather than for solving each problem as a specific instance as it arose.

The optical controversy, of course, occurred right at the beginning of Newton's career, long before he had worked out how to reformulate the whole of physics and many years before he had formulated his most powerful abstract concepts or had systematised his practices, but it came about because Newton had, almost inadvertently, introduced a method which came naturally to him but which was incomprehensible to most of his contemporaries. For Newton, all that was necessary to make sense of his experiments was to accept the fact that light rays which produced sensations of different colours were different in the abstract and mathematically

---

[8]For the benefit of readers in future times, in addition to its use as an index of the perceived importance of various scientific journals, 'impact' was an arbitrary and undefinable parameter invented by early twenty-first century politicians in the UK to measure the supposed value of research funded by tax-payers, and was largely aimed at identifying projects which seemingly provided an immediate return on investment, despite the overwhelming evidence that such returns nearly always resulted from discoveries that could not have been predicted — a classic case of a misunderstanding of the scientific process through a lack of knowledge of scientific history.

measurable property of refrangibility, the quality defined as $\sin i/\sin r$ in Descartes' law of refraction. This was true whatever mechanical hypothesis was used to explain the fact, if any were necessary at all, and he would rather jettison everything than concede this fundamental truth. The method he was in the process of discovering mattered more to him than any of the specific conclusions it uncovered. So, although it was his rejection of the hypothetical method which had triggered the long series of optical controversies, and, although he realised that hypotheses had a role to play in scientific theorising, most notably when he wrote his *Hypothesis Explaining the Properties of Light* in 1675, he was not prepared to concede that hypotheses should take precedence when a more secure abstract truth had been established.[9]

The experiments reported in 1672 required a very high degree of experimental technique, which Newton was forced to spell out in detail in subsequent replies in the *Philosophical Transactions* to criticisms by the Jesuits of Liège.[10] Newton took a great deal of trouble to secure the right kind of prisms and even constructed his own water-filled ones. A skilled experimenter like Hooke could reproduce the experimental results after a number of trials, but an amateur such as Anthony Lucas could not. Approaching the work with a preconceived attitude was likely to lead to a negative outcome. Even after the main controversy had died down, when Newton refused to prolong the correspondence, continental Cartesians, such as Edmé Mariotte continued to publish papers saying that both theory and experiments were wrong,[11] and this became perfect ammunition for a convinced Cartesian such as Leibniz, whose attack on Newton, though largely covert, was mounted simultaneously on all possible fronts: dynamics, gravity, optics and metaphysics, as well as mathematics.

Newtonian optics only became accepted after its author became celebrated for his work in celestial dynamics, and then by no means immediately.

---

[9] In a draft Query Newton makes clear that his early work in optics had led to seemingly irreducible abstract or semi-abstract ideas which included 'the original differences of the rays of light in respect of refrangibility reflexibility & colour & their alternate fits of easy reflexion & easy transmission & the properties of bodies both opake & pellucid on which their colours depend & these discoverys being proved may be assumed as Principles in the Method of Composition for explaining the phænomena arising from them' (CUL Add. 3970.3, f. 242^r, NP).

[10] Linus (3); Gascoignes (1); Lucas (1). Schaffer (1989) has a lengthy discussion of the controversies in relation to the detailed claims of the experiments and the experimental techniques used. Newton, like Faraday after him, was an experimenter who left nothing to chance, working through many possible variations on his experiments to close down every possible loophole (Shapiro, 2002, 230).

[11] Mariotte (1681).

The *Opticks* was not published until the *Principia* had been in print for 17 years. Hooke had conceded that some of the experiments were correct as early as 1675, but the exact conditions specified by Newton were not always easy to match, so controversy persisted and Desaguliers was still doing Newtonian demonstration experiments to prove their validity as late as 1722.[12] The immediate and comprehensive rejection of the early paper on the analysis of white light became a fact of great importance to the way Newton subsequently chose to present his work. He was already inclined to shun publicity, either because of his natural temperament, or because he wanted to avoid drawing attention to his more esoteric and clandestine investigations; but he now became even more cautious and secretive about his ideas.

Whether or not Frank Manuel's representation of Newton as a man with a Messianic complex is a genuine representation of his character,[13] it seems to be an accurate representation of the position in which he now found himself. As a theologian, Newton was comfortable with a mind-set based on articles of faith. He probably also knew that he had special powers in uncovering truths in a holistic way, which could seem almost like divine revelation. What he couldn't explain was how he had arrived at his abstract and generalising ideas. In this position, a scientist has to find a strategy which will convince people who work in an entirely different way. To say that the ideas only have to be accepted to become obvious won't convince people who want to know why they should take the initial step of accepting them in the first place, especially if it means jettisoning their own pet theories. This is why many ideas that in retrospect seem obvious have been met with hostility at the time. Newton had a strategy. He decided to present his conclusions as the *direct* results of the kind of experimental inquiries popular with his mechanically-minded contemporaries. How could this fail? He produced a powerfully-argued case in his brilliant report on 'A New Theory of Light and Colours';[14] but his strategy still failed — because his then more powerful opponents already had model-dependent theories available which they could modify to accommodate any new set of data.

The strength of Newton's inductive method in producing powerful results of great generality had put him in a vulnerable position. The process by which he gained his greatest insights was inexplicable — it doesn't follow the usual rules of deductive logic, and it could not be duplicated by an

---

[12] Desaguliers (1722).
[13] Manuel (1968).
[14] 'A New Theory', *Corr.* 1: 92–102.

iterative chain of reason. So he was unable to explain the source of his superior insights to the satisfaction of other philosophers, and they weren't willing to concede that they should use his starting point. He was only able to convince them of his superiority at a later date by sheer mathematical power. The long drawn out battle over his early work on optics, so obvious to him, incomprehensible to his contemporaries in the abstract form he stated it, eventually tried his patience beyond endurance. It was a harsh lesson for a researcher at the beginning of his career. For a long time, he made an effort to try to accommodate his vision to encompass other people's theories, ideas and questions, and we see this in polite and courteous letters to many correspondents — for example, Michael Dary, Francis North, William Briggs and John Harington, and perhaps Joshua Maddock.[15] In the end, however, the endless criticism of the experiments on which he based his theory took its toll, and he found it impossible to tolerate the sheer refusal of correspondents to see the fact that to him was blindingly obvious but could not be derived from anything else: no matter what hypothesis explained the nature of light, it was composed of rays 'differently refrangible'. 'This I believe', he wrote, 'hath seemed the most Paradoxicall of all my assertions, & met with the most universall & obstinate Prejudice. But to me it appears as infallibly true & certaine, as it can seem extravagant to others'.[16]

From a later vantage point, we see that Newton's theory of light would have been completely obvious *once accepted* and then a powerful source of new development. However, to accept such a theory, his contemporaries would have had to adopt a new type of thinking, beyond that to which they had become accustomed. The same pattern has repeated itself many times since in the history of physics. Though nature is grounded in ideas that are fundamentally simple, this simplicity is of an abstract kind and very different from the supposed 'simplicity' of 'concrete' everyday experience, which is actually complex. To adopt nature's mode of simplicity is a very difficult exercise, because it runs counter to normal human experience. So, simple and 'obvious' facts are very hard won and always strongly resisted. Even now, the scientific community has never been fully won over to the fact that simplicity and the concrete are at opposite poles of human experience, and it has certainly never fully accepted its consequences. 'Fundamental'

---

[15] Newton to Dary, 6 October 1674, *Corr.* I, 319–320, and 22 January 1675, *Corr.* I, 332–333; to North, 21 April 1677, *Corr.* II, 205–207; to Maddock, 7 February 1679, *Corr.* II, 287–288; to Briggs, 20 June 1682, *Corr.* II, 377–378, and 12 September 1682, *Corr.* II, 381–385; to Harington, 30 May 1698, *Corr.* IV, 274–375.
[16] Newton to Oldenburg, 5 December 1674, *Corr.* I, 328.

theories are still being put forward in a style which is incompatible with producing lasting results. The difficulty is increased because most of physics actually concerns the concrete and complex; only the fundamentals are simple, and most physicists have no occasion to concern themselves with these. Because concrete models work so well in other areas of physics, the tendency to describe fundamentals using them can be almost irresistible, but long experience of the failure of this strategy suggests that some aspects of physics can never be expressed in this way.

1.4. Gravity Established as a Universal Theory

It may have been the earlier criticisms engendered by his optical work which made Newton compose the *Principia* very largely in the geometrical style rather than in the much more revolutionary analytic. Had he used revolutionary mathematics alongside revolutionary physics his work would have had much less immediate impact. He was determined that this time he would not fail in his public message. Even after he gained fame as the author of the *Principia*, the experience over the optical paper had a profound effect on him. It is well known that he felt obliged to overstate the accuracy of his work in a number of specific instances, in order to avoid giving ammunition to the continental Cartesians, who subjected his work in all areas to a never-ending barrage of hostile criticism. Leibniz, the most prominent of the Cartesians, conceded nothing to Newton in physics, recognising only that he had made advances in mathematics. Even there, he thought nothing of appropriating Newtonian results in the physical application of mathematics if he could provide them with a Cartesian interpretation which would undermine the one given by Newton in the *Principia*. The priority battle over the calculus took place in the context of a much more extensive all-out war concerning the fundamentals of metaphysics and natural philosophy.

Gravity was a particular problem. How could one invoke a force of gravity, his opponents argued, if one had no physical explanation for it? Newton would have none of this kind of argument. The physical explanation, if any existed, had no bearing on the existence of the gravitational force: 'But hitherto I have not been able to discover the cause of those properties of gravity from phænomena, and I feign no hypotheses; for whatever is not deduced from the phænomena is to be called an hypothesis; and hypotheses, whether metaphysical or physical, whether of occult qualities or mechanical, have no place in experimental philosophy'.[17] 'Gravity must be caused by an

---

[17] General Scholium (with the usual amendment of 'frame' to 'feign').

agent acting constantly according to certain laws, but whether this agent be material or immaterial is a question, I have left to the consideration of my readers'.[18] 'And to us it is enough that gravity does really exist, and act according to the laws which we have explained, and abundantly serves to account for all the motions of the celestial bodies, and of our sea'.[19]

The catastrophic failure of the embryonic peer review system in 1672 meant that the truly revolutionary nature of Newton's work was not understood, and, to a large extent, it was only vaguely apprehended long afterwards. His powerful methodology was never fully accepted, even by the people who approved of its results. There seems to have been two reasons for this. The first was that Newton was a powerful analyst. The ability to think in this way seems to be extremely uncommon, and the ability to recognize such thinking does not seem to be particularly widespread either. The Greeks probably produced only two genuinely analytical thinkers — Parmenides and Zeno — both greatly misunderstood by their contemporaries and even down to this day. The second was that Newton was primarily a theologian, and his contemporaries knew it. As he himself wrote to Bentley after he had published the *Principia*: 'When I wrote my treatise about our Systeme I had an eye upon such Principles as might work with considering men for the beliefe of a Deity & nothing can rejoyce me more then to find it usefull for that purpose'.[20] Newton had introduced or reintroduced a whole series of theological concepts into physics, just when it was establishing itself as a modern, 'secular' subject, based on good materialist principles.

Though it was certainly true that Newton had derived his theory directly from experiments, it was also true that this was only possible because he had introduced into science a new and non-mechanistic style of thinking which had more in common with late mediaeval modes of thought than anything fashionable in his day. His contemporaries rejected his abstractions because they thought that he was not being 'modern'. The mechanistic style — the invention of mechanistic hypotheses or mechanical models — was modern; abstract categories were 'occult', and had been banished by Descartes, once and for all, from the study of nature. But, as Newton protested for the benefit of Leibniz:

> To tell us that every Species of Things is endow'd with an occult specifick Quality by which it acts and produces manifest Effects, is to tell us nothing:

---

[18] Newton to Bentley, 25 February 1693; *Corr.* III, 253–356, 254, Cohen, *Papers and Letters*, 302–303.
[19] General Scholium.
[20] Newton to Bentley, 10 December 1692; *Corr.* III, 233.

but to derive two or three general Principles of Motion from Phænomena, and afterwards to tell us how the Properties and Actions of all corporeal Things follow from those manifest Principles, would be a very great step in Philosophy, though the Causes of those Principles were not yet discover'd: And therefore I scruple not to propose the Principles of Motion abovemention'd, they being of very general extent, and leave their Causes to be found out.[21]

Perhaps significantly, he used his opponents' own favourite metaphor of clockwork to illustrate his point, reminding them that the gravity that drove clockwork was just as much a mystery as the gravity that drove the world system:

> ... certainly God could create Planets that should move round of themselves without any other cause then gravity... And to understand this without knowing the cause of gravity, is as good a progress in philosophy as to understand the frame of a clock & the dependance of ye wheels upon one another without knowing the cause of the gravity of the weight which moves the machine is in the philosophy of clockwork...
>
> And therefore if any man should say that bodies attract one another by a power whose cause is unknown to us or by a power seated in the frame of nature by the will of God, or by a power seated in a substance in which bodies move & flote without resistance & which has therefore no vis inertiæ but acts by other laws then those that are mechanical: I know not why he should be said to introduce miracles & occult qualities & fictions into ye world...[22]

There were many other things that could not be explained mechanically, yet they were accepted by the mechanists and not considered occult:

> The same ought to be said of hardness. So then gravity and hardness must go for occult qualitys unless they be explained mechanically. And why may not the same be said of the vis inertiæ & the extension the duration & mobility of bodies, & yet no man ever attempted to explain these qualities mechanically, or took them for miracles or supernatural things or fictions or occult qualities. They are the natural real reasonable manifest qualities of all bodies seated in them by the will of God from the beginning of the creation & perfectly uncapable of being explained mechanically, & so may be the hardness of primitive particles of bodies.[23]

---

[21] Q 31.
[22] Draft letter to editor of *Memoirs of Literature*, after 5 May 1712; *Corr.* V, 298–300, quoting 300. The editors of the *Correspondence* say that this 'is an early appearance' of 'a non-Cartesian aether, here invoked as a mere conjecture' (*op. cit.*, 301).
[23] *ibid.*

As Newton wrote in a letter to Cotes: '... Experimental philosophy proceeds only upon phenomena and deduces general propositions from them only by induction. And such is the proof of mutual attraction. And the arguments for the impenetrability, mobility and force of all bodies and for the laws of motion are no better...'[24]

Samuel Clarke expressed the Newtonian position in another letter to Leibniz:

> That any body should attract another without any intermediate means, is indeed not a miracle, but a contradiction: for 'tis supposing something to act where it is not. But the means by which two bodies attract each other, may be invisible and intangible, and of a different nature from mechanism; and yet, acting regularly and constantly, may well be called natural; being much less wonderful than animal motion, which is never yet called a miracle.
>
> If the word natural *forces*, means here mechanical; then all animals, and even men, are as mere machines as a clock. But if the word does not mean, mechanical forces; then gravitation may be affected by regular and natural powers, though they be not mechanical.[25]

Newton was accused in his own day of *not* being a mechanist — of believing in 'occult' or 'magical' qualities, and his thoughts on the microphysics of matter, not to mention his work in alchemy, suggest that he was in no sense a believer in a clockwork universe. Ultimate causes for him were not mechanical. Paradoxically, Newton's theological concerns made him a more modern thinker than the mechanists of his own day and of ours.

Though Newton used rational thinking, he was not a rationalist. He did not believe that fundamental truths could be discovered by reason. Also, it would have been unthinkable for Newton to have created an image of God as a 'watchmaker', as is often alleged. The phrase comes from a letter by Leibniz, who interprets a passage from Query 31 as suggesting that God had actually to interfere from time to time with the motions of planets and comets to keep them in their orbits.[26] But Clarke replied to this, most certainly with Newton's approval: 'The notion of the world's being a great machine, going on without the interposition of God, as a clock continues to go without the assistance of a clockmaker; is the notion of materialism and fate, and tends (under pretence of making God a *supra-mundane*

---

[24] Newton to Cotes, 31 May 1713; *Corr.* V, 400.
[25] Clarke, Fourth Reply to Leibniz, 26 June 1716; Alexander (1956, 53).
[26] Leibniz to Princess Caroline, November 1715; Alexander (1956, 11).

intelligence) to exclude providence and God's government in reality out of the world'.[27]

The clockwork universe idea, of course, long predated the seventeenth century, and has nothing to do with the introduction of the Newtonian system. Nicholas Oresme, for example, was already using the metaphor as early as the fourteenth century, that period when mechanical clocks of the highest quality, and often of great sophistication, first began to be constructed. A version of it is found even earlier in that great thirteenth century textbook, the *De Sphaera* of Sacrobosco.[28] By the seventeenth century, it had become as much a part of general culture as it was of science. While Kepler, Boyle and others used the idea in their scientific works, so also did poets such as Francis Quarles and Thomas Traherne.[29] Newton was, in this sense — and as on other occasions — being accused of being out of line with the general consensus; his ideas did *not* fit in with clockwork regularity, and he made no attempt to claim that they did.

For Newton, as he stated explicitly in his works, God was a 'governor' ($\pi\alpha\nu\tau o\kappa\rho\alpha\tau o\rho$, in the General Scholium), who moved everything by his Will alone.[30] David Gregory recorded in a memorandum of 21 December 1705: 'The plain truth is, that he believes God to be omnipresent in the literal sense... What cause did the Ancients assign of Gravity. He believes that they reckoned God the cause of it, nothing els, that is no body being the cause; since every body is heavy'.[31]

Newton knew that the world would not instantly recognise the meaning of the *Principia*. The battles over the invention of the calculus were part of a much bigger struggle over the explanation of the great world system, and winning one battle was a necessary prelude to winning the other. That Newton ultimately succeeded is history, but it was not inevitable. Had anyone else solved the problem of planetary motions — and several people

---

[27] Clarke, First Reply to Leibniz, 26 November 1715; Alexander (1956), Smith (2007) states that the clockwork aspect (a common enough metaphor elsewhere) is nowhere apparent in the *Principia*, and appears to have been grafted onto the understanding of the Newtonian system by Laplace.
[28] Oresme (1968, 289); Sacrobosco (1949).
[29] Kepler to J. G. Herwart von Hohenburg, 16 February 1605, quoted Koestler (1967), IV, 331: 'My aim is to show that the heavenly machine is not a kind of divine, live being, but a kind of clockwork, insofar as nearly all the manifold motions are caused by a most simple, magnetic, and material force, just as all motions of the clock are caused by a simple weight. And I also show how these physical causes are to be given numerical and geometrical expression'. Boyle (1690); Quarles (1624); Traherne (1675).
[30] Q 31.
[31] David Gregory, memorandum, 21 December 1705; Hiscock (1937, 30).

came close — the Newtonian universal method, which was a much more significant innovation, might never have seen the light of day. The early version of Newton's method had already been universally rejected in 1672, when he had applied it to optics, and the underlying philosophy of the *Principia* was again opposed on almost all sides; only the mathematical results were accepted, and that part of the philosophy they necessarily carried with them, but the whole basis on which they were constructed was considered 'unphysical' and 'occult'. The story of modern physics is, from one point of view, the gradual coming to terms with the Newtonian legacy.

Chapter 2

Analysis and Synthesis

2.1. Newton and Leibniz

The controversy that inevitably followed on from Newton's radically innovative work in optical theory and gravity also engulfed his less problematic work in mathematics, though here it became a personal battle to secure his title to the authorship of the only part of his work that remained uncontested on intrinsic grounds. The mere listing of Newton's discoveries in mathematics shows how powerful the toolkit was that was available to him in his work in physics.[1] We also know that he made good use of it in the years to come. However, the attacks that were made on his optics only confirmed his natural reluctance to lay himself open to public dispute, and, though he was persuaded to send his early tract on calculus *De analysi*, to the mathematical impresario, John Collins, in London, he could not be persuaded to publish his mathematical work in the Royal Society's *Philosophical Transactions*. Ultimately, this led to the most notorious of his major disputes with his contemporaries, the argument with Leibniz over priority in the invention of the calculus as a general method.

    The quarrel with Leibniz was one of a kind that has affected science at frequent intervals down to the present day. Universally deplored by the worthy historians who consider science a purely objective search for truth, the priority dispute is unlikely to disappear in what will always be a competitive enterprise between individuals or groups of individuals. In addition, though the dispute between Newton and Leibniz was mainly about the invention of the calculus, it also spilled over into other things. Both men were outsiders, obliged to force their way into the scientific establishment, and never completely secure in their positions. Both were highly original thinkers, with wide interests, each determined, for different

---

[1] This is discussed in detail in *Newton and the Great World System*, Chapter 2.

reasons, to be number one. Leibniz, much more than Newton, was an urbane and sophisticated type of person, who has always had his supporters. But he could, nevertheless, also be vain and arrogant, and a master of subtle duplicity.[2] It was probably inevitable that two such talented and egotistical men would become major rivals, but it may have been the supporters rather than the principals who turned the rivalry into an open conflict.

In addition, priority was possibly less important to the participants, at least to Newton, than *credibility*. Newton seems to have been strangely unconcerned about claiming priority in many cases where it would have been relatively easy for him to do so. He generally cites his earlier work on the basis of *to whom it was communicated* and *when*, rather than when it was actually discovered. His greatest anxiety was the possibility of being accused of plagiarism and so not being regarded as a credible witness. The use of an encrypted anagram in his correspondence with Leibniz may have been intended to prevent the kind of retrospective claim on his most important ideas that he had discovered was inevitable in any correspondence with Robert Hooke.

The basic facts are now well known. Although some individual cases that we can now designate as differentiation and integration (then known as the methods of tangents and quadrature) had been known previously, no general method had been devised, in particular the fact that the methods were inverse operations, before Newton invented his version of the calculus between 1664 and 1666, and wrote his first tract on it, *To resolve problems by motion*, in the latter year.[3] It was followed rapidly by a number of others, including *De analysi*, written in 1669,[4] and the more algorithmically-structured *De methodis*, written in 1670–1671.[5] Leibniz's invention came

---

[2] Richard C. Brown (2012), 118, says of his early correspondence that 'one detects a feeling that he promised more than he actually had achieved, since his announcements of grandiose discoveries often evaporated upon inspection. There is certainly sometimes a tone of boastfulness in Leibniz's correspondence that contrasts unfavourably with the businesslike directness of Newton'. Leibniz was also a master at the art of ostentatiously praising his rivals in public while attacking them in private or condemning them in anonymous reviews.

[3] *MP* I, 400–448. The earlier methods of tangents and quadrature were also conceived almost exclusively in geometric terms, differing significantly in structure from the calculus methods which followed. Though, with hindsight, we can see that the individual results were equivalent to some of those later achieved by calculus methods, the calculus was by no means an inevitable consequence waiting to happen. Brown (2012) discusses this in relation to mathematicians like Barrow and the early Leibniz.

[4] *MP* II, 206–247.

[5] *MP* III, 322–353. Guicciardini (2009a) considers this as Newton's 'masterpiece' in algebraic calculus (179, 341, 368).

later, seemingly in October 1675.[6] The two methods were broadly similar but used quite different notations, with Leibniz's being considered particularly versatile and still in use today.

Like many major discoveries, the calculus can seem to be an inevitable consequence of the work preceding it, which produced numerous results which could be subsequently incorporated into the new system. So it is easily assumed that Leibniz, once he had started on his quest for a universal 'calculus', would inevitably arrive at the structure we now call the differential and integral calculus in the same way as Newton had a decade earlier. However, this is not really how truly ground-breaking discoveries actually occur. They are ground-breaking because *unexpected*, and are often opposed by the discoverer's contemporaries. Because Leibniz had for several years before 1676 produced results that seem now to be part of the algorithmic calculus does not necessarily mean that he was on an inevitable path towards discovering it. Richard C. Brown's extensive study of Leibniz's manuscripts of this period has left the historical picture of the development of the new analysis more problematic than it had previously seemed to be. In particular, says Brown, 'the exact nature of the psychological relation of Leibniz's work to Newton's may never be known'. Brown believes that the judgements of earlier scholars that Leibniz's achievement of what we now call the calculus came relatively early, in 1675, 'seem a bit too generous and are not supported by the manuscripts'. They are the result of reading back into earlier manuscripts insights that were only achieved later.[7]

It is something of a quirk of history that Newton and Leibniz came to be involved in the same area of innovatory mathematics at close to the same time. Both men were engaged on grand intellectual programmes that extended far beyond mathematics, but they were utterly different in conception. Newton was seeking to understand the secrets of the natural world. Calculus was forced upon him because it was the only way of understanding how space, the single physical quantity that can be directly observed and measured, varied with time, the only other quantity whose variation we can apprehend. A polymath, but not a natural mathematician like Newton, Leibniz was looking for a universal language, a universal system of logic beyond specific areas of knowledge. Calculus came to him because change is a universal component of human knowledge and calculus gives us a way of symbolising change. The symbols were, for Leibniz, in some

---

[6] See, for example, Hofmann (1974) and nearly all biographies of Newton.
[7] Brown (2012, 118, 122).

ways more important than the process. For Newton, symbols were merely tools, for Leibniz they were part of the meaning of what he was trying to express. Ultimately, however, the symbols would only have meaning if they represented a process. When he found that the symbols could represent a universal process Leibniz believed he had achieved a major step towards his ultimate goal. However, it was only a universal process because space and time are universal components of human constructions. The physical description had become a model for the whole structure of human thought.

In October 1676 Leibniz visited London for a second time, after receiving the first of Newton's two letters sent to him through Oldenburg. During the 10 days he spent in the city, he met John Collins, who showed him various mathematical tracts, including Newton's *De analysi*, on which he took some notes, though mainly on the section on infinite series. Although Collins had copies of *To Resolve Problems by Motion* and passages from Newton's mathematical notebook, CUL Add MS 4000, it is believed that Leibniz never saw them. In three places in *De analysi*, however, there is direct use of the inverse relationship between fluxions and fluents. One of these is 'virtually identical to the modern proof that the derivative of the area function is the ordinate y'. Though most of the tract is concerned with quadrature or integration, the method of tangents or differentiation via vanishing infinitesimal quantities (symbolised by o) is used at the end to prove the basic rule of integration for a term with a rational power which had been introduced at the beginning.

Only a handful of pages out of about forty are of this kind, and it is, of course, conceivable that Leibniz could have missed their significance in a hurried examination. There are certainly no notes by Leibniz on these pages, an indication taken by earlier scholars as meaning that he was already familiar with what they contained. Nevertheless, the question of the relation between the work of Newton and Leibniz is now no longer so easily resolved in the way it once may have seemed to be, for Brown, whose work is not mainly focused on the priority dispute but rather on the uniqueness of Leibniz's overall programme, is of the opinion that Leibniz's characteristic calculus algorithms were not developed until 'after the second visit to London'. In fact, Brown says that, 'although Leibniz had been thinking of a 'calculus' ever since 1672/73, the parts of it for which he is remembered today were only formulated *after* the London trip, and some respects were not perfected until the 1680s'.[8] So, although Leibniz had already introduced versions of

---

[8]Brown (2012, 117, 124, xvii) says that this opinion was 'gradually forced upon' him on the basis of studies undertaken over a long period of years.

what we now consider the 'differential' and 'integral' symbols by 1675, he had not yet defined their use as aspects of the calculus algorithm.[9] He could certainly have found such an algorithm in *De analysi*, even if this work was not written with the more logical structure and greater technical precision of *De methodis*.[10]

Whatever the date of Leibniz's breakthrough moment, Newton wrote his second letter on 24 October, immediately after this second visit, revealing his discovery of the inverse connection between tangents and quadratures, or fluxions and fluents, although encrypted in an anagram. The letter also refers to the *De methodis*, written 5 years earlier, and gives some indication of Newton's higher methods of quadrature, by antidifferentiation, and by substitution of conic sections, though without being totally explicit about the details. The 'prime theorem', used in the first of these methods, however, is given in full. Newton's letter was held back by Oldenburg on account of Leibniz's travels, and only reached him many months after he had returned home. Though he probably couldn't decipher the anagram, Leibniz must have known or quickly guessed, by this time, what it contained. Even if he had not been impressed by what Collins had written in April 1675, or by *De analysi* in October 1676, he cannot have failed to notice with the accumulated evidence that Newton had devised a method with remarkable similarity to his own. Immediately on receiving the letter, Leibniz sent one of his own with a full account of his new algorithm, which Newton received on 30 August 1677. Four days later, the intermediary, Oldenburg, was dead and the correspondence lapsed for another 16 years.

Nothing much happened for several years. Collins died in 1683 without persuading Newton to let him release his manuscripts. Leibniz eventually published the first paper on the differential aspect of his calculus in *Acta eruditorum* on 10 October 1684, without indicating that he was aware of Newton's work, which he had seen in *De analysi*, as well as in the two lengthy

---

[9] Brown says, for example, that dx and dy only became infinitesimal, rather than constant, differences in the *Calculus tangentium differentialis*, written in Amsterdam in November 1676 (2012, 128).

[10] As Brown says (2012, 224–227), it is virtually impossible to pin down exactly what Leibniz may have learned from *De analysi*, or from his discussions with John Collins, as his own work was conceptually different from that of Newton. Leibniz, unlike Newton in his early algebraic work, had 'little interest' in the differential 'as an independent object, function, or rate of change', seeing it rather as a device which allowed a tangent to be constructed from the subtangent (226); his work, unlike Newton's, was limited by the 'homogeneity requirements of the traditional geometric treatment' (150). Brown believes that 'conscious plagiarism' by Leibniz is 'unlikely' (226) and seems to imply a more subliminal kind of influence. Whatever may have triggered Leibniz's 'epiphany', however, seems to have occurred soon after his second London visit.

epistolae of 1676, both of which extracted results from *De methodis*.[11] He also made no mention of the tangent letter, which had indicated that Newton had developed an advanced fluxional method as early as 1672. He further avoided any reference to Collins's reports of 1673 and 1675, of the *Historiola*, which Collins had finished in mid-June 1676, of Oldenburg's letter of December 1674, and of the works of Barrow whose *Lectiones opticæ* and *Lectiones geometricæ* he had acquired on his first visit to England in 1673. The work was presented as though it was entirely *sui generis*.[12]

Newton made no attempt to rush into print with any of his early mathematical tracts. He was, in any case, heavily involved in preparing the precursor tracts to his *Principia*, the first of which, *De motu corporum in gyrum*, with Newton's first application of calculus to physics, with 'one or more orders of the infinitesimally small', was received by Halley on behalf of the Royal Society in November 1684.[13] However, as early as 1685, one of Newton's disciples, John Craig, published some calculus results by Newton, using Leibnizian notation, and naming him alongside Leibniz as inventor,[14] while David Gregory had included work from *De analysi*, in his *Exercitatio geometrica de dimensione figurarum*, which was published in June 1684, and named Newton as the source, even before Leibniz's paper appeared in the journal.[15] Newton responded to Gregory by drafting his *Mathesos Universalis Specimina* between late June and mid-July 1684, and then *De computo serierum*, a work on infinite series, between July and August.[16] Neither of these works was published or even completed, but a summary of his two *Epistolae* appeared in Wallis's *Algebra* in 1685, while, in the same year, Gregory obtained from Craig Newton's prime theorem on quadrature

---

[11] Leibniz (1684).

[12] Unlike Newton, Leibniz frequently denied any indebtedness to Barrow, but it has recently been argued that his work on the fundamental theorem of the calculus of 1693 contained unacknowledged borrowing from this predecessor's pre-calculus geometric work, outlined in lectures delivered in 1667, edited by Newton, and published in 1670 (Nauenberg, 2011). Blank (2009, 608) quotes Ehrenfried Walter von Tschirnhaus, Jakob Bernoulli and some modern authorities as believing that Leibniz borrowed significantly from Barrow.

[13] Herivel (1965), 257–289.

[14] Craig (1685).

[15] Gregory (1684). Gregory was mostly concerned with publishing results on series by his uncle, James, which in some instances coincided with similar ones obtained by Newton in *De analysi*.

[16] Newton, *MP* IV, 526–588 and 590–616. E. W. von Tschirnhaus, who had travelled to London and Paris in 1675 and become acquainted with the work of Newton and Barrow, as well as that of Leibniz, had published a work in 1682, which included calculus results taken from Leibniz, while acknowledging only Barrow (Blank, 2009, 606, referring to Kracht and Kreyszig, 1990).

from the *Epistola posterior*. He published it as his own work in Archibald Pitcairne's *Solutio Problematis de Historicis seu Inventoribus* three years later, although he subsequently became a genuine admirer of Newton and recruited other Scottish associates such as George Cheyne and John Keill to the Newtonian cause. Certainly, some of Newton's work was finding its way into print, despite his own intentions, and it could even be argued that Craig's work of 1685, produced, we are told, with the active help and encouragement of Newton, was the first extensive published account of the inverse method or integral calculus.

As Guicciardini has shown,[17] Newton was actually strangely unconcerned about making claims on the direct method, that is, the method of tangents, or differentiation in Leibniz's language, even though he was the first to discover the general algorithm which generated the fluxion for virtually any curve or algebraic or transcendental expression, and had told Collins what it could accomplish in his letter of 10 December 1672. Even at the height of the subsequent priority dispute, Newton was quite prepared to concede the discovery to other early workers such as Barrow, de Sluse and James Gregory, who had either not reached this stage in their work or did so after him. The work whose intellectual ownership he was particularly concerned about involved the *inverse* method, the method of quadrature or finding the fluents, the method that Leibniz called integration. Newton regarded the panoply of integration techniques he had discovered, which included the antidifferentiation method which emerged from the fundamental theorem of calculus, as powerful steps towards the ultimate goal of geometrical constructions which would alone represent verifiable physical truth. Leibniz, by contrast, thought that the direct method or differential calculus, where he had codified the rules and introduced a versatile symbolism, was his special contribution, and it became the subject of his first paper.

Leibniz's first paper, as it happens, was a far from lucid exposition, and would have been unlikely, on its own, to have provided a basis for the rapid development of the subject that followed. It was, in fact, partly on the basis of Craig's complaints about its impenetrability that Leibniz produced a second paper in 1686, which expanded on and augmented the previous work, and made his exposition clearer.[18] By introducing the fundamental theorem of the calculus, it also ventured into integration or the inverse

---

[17]Guicciardini (2009a), where the issue is discussed extensively.
[18]Leibniz (1686). Hall (1992, 38) says that Newton's early 'To Resolve Problems by Motion', of 1666, was actually a clearer exposition than Leibniz's paper of 1684. It

process. Following this, Leibniz's work was almost immediately taken up on the Continent, and developed very effectively by other mathematicians, in particular by the brothers Jakob and Johann Bernoulli. It was Johann Bernoulli who, among other things, replaced Leibniz's original idea of the differential as 'a minute fixed quantity'.[19] He would actually also later lay claim to being the inventor of the integral calculus, and he certainly did far more than Leibniz to develop its extensive possibilities. Whatever the real origins of the calculus algorithm, this is the point at which it began to take off, and Brown considered this to constitute a Kuhnian paradigm shift in human knowledge.

Newton showed no immediate response to Leibniz's publications — which he wouldn't have known about for some time anyway. It was not until 1693 that the first purely *mathematical* account of his fluxions appeared in the second volume of Wallis's *Opera*, with material from the *De quadratura* which he was then writing to counter the danger of anticipation — from Gregory and his friends rather than from Leibniz. Newton, however, in addition to including a number of results and methods of the new mathematics, made a reference to the early nature of his discoveries in calculus when the *Principia* was published in 1687, along with a statement that he had communicated his method to 'that most skilled geometer ten years earlier', which he had concealed in an anagram, now correctly decoded as: 'Given any equation involving flowing quantities, to find the fluxions, and vice-versa'. Leibniz, he said, had revealed that he had also come across a similar method, 'which hardly differed from mine except in words and notation'. His words were quite reasonable and perfectly nonconfrontational, in line with Leibniz's similar comments, made subsequently, but he later changed this to a statement that the work had been communicated to John Collins as early as 1672.[20] In the context of the time, before the establishment of scientific journals, communication to a correspondent like Collins was considered a form of 'publication', as indeed it was, if someone like David Gregory, whom Newton didn't know, could quote his results in a printed book.[21] The earliest journals in fact were merely records of such

---

also included results of integrations that Leibniz could not have achieved with his more geometrically-based conceptions and assumptions of homogeneity.

[19] Hall (1992, 113). Leibniz's 1684 paper had generally avoided infinitesimals (Brown, 2012, 135).

[20] II, Lemma 2, Scholium, first edition, 1687, translated in Hall (1980, 33) and (1992, 251).

[21] *De analysi* had been passed to David Gregory's uncle, James, by Collins. The work had even found its way abroad, being sent to René de Sluse in Belgium and Jean Bertet and Francis Vernon in France. Oldenburg was pretty explicit about its methods of quadrature

correspondence, printed to make them available to more recipients, and the earliest of all was only four years old at the time when Collins received *De analysi*.[22]

2.2. Mathematicians at War

After twenty years of shadow boxing between the participants, which included a seemingly cordial exchange of letters in 1693, the dispute came to a head in the period 1697–1716. On New Year's Day 1697 Bernoulli, who was genuinely convinced that Leibniz's work stood out on its own, decided to test the powers of the Newtonians with two challenge problems requiring advanced calculus issued to the mathematicians of Europe. Newton received them after coming home from a day's work at the Mint and, as has often been recounted, solved them the same night.[23] They were published anonymously but Bernoulli recognised 'the lion from his claw'.[24] It is not yet definitely established whether there was, at this stage, any ulterior design in the challenge with regard to Newton. It could equally well have had its origin

---

and tangents in letters to de Sluse of 1669 and 1673 (1965–1986, VI, 227; IX, 427–430). Information about the work also reached G. A. Borelli in Italy, and several important people in England. One of the important aspects of scientific history, though it can never be written down, is how much information has passed through invisible channels of communication. News of mathematical discoveries would certainly have been passed around in seventeenth century Europe by channels we cannot now recover.

[22] The *Philosophical Transactions*, which effectively became the first fully scientific journal, had only been created in March 1665 and was originally Henry Oldenburg's private account of exactly such correspondence.

[23] *De ratione temporis*, *Philosophical Transactions*, January 1697. Letter to Montague, 1697, *MP* VIII, 72–79. There is a genuinely contemporary account for this story (Catherine Barton, Keynes MS 130.5, Westfall, 1980, 582–583), but, as with all such accounts, there is always the possibility of it shading into hagiography. Nevertheless, it seems that the work that Newton had previously done on the solid of least resistance and in *Principia*, Book I, Proposition 50 had prepared him specially to solve this problem relatively quickly.

[24] Chandrasekhar (1995, 573), offers a translation by Raghavan Narasimhan of Bernoulli's letter to Basange de Beauval: 'Thus it is, Dear Sir, that my problem remained unsolved after it had been examined by several people in Holland. It was then sent to England where I had great hopes that it would have a happier end, since in England there are some excellent geometers skilled in using our methods or similar methods. In fact, the January issue of *Philosophical Transactions*, which you were kind enough to send me, shows that I was not mistaken, since it includes a construction of the curve of quickest descent that deals with the problem perfectly. Although its author, in excessive modesty, does not reveal his name, we can be certain beyond any doubt that the author is the celebrated Mr. Newton: for, even if we had no information other than this sample, we should have recognized him by his style even as the lion by its paw ... It is only to be wished that Mr. Newton had published the solution and the method by which he found the desired curve. ...'

in the ongoing feud between Johann Bernoulli and his brother, Jakob,[25] and Leibniz would refer to Newton's success favourably in a reply to the Queen of Prussia in 1701.

Two years later, however, an erstwhile follower of Newton, Nicolas Fatio de Duillier launched an attack on Leibniz as a plagiarist; Fatio may have been influenced in his antipathy by Leibniz's rejection of his *own* claim to be an independent discoverer of the calculus in 1686. He seems to have acted independently of Newton, with whom he was now no longer closely associated, but his views reflected ones which he had privately given in correspondence with Huygens as early as 1691 and 1692, when he had specifically expressed his surprise that Leibniz had made no mention of Newton in his articles for *Acta eruditorum*.[26] Fatio's published accusation was not as wild as is sometimes claimed. He stayed close to the truth as he saw it: Newton's invention had preceded Leibniz's by many years. Whether Leibniz's invention was independent or not would be left to the judgement of those who had seen the relevant documents.

Later in the same year, Wallis published the two extensive mathematical letters Newton had written to Leibniz in 1676, the *Epistola prior* and the *Epistola posterior*. Newton's early tracts started being recovered by his followers and being printed for the first time in the early 1700s. It was then that Newton began to realise for the first time how extensively Collins had promoted his work. He began to suspect that Leibniz might really have been a plagiarist, especially as his almost transparent attempt to appropriate Newton's dynamics in his *Tentamen* of 1689 was already pointing in that direction.[27] Though the *Tentamen* was presented as if containing earlier work, Leibniz's own drafts, now discovered, show that it was based on notes made directly from the *Principia*;[28] even so, Leibniz managed to imply that Newton had *now* found the law of gravity that he had already discovered!

---

[25] Ferguson (2004). Bernoulli had described his brother as 'incompetent' and wished to demonstrate it publicly.

[26] Fatio (1699). Fatio to Huygens, 18 December 1691, Huygens, *Oeuvres complètes*, 1888–1950, X, 214; 1692, *ibid.*, 257–258.

[27] Leibniz (1689). There is a pointed reference to the *Tentamen* as an unsuccessful 'imitation' of the *Principia* in a draft for the *Recensio libri* (1722), quoted Cohen (1999, 281). Blank (2009, 609) believes it 'likely' that Newton was unaware of the *Tentamen* until after 1710. If he had seen the *Tentamen* when it was first published in 1689 his letter to Leibniz of 1693 would have been a good deal less cordial!

[28] Meli (1993). There is no mention of Newton in the *Tentamen*, just as there is none in the calculus paper of 1684. Leibniz, however, openly protested on more than one occasion that he had written it before seeing Newton's work.

An even more blatant case was a list of corrections to the *Principia* which Newton had written out in 1690 and sent to Huygens, with Fatio, in 1691, which were printed in the *Acta eruditorum* of 1701 as a list of errors detected by Leibniz and his supporters.[29]

Matters simmered for several years, with covert (anonymous) attacks by Leibniz and Bernoulli — a 'lion by day, jackal by night',[30] who made a show of being Newton's friend while sticking in the knife with publications which the latter took to be the work of Leibniz — and an open and no-holds-barred challenge by Newton's rough-tongued lieutenant, John Keill, written in 1708 and published in 1710. The war of words reached a new pitch of intensity with a statement by Leibniz, in a supposedly anonymous review of Newton's first two published mathematical tracts, *De quadratura* and *Enumeratio linearum tertii ordinis*, in the *Acta Eruditorum* of January 1705, which could be read as tantamount to an accusation of plagiarism by Newton, the first made directly by either participant. Though it went unnoticed by Newton at the time, as he seems to have made no attempt to keep up with the journal literature and had no close connection with Keill, it was a statement that Leibniz came to regret, as a public attack on his character was the one thing guaranteed to goad Newton into direct action. In more immediate terms, it led to Keill's response in which he asserted, in a direct reversal of Leibniz's prior insinuation, that the latter had published Newton's method of fluxions, simply changing 'the name and symbolism'.[31]

The evidence was now available in the papers of John Collins that William Jones had acquired in 1708, which proved beyond doubt that *De analysi*, had been written in 1669. Newton, for once, overcame his reluctance to publish and allowed Jones to issue four of his mathematical tracts in a volume which was presented to the Royal Society on 31 January 1711. The volume included the 'smoking gun' text, *De analysi*, which Jones was able to date to 1669 by quoting the letters of Collins and Barrow, together with a mass of other documentation relating to Newton's discoveries. Westfall believes that Newton may have known about Leibniz's

---

[29] Johannes Gröning, *Historia Cycloeidis*, Hamburg, 1701. We may also mention that, in 1702, Jakob Bernoulli, Johann's brother, published series for sin x and cos x, as original results, though de Moivre had previously published them, with proof, as Newton's in the *Philosophical Transactions* for May 1698 (20, 190–193), while, as we have previously noted, it seems highly possible that Leibniz's work on the fundamental theorem of the calculus borrowed from Barrow's pre-calculus *Lectiones geometricae* in a way that he never acknowledged (Nauenberg, 2011).
[30] Hall in *Corr*. VII, xxi.
[31] Leibniz (1705, 35); Keill (1708, 186).

accusation as early as 1708 and may even have contributed to Keill's 'riposte'.[32] There is no direct evidence, but, ultimately, it is immaterial. Leibniz's attack was of the kind which he would never have left unchallenged at whatever date he found out about it.

Newton, who had grown much more worldly-wise during his years of public service, and had learned to read men's characters from what they said or what they avoided saying in their documents, was suspicious of the long time gap between the *Epistola posterior* and Leibniz's announcement of his discovery in June 1677, which included the second visit to England and a probable reading of *De analysi*. He didn't know that Oldenburg had only sent the *Epistola* after a long delay. However, the reply of 1677 came at an extremely convenient moment, when Leibniz (who, in fact, seldom held anything back which he had discovered) could be seen as reacting to Newton's *Epistola* with the full algorithm which he had discovered a long time before and was now presenting in epistolary form for the convenience of his correspondent. Newton noted that Leibniz, like Hooke, had a habit of making retrospective claims on results that had been sent to him, even when he clearly hadn't understood the material at the time. The 1677 letter was the first time he had ever put forward a general differential method. Leibniz had too much previous 'form' to be trusted on his own account, and, for Newton, the accusation of 1705 could be seen as a measure of his opponent's intrinsic dishonesty. Whatever Leibniz discovered and at whatever time, and with whatever degree of independence, it cannot be denied that Newton had a case, and that his actions were provoked by an attack on his personal integrity which moved the unspoken conflict into completely new territory.

When Leibniz protested about Keill's paper, writing to Hans Sloane, the Secretary of the Royal Society, on 21 February 1711, and Keill was asked to explain his actions, he supposedly drew the attention of the President to the offending review. Even then, Newton did not respond directly, but encouraged Keill to send a less confrontational letter to Leibniz, which pointed out unequivocally that Newton was the 'first discoverer' of the calculus, as was evident from his correspondence of 1676 with Oldenburg. The letter was certainly authorised by Newton as President, and may have been drafted by him.[33] The stage was now set for the ultimate confrontation, and the trigger was another letter from Leibniz, dated 29 December, which

---

[32] Jones (1711) and Westfall (1980, 716).
[33] Leibniz to Sloane, 21 February 1711, *Corr.* V, 96–97; Keill to Newton, 3 April 1711, *Corr.* V, 115. Keill to Leibniz, May 1711, *Corr.* V, 133–149. Westfall (1980), 723, believed the letter of May 1711 was drafted by Newton, though once again direct evidence is lacking.

arrived at the Royal Society on 31 January 1712, and which attempted to separate Keill's 'unjust braying' from the views of the 'illustrious' President, whose fluxions he now recognised as an independent discovery, founded on 'principles similar to our own'. At first, Newton thought once again about drawing back from the dispute, but something persuaded him to change tack in a rather dramatic way. It may have been another anonymous review, this time of Jones's volume, which appeared in the *Acta eruditorum* for 1712, which avoided discussion of the date of *De analysi*, but contrived to imply that it contained nothing that went beyond Archimedes and Fermat, unlike the differential calculus of the 'illustrious Leibniz'.[34]

Leibniz had demanded an impartial inquiry from the Royal Society into the whole matter. Newton, by a brilliant stroke of political manoeuvring, gave them one, by setting up a committee on 6 March, ostensibly to adjudicate between Keill and Leibniz. On 24 April, just six weeks later and a week after appointing its final three members, the committee reported in Keill's, and so in Newton's, favour. The report was made public in a volume with the title *Commercium Epistolicum D. Johannis Collins, et aliorum de analysi promota*, or 'The Correspondence of the Learned John Collins and Others Relating to the Progress of Analysis', published early in 1713 as the work of the Royal Society but actually written by Newton himself, which stated in unequivocal terms that, not only had Newton preceded Leibniz as inventor, but that the latter had published his calculus without acknowledging the work of his predecessor, with which he was familiar.[35] Newton used the techniques he had employed against the likes of the master-forger William Chaloner, compiling and manipulating documentary evidence, to stack up the legal case against Leibniz. He compounded this two years later by publishing an anonymous review or 'Account' in the *Philosophical Transactions*, again written by himself.[36] The 'Account' claimed that the report was the work of an international committee, 'composed of Gentlemen of several Nations', on the basis that it included the Huguenot mathematician Abraham de Moivre and the King of Prussia's minister in London, Frederick Bonet, among its members. While strictly 'true', this claim has been criticized as creating a false impression

---

[34] Leibniz to Sloane, December 1711, *Corr.* V, 207 (Latin), translation (208). Leibniz (1712). Westfall (1980, 724, 720).

[35] *Corr.* V, xxvi–xxvii. *Commercium Epistolicum*, 1712, issued February 1713. Westfall (1980), 724–727.

[36] 'Account', *Phil. Transactions*, January-February 1715, 173–225.

of the committee's membership, which was not publicly known until the nineteenth century.

The 'Account' is not a pleasant document. It is ferocious and uncompromising, but, like the *Commercium Epistolicum*, it nowhere directly accuses Leibniz of plagiarising Newton's calculus. It challenges Leibniz, using the manner of the most effective kind of legal advocacy, to produce the evidence to support the claims made in the review of 1705 that he had preceded Newton in the invention, and that Newton had borrowed from *him*. It denies that the Royal Society has passed judgement on either side in the quarrel, only that it has assembled the evidence from which readers can make their own judgement. If Leibniz has any evidence to counter that in the *Commercium Epistolicum*, he is invited to come forward and make it available to the scientific community. Cases where Leibniz had seemingly appropriated the work of others, whether inadvertenly or otherwise, are catalogued in both documents, in the classic mode of damaging the credibility of the witness. Chief among these in the 'Account' is the appropriation (somewhat inexpert, as Newton was able to show) of Newton's dynamics and gravitational theory in the *Tentamen*. Clearly, this rankled with Newton as much as the 1705 accusation of plagiarism.

One thing that stands out in the 'Account', however, is the exact assignment of credit to the work of the predecessors of both men — Wallis, Descartes, Fermat, Pascal, Cavalieri, Hudde, Mercator, Sluse, Barrow and Gregory are all named as significant contributors. Newton was clearly determined to avoid Leibniz's quite frequent fault of claiming work already accomplished as a new discovery. Newton, as always, was concerned not primarily with claiming priority, but rather with maintaining his own credibility. It is hard to avoid the impression that the entire affair could have been avoided had Leibniz not published the offending review of 1705, forcing Newton into a defensive strategy where the more ferocious side of his character necessarily came to the fore.

There was a dicey moment in October 1712, before the *Commercium Epistolicum* came out, when Bernoulli's nephew, Nicolaus, told Newton that his uncle had found an error involving a numerical factor in the first example illustrating Proposition 10 from Book II of the *Principia*; Newton, after expending an enormous amount of effort in locating the source of the error, had just enough time to paste over the relevant pages in the second edition of the book, which came out at the end of June 1713.[37] Bernoulli, who, as

---

[37] *MP* VIII, 48–61, 312–420. A. de Moivre to Johann Bernoulli, 18 October 1712, *Corr.* V, 347. Whiteside (1970), 126–129; Westfall (1980), 740–743. Bernoulli found the error

Newton's 'friend', was elected a fellow of the Royal Society as a result (!), along with Nicolaus, must have been furious with the leak, for the error (which was actually geometrical and did not directly involve calculus) was such that it gave the impression that Newton didn't understand second differentiation, a really damning indictment. It was only because Nicolaus, in characteristic Bernoulli style, wanted originally to claim credit for the work himself that Newton found out about it at all. Bernoulli didn't get an acknowledgement in the edition because, by the time it was published, Newton felt that his supposed friend had tried to trick him by publishing corrections timed to come out immediately after what he thought would be a still-uncorrected text of the *Principia*. *Acta Eruditorum* howled with rage about the matter in 1713/14 because, with the delay in publication of Newton's work, and the subsequent pasting in of the corrections, the plot had failed.[38] Later in life, Bernoulli competed with his own son in publishing new results, sometimes predating his work by as much as two years in order to claim priority.

As a collection of documents, the *Commercium epistolicum* had been put together with the skill of a historian, as well as the cunning of a lawyer, and Leibniz found it difficult to organize a counter-case, which would unequivocally demonstrate his priority and independence. Westfall points out that Christian Wolf, who clearly had not read *De analysi*, and had no idea that Leibniz had done so as early as 1676, saw in 1715 that Leibniz had neglected to demonstrate that the tract didn't contain his calculus algorithm and urged him to remedy this defect.[39] The stark and uncompromising presentation of the documentary evidence in the *Commercium*, with the supporting correspondence of the long-dead Collins and Barrow, made it virtually impossible for him to do so. Instead, he took the diversionary tactic of attacking the basis of the Newtonian system of natural philosophy, which had still to make headway on the Continent.

Though Leibniz (under the cloak of anonymity) and Bernoulli launched quite brutal assaults on the second edition of the *Principia*, the tide now turned in Newton's favour. Bernoulli's cover was blown by a document of 29 July 1713 known as the *Charta Volans*, which though published anonymously was clearly the work of Leibniz, and referred to an 'eminent mathematician',

---

by producing his own calculation; he was unable to locate its source, which was found by Newton himself.

[38] Cohen (1971), 254–256. *Acta eruditorum*, March 1714, 134. Bernoulli had published his account of the errors in the *Principia* in the *Acta eruditorum* for February and March 1713, 77–95, 115–132.

[39] Westfall (1980, 720).

who was clearly Newton's ostensible 'friend'.[40] Leibniz himself had few overt friends left when he died soon afterwards in November 1716, while contemplating the unrewarding prospect of compiling a history of the House of Brunswick — two years after they had abandoned him in Hanover to become the new royal family of England. Newton allowed Bernoulli to think that he still had some credibility with him for a few more years, but a final attempt at self-justification by the latter in February 1723 proved one too many and Newton abruptly broke off the correspondence.[41]

Newton's manipulation of the inquiry has been much criticised but the *Commercium Epistolicum*, though certainly biased, shows one interestingly positive aspect of his character with respect to those of such opponents as Leibniz, Bernoulli or Robert Hooke. There is no significant evidence of outright falsehood, plagiarism or use of spurious data in the work.[42] The *Commercium Epistolicum* may make what have been taken by later readers to be unfair inferences about the independence of Leibniz's work in the calculus, but the documents quoted are genuine, and Newton believed that the documentary evidence, recovered mostly from the papers of Collins which he had now seen for the first time, indicated that his allegations must be true. This is a pattern that shows up in nearly all of Newton's disputes, and it gives us confidence that his recollections of events where

---

[40] *Corr.* VI, 15–19. The identity of the 'eminent mathematician' was inadvertently revealed by Bernoulli himself in 1716. Prior to this (in 1713), Bernoulli had told Leibniz that he wished to remain anonymous because Newton 'has heaped many testimonies of his goodwill upon me' (*Corr.* VI, 5).

[41] *Corr.* VII, 218–223.

[42] Newton certainly used trickery and sleight of hand in his handling of the priority dispute, but he generally avoided outright falsehood. According to Hall: 'The *Commercium*, though advocating a case, contained no falsehoods that Leibniz was ever able to demonstrate; its fault is that of omission of Leibniz's side of the story'. The only slight deception in the entire work that Hall found (one which he considered 'rather unimportant') was the implication that *De Quadratura* was an early work, and so the dot notation was early too. (1992, 313–314) (In fact, there is an early version of a dot notation for partial derivatives dating from the 1660s though it was originally used in a different way, and would be completely transformed in the later work.) Another authority, Westfall (1980), 728, quoted Newton's dating of the principal propositions of the *Principia* in one of the notes to 1683, rather than 1684. Newton's use of experimental data is discussed below. Concerning plagiarism, there isn't a single example of him appropriating the work of another author, in stark contrast to Hooke, Leibniz and Bernoulli, and, contrary to what has been claimed surprisingly often, his works are full of attributions.

there is no documentation are generally valid. Certainly, more than one conspiracy theory concerning what Newton had or had not discovered has been disproved by the subsequent discovery of documentation backing his claims.[43]

Of course, printing in what we would now regard as a 'scientific journal' was, as we have seen, only about ten years old, back in 1675, and by no means the only method of 'publication' at the time.[44] There was more publication of the work than is usually claimed. According to Niccoló Guiccardini:

> In the period preceding the printing of the *Principia* most of Newton's mathematical discoveries were rendered available to the mathematical community through rather oblique ways. Newton engineered a complex publication strategy. He allowed some of his mathematical discoveries to be divulged through letters and manuscript circulation. Manuscripts were shown to a selected group of experts in the field (such as John Collins, John Craig, Edmond Halley, David Gregory, and Fatio de Duillier) who visited Newton in Cambridge, they were deposited at the Royal Society in London or as Lucasian Lectures in the University Library at Cambridge, and they were even copied (sometimes in mutilated form).[45]

Guiccardini refers to this as 'scribal publication', and it was a method used in Restoration England for other (e.g. literary) purposes.

Even now there is no absolute consensus on 'publication'. Different interest groups push different interpretations. Does it mean 'peer reviewed journal' as opposed to news conference, website, conference proceedings, etc. What does 'peer reviewed' mean? Is it submission prior to the work existing in a printed form? Is submission of a rejected paper to be taken into account? Theodore Maiman, for example, published his discovery of the laser in a newspaper, available to all, because he knew he would have no success in

---

[43] The classic example is the discovery of the manuscript of the lunar gravitation test of the 1660s, which completely refuted a long-held conspiracy theory, though that theory continues to be resurrected. Brown (2012), xvii, says that the evidence revealed by the manuscripts he has examined means that 'Newton's suspicions' cannot now be 'dismissed as 'paranoia'; he had some rational grounds for them and his suspicions were fuelled in part by aspects of Leibniz's character as well as by the Iago-like insinuations of Newton's colleagues'.
[44] Gleick (2003, 221), considers Newton's own heavily-attacked 'New Theory' of 1672 to be 'the first major scientific work' ever 'published in a journal'.
[45] Guicciardini (2003b).

submitting to a peer-reviewed journal. It was not officially approved of, but it couldn't be denied.[46] Then again, does 'publication' mean publication of an announcement or publication of a method of discovery?

It is little known that, however long it took Newton to publish his new mathematical system, a published *statement* of its content came at a very early date. The poet and Fellow of the Royal Society, Sir Edward Sherburne, early in 1675, published a translation of an ancient astronomical poem, the *Sphere of Marcus Manilius*, which included an appendix constituting a history of science almost up to the time of writing. At the end of this survey, which is dated 1673, Sherburne wrote:

> *Mr Isaac Newton* Lucasian Professor of Mathematics in the university of Cambridge and Fellow of Trinity college has lately published his reflecting telescope; new theories of light and colours; hath ready for the press a treatise of *Dioptrics* and divers astronomical exercises, which are to be subjoined to *Mr Nicholas Mercator's Epitome of Astronomy*, and to be printed at Cambridge. From him besides is to be expected a *New General Analytical Method* by him for the *Quadrature of Curvilinear Figures*, the *Finding of Centres of Gravity*, their *Round Solids*, and the *Surfaces* thereof, the straitening of curved lines; so that giving an ordinate in any figure as well such as *Des Cartes* calls geometrical, as others, to find the *Length of the Arch Line*, and the *Converse*. Such an invention, to wit, but in one particular figure the circle, the learned *Snellius* thinks transcendent to any thing yet published; and how much conducing to the benefit of astronomy, and the mathematical sciences in general, such an universal method is, I leave to others, together with myself to admire and earnestly expect.[47]

---

[46] Maiman was right. He produced the first working laser on 16 May 1960, seven years after Townes, Gordon and Zeiger had produced the microwave equivalent (the maser), and demonstrated it at a news conference on 7 July. There was at the time massive competition between laboratories to produce the first working device, amid a large degree of scepticism that it could be done at all, and there were conflicting claims on ownership of the theoretical basis. In a classic example of the detrimental effect of power structures on the development of physics, *Physical Review Letters* took just two days to reject Maiman's paper on the basis that the work wasn't novel. Maiman was fortunate that *Nature* chose to accept it. Even after that it took a long time for Maiman's part in the story to become fully accepted as more than an example of opportunism, and he was not a recipient of the Nobel Prize which, in 1964, was shared between Townes and two other early pioneers, Basov and Prokhorov.

[47] Sherburne (1675). It was first drawn to my attention by Dr. Patrick Gavin from the copy in Stonyhurst Library. The examples quoted are similar to those mentioned in Oldenburg's (later) letter to Leibniz of December 1674.

This was the first published claim that Newton had devised a new general method of calculus techniques, involving an inverse process. It is pretty comprehensive and the information can only have come from Collins or Oldenburg. It also indicates, what we know from other sources, that, whether printed or not, copies of mathematical work by Newton were certainly quite well known among a group of mathematical scholars, and some were even sent abroad. The book was reviewed in the *Philosophical Transactions*, in March 1675, before Leibniz made his apparently independent discovery. It seems never to have been referred to during the dispute, as it would almost certainly have been *prima facie* evidence, far superior to the anagram that Newton had concealed in his letter to Leibniz.[48] But Oldenburg, Collins and Sherburne were already dead when the dispute reached its climax. Whiteside, in his edition, of Newton's *Mathematical Papers*, published Sherburne's statement in his commentary, but seems to have missed its real significance.[49] Westfall, perhaps influenced by Whiteside, also refers to it without giving it the special emphasis it clearly deserves.[50] Whiteside's emphases, in fact, seem to have significantly affected subsequent comment, for, in addition, to stressing that Newton's work remained unpublished, he was also categorical in his opinion that the bulk of the *Principia* was not done by analysis, though its 'geometrical' theorems were often equivalent to the procedures of analytical calculus. This issue has become something of a separate area of dispute among modern historians although it would not have been considered so important among the warring seventeenth century mathematicians.

2.3. Analytic and Synthetic

As his later geometric writings suggest, Newton may well have considered his early analytical researches as less well-founded and less elegant than the synthetic methods he adopted for most of the *Principia*. He claimed that he had found many of his propositions first by analysis and then translated them to the more elegant synthetic form in writing the book, and he further claimed that this had also been an ancient practice. Many historians have assumed that this was a polemical point used in making a case against Leibniz and have stated that he almost certainly derived the results exactly as they are written down in the book. However: 'A study

---

[48] *The SPHERE of M. Manilius made an English Poem, with Annotations, and an Astronomical Appendix: By* Edward Sherburn, *Esq., Philosophical Transactions*, vol. 9, no. 110, March 1675; abridgment, 1809, ii. 185–625 January 1674/75.
[49] *MP* III, 9–10.
[50] Westfall (1980, 258).

of the circulation of the mathematical manuscripts within Newton's circle shows that he revealed the fluxional analysis which stays behind some of the *Principia*'s demonstrations only to a selected group of acolytes'.[51] There are enough traces of the analytical method in the *Principia* to show that at least some (though very likely a minority) of the propositions were proved this way, and there are also cases of propositions (especially those relating to the Moon's motion) which exist in manuscript in analytical form, but which are represented synthetically in the published text. For Guiccardini: 'It was only to his disciples, who came to visit him in his rooms in Trinity, that he revealed the hidden fluxional analysis'.[52]

In fact, the *Principia* is almost entirely a work of calculus (however defined) and *was recognised as such*, with the Marquis de l'Hôpital declaring in 1696 that, 'lequel est presque tout de ce calcul' ('nearly all of it is of this calculus').[53] The modern scholar Clifford Truesdell, though in some respects hostile to Newton, calls it 'a book dense with the theory and application of the infinitesimal calculus'.[54] Its 'geometry' is certainly not classical geometry, and nearly always incorporates a limiting process, and this is the method used in most of the theorems. It is not 'calculus' in the algebraic sense, but it is calculus in the sense that all processes in which a tangent or area becomes exact as a limit is reached can be so defined, and it links directly to Newton's geometrical method of first and last ratios, given in a detailed series of lemmas in the very opening section of the book. It is also calculus in the sense that it concerns the attempt to find positions of objects at different times from laws which are only concerned with the *rates of change* of positions with respect to time. Most of its propositions are directly translatable into algebraic calculus, as numerous subsequent commentators have found. Certainly, none of Newton's most important contemporaries and successors, including Bernoulli, Euler, Lagrange and Laplace, had the slightest doubt that Newton's method used calculus even if they thought Leibniz's version was superior.[55]

---

[51] Guicciardini (2003b, 413).
[52] Guicciardini (2003b, 415).
[53] Marquis de l'Hôpital, *Analyse des Infiniment Petits*, Paris, 1696, preface. Guicciardini (2009b) says that the passage is now believed to have been written by Bernard de Fontenelle.
[54] Truesdell (1968a, 99).
[55] Bernoulli's challenge to *Principia*, II, Proposition 10 wasn't based on the idea that it didn't use calculus but that Newton's calculus was inferior to that of Leibniz. For Smeenk and Schliesser (2013, 149), 'The contrast between 'the calculus' and Newton's geometrical

But there are also more conventional approaches to calculus present in the *Principia*, some quite explicit. In Book I, Proposition 41 uses the 'instantaneous velocity quotient of an infinitesimal increment of arc and an infinitesimal element of time', and states that quadrature is required, while Section I, as we have indicated, uses first and ultimate ratios, outlined in a series of eleven lemmas, while stating in the accompanying Scholium, 'hereby the same thing is performed as by the method of indivisibles', which he considers 'somewhat harsh'. Quadrature is also in Book I, Propositions 39, 46, 53, 54, 56 and 93, and Corollary 1 to Proposition 91, and in Book II, Proposition 51. Proposition 45 of Book I uses the binomial expansion in considering the motion of apsides in perturbed elliptical orbits, while the Scholium to Proposition 93, in the converging series expansion of $(A+o)^{m/n}$, uses a combination of fluxions ('moments') and infinite series.

In Book II, Lemma 2 is an exposition of fluxions, the use of fluxions and infinite series, and the attached Scholium refers to the general method, besides claiming priority for the invention of calculus by quoting a letter to John Collins of 10 December 1672, by which he had made it known, and a treatise he had written in 1671, *De methodis serierum et fluxionum*. Proposition 11 uses fluxions, while the conclusion following Proposition 18 in Section IV refers to the fluxionic methods of Proposition 10 and Lemma 2 as the source of 'managing' the immediately preceding problems.[56] Proposition 14 makes explicit use of moments as well as limits, fluxions and infinite series, while infinite series also occur in Propositions 18, 21, 22 and the Scholium. In Book III, Lemma 2 (just before Proposition 39) again uses fluxions; Proposition 28 calculates a radius of curvature, while quadratures are required to complete the analysis in many places, including Propositions 26–35.

Much has been made of the fact that the *Principia*'s propositions were probably found in the form there published. But we don't actually have the evidence to say that Newton worked out his results exactly as they are written down. Hardly anyone ever does this. Newton himself claimed: 'I wrote the Book of Principles in the years 1684, 1685, & 1686, & in writing it made much use of the method of fluxions direct [derivatives] & inverse

---

reasoning is … not as sharp as the question' ['Did Newton use calculus in the *Principia*?'] 'presumes'.

[56] Guicciardini (2009a, 225) sees it as 'a notable attempt to summarize the rules of the direct method of fluxions in a compact and general form', probably in response to Leibniz, although there is a 'faulty demonstration' in Case 1 which is not found in *Geometria curvilinea* or the Addendum to *De Methodis*.

[integrals], but did not set down the calculations in the Book it self because the Book was written by the method of composition, as all Geometry ought to be ...'[57] Though this has been disputed ever since, it doesn't actually say that there was an analytical *version* of the *Principia* which was translated into synthetic form for publication. It says that Newton made his *discoveries* by analysis, possibly never writing them down in this form, which is a much more plausible, if greatly exaggerated, claim.[58] Various scholars have seemed to detect an analytical background to individual sections, as Ehrlichson has done for Sections I and II of Book II, which were originally part of Book I.[59] Problems that seem to have been particularly difficult, such as those connected with the lunar theory and the solid of least resistance (Scholium to Book II, Proposition 34), nearly always involve the direct use of analysis, suggesting how much value it was to Newton as a discovery method.

Certainly, some of the propositions did have hidden analytical derivations and some are referred to in *Principia*, but others aren't (for example, the lunar ones). A great many of Newton's demonstrations in the *Principia* have missing steps which can only be supplied by analysis. Sometimes we have documentary evidence of these, but at other times we don't. Because of these missing steps, disputes have arisen among scholars and mathematicians about how far Newton's analyses went, but some of the best informed interpreters have detected their presence in Newton's work, and it would be a mistake to assume that they never existed. His mixed methods show a degree of pragmatism under the exterior of rigorous geometry, in line with his special skills as a natural problem solver. The outright denial of Newton's claims of an analytical background doesn't take into account how much creative mathematicians work by instinct. The creative mathematician-physicist commentators can see the connections — the 'scholars' sometimes cannot. The creative interpreters don't always need to see the details to see how another creative thinker operates. A creative mathematician can often

---

[57] Draft letter to Des Maizeaux, c 1718, Cohen (1971, 295).
[58] The spurious 'lost' analytical version of *Principia* has become something of a 'straw man' based on a misinterpretation of what Newton actually said about his methods. Guicciardini (1998) refers to the 'Costabel thesis' (Costabel, 1967) as implying that Newton did not use the formulation and solution of differential equations in the *Principia*, an extreme view that became widely accepted, but was countered by Whiteside (1970). Guicciardini's article gives strong support to the Whiteside view. He states quite clearly that: 'The analysis of Newton's manuscripts and of the texts of some of his disciples show that the Costabel thesis is wrong' (333).
[59] Ehrlichson (1991a, 1991b, 1996).

see how another one works even when steps are missing. Almost certainly, there is no missing 'analytical text' of the *Principia*, but this does not mean that Newton must only have used geometrical methods, when that is all that survives. We should also avoid the argument that says that Newton's propositions should not be explained by using analytical forms which are more familiar to us today than the synthetic ones in the published text, as the analytical mathematics was always available to Newton and he would certainly have seen the connections.

In general, besides the issues relating to its status as analysis, the *purely* geometric nature of the *Principia* has been consistently over-emphasized.[60] A glance through the book will reveal that it is also full of algebraic expressions. Contrary to claims that are sometimes made, Newton, from the beginning, wrote down dynamical equations in algebraic form. His earliest texts on algebraic calculus, some of which were published in his lifetime, begin with the attempt to express dynamical motion.[61] The dynamical writings in the Waste Book (CUL Add MS 4004) from the same period are largely couched in algebraic form.[62] The algebraic lectures of 1673–1683, published as *Arithmetica Universalis* in 1707, contain a distinctly algebraic treatment of the collisions of elastic bodies, by contrast to those of Wren and Huygens, which are based on diagrams.[63] A Scholium intended to follow Proposition 17 in a revised version of *Principia*, Book I, which may date to the years from soon after the publication of the first edition,[64] has a seemingly geometric treatment of a curve of the form $ax = v^n$, where a is a constant, x is an abscissa representing a freely flowing time and v is a varying ordinate, in which Newton explicitly states that 'the method of fluxions will give' a calculation which, after two differentiations of the function, leads to an expression for the acceleration:

$$\ddot{v} = -\frac{(n-1)}{v}\dot{v}\dot{v}.$$

---

[60] A book review by Palmieri (2004), in order to emphasize the supposed superiority of Euler's algebraic methods, makes the incredible claim that the *Principia*, is 'a work of geometry wholly in the style of Euclid's *Elements*'. Palmieri also uses the phrase 'Leibnizian calculus', as though calculus, as applied to physics, must necessarily be 'Leibnizian', a fact contradicted by the work of Maclaurin and many others, as well as by that of Newton.
[61] This is especially true of the first extensive calculus tract, the unpublished, 'To Resolve Problems by Motion', October 1666, *MP* I, 400–448.
[62] Herivel (1965), 128–182.
[63] See, for example, Problem 12, *MP* V, 148 and 150.
[64] CUL MS Add 3965 ff, 4^v–5^r; 5^r–6^r. *MP* VI, 588–593. As 'Calculations of Centripetal Forces', *Unp.*, 65–68, translation, 68–71.

For the particular case in which $n = 1/2$ and the acceleration is constant, this leads to the familiar kinematical equation for uniform acceleration:

$$\text{velocity}^2 = 2 \times \text{acceleration} \times \text{distance}.$$

Whiteside points out that the Scholium incorporates a statement 'in all its generality' of 'the fluxional measure of a central force', 'long before its rediscovery in equivalent form by Varignon, de Moivre and Johann Bernoulli'.[65]

The manuscript interestingly shows Newton switching over to algebra and fluxions for his convenience at a crucial point in the calculation, just as he would very likely have done on numerous occasions when composing the *Principia*. Such a switch does not seem to have been considered a big issue by Newton's contemporaries and immediate successors, who appear to have performed such operations on a regular basis. The attempt to separate out 'analytical' and 'geometrical' approaches to dynamics is anachronistic in the late seventeenth century context, as they were widely recognised as interchangeable and results were regularly translated from one to the other. The purely analytical method certainly won out in the eighteenth century, in the same way as the Leibnizian version of calculus, but it is a modern legend that this was necessary to develop a modern version of mechanics, and that anyone thought so before that time.

The idea that verbal, geometric and algebraic expressions somehow have a different status in representing physical relationships is a modern prejudice based partly on later ideas about the integrity of algebra as a description of an operation on symbols irrespective of meaning and on the fact that the algebraic style later became the dominant one. It was not one which dominated mathematical physics in the late seventeenth century. Consequently, the *Principia*, though strongly expressing Newton's preference for a geometrical approach, contains mathematics in a wide variety of styles, including algebra, several forms of calculus and even transcendental mathematics, and its readers would not have seen any great distinction between arguments which variously used expressions such as 'squared radius', OP^2 or r^2, or which used a verbal rather than an algebraic definition of force.

Again, the idea that 'modern dynamics' or even 'modern physics' begins with purely algebraic statements of dynamical equations has no basis in historical reality. This way of thinking seems to have been initiated by

---

[65] *MP* VI, 28.

Clifford Truesdell, one of Newton's more prominent and persistent critics, and recurs regularly among those authors who wish to emphasize the importance of Newton's successors in creating mechanics as we know it today.[66] However, the Newtonian origin of eighteenth century mechanics was proclaimed by its own proponents, and was not a product of nineteenth century British historical writing, as the more extreme revisionists have occasionally tried to imply. The lineage is clear in the works of Varignon, Bernoulli, Clairaut, Maupertuis, Euler, Lagrange and Laplace. There was no system of mechanics produced independently of Newton, however much it was subsequently developed or modified. In the eighteenth century it was impossible to ignore the *Principia*, and it would have been self-defeating to do so.

In addition, as we have said, Newton could and did write dynamical equations in an algebraic form at an early stage in his career, but the contemporaries who later translated his later geometrical and verbal statements into algebraic constructions did not consider this in itself to be a major feat, especially as the algebraic expressions in the *Principia* often made the connections fairly obvious.[67] They were merely translating the theorems into the form that their own mathematical work made most convenient for them. Of course, in the hands of Euler, especially, this led to major developments, but this simply meant that Euler, having absorbed the basic physical structure provided by Newton and his immediate successors, continued the process by selecting the style of mathematics from those available from the Newtonian period that he felt most suited his own post-Newtonian development.

Whether or not they quoted 'Newton's laws' in their works, as later writers on mechanics would do, Varignon, Johann and Jakob Bernoulli, Euler and many others were fully conscious of the fact that they were working in

---

[66] Truesdell (1960, 1967, 1968a, 1968b). It is important to record that Trusedell was writing before the significance of the 'conservation of energy' theorems, Propositions 39–41, was discovered, and before the true origin of Leibniz's *Tentamen* had been revealed. Truesdell did not deny Newton's historical importance, but he was over-zealous in denying him credit for certain discoveries and developments which later research has shown can be traced back to his work, and some people quote such relatively early writings as though they still have the authority that they seemed to have before such research was published. Some of Truesdell's equally negative comments on *Principia*, Book II are actually refuted by modern scientific and engineering practice.

[67] According to E. A. Fellmann, for instance (1988, 20, and 1973), Leibniz's annotations on his own copy of the *Principia*, discovered before his notes for the *Tentamen*, 'show that Leibniz did not have the slightest difficulty here in mastering Newton's arguments in detail'.

a line of development that stemmed directly from Newton's first systematic formulation of mechanics and its application to the study of the System of the World. And Newtonian mechanics was not only defined by Newton's laws. Many Continental mathematicians realised at an early date the significance of the new opportunities offered by Propositions 39–41 of the first book of the *Principia*, which, in many ways, presented an alternative and more powerful approach to mechanics, based on the conservation of energy. Whether acknowledged or not, many of Newton's results entered into the mainstream of Continental dynamical theory, in addition to results in the Newtonian tradition obtained by such successors as Cotes, Taylor and Maclaurin. Leibniz consciously appropriated major aspects of Newtonian mechanics in his *Tentamen*. Varignon quoted Newton in producing an algebraic version of his second law. Johann Bernoulli had the full precedent of Newton's work in developing his *vis viva* theorem from Newton's conservation of energy propositions, as well as corresponding with Newton on dynamics as his ostensible 'friend'. Maupertuis was purposely schooled by him in the Newtonian system. Clairaut was never anything but a Newtonian. Daniel Bernoulli, Johann's son, and Euler, another of Johann's pupils, defected from the Cartesian camp at a relatively early stage.

Laplace, in whom this tradition culminated, emphasized the significance of the *Principia* in 1823 as including, in particular, the many developments that had followed from it in the succeeding century and a half: 'This admirable masterpiece contains the germ of all the great discoveries that have since been made concerning the system of the world: The history of the developments by the successors of this great geometer would at the same time be the most useful commentary upon the Work and the best guide to the attainment of new discoveries'.[68] Laplace notably refers to Newton as a 'geometer', a class in which he clearly included both himself and his immediate predecessors in the Newtonian tradition.

2.4. The Dual Methods

There is, of course, a very definite sense in which Newton was a geometer much more than he ever was an algebraist. As Guicciardini has shown,[69] Newton, in his mature years, regarded algebraic analysis merely as a

---

[68] Laplace (1823), as quoted and translated by Fellmann (1988, 32). To this we may add Lagrange's often-quoted remark that 'Newton was the greatest genius who ever lived, and the most fortunate; for we cannot find more than once a system of the world to establish'. (Shank, 2008, 4).
[69] Guicciardini (2009a).

heuristic analytical tool whose purpose was to lead to the true geometrical construction, which represented the only certainty in nature. In his more extreme moments, as in the *Account*, he even alleged that Leibniz's calculus was incomplete as it showed the analysis but not the synthetic construction. He also only allowed his algebraical lectures to be published, as *Arithmetica Universalis* (1707), as part of an attempt to be re-elected as MP for Cambridge in 1705, and made absolutely no attempt to cooperate with the editor, William Whiston, in the process. As a result of his views on algebra, he downplayed some of the work on the direct method of calculus which might have been considered crucial in the dispute with his opponent, and, in the *Principia* and elsewhere, was reluctant to give details of the analytical methods he had used in deriving his results, even when no synthetic reconstruction was possible, thus leaving many gaps in the arguments to be filled in by his disciples or successors. One consequence of this was his conviction that a truer, geometrical method of analysis had been discovered and used by the ancients. This was a belief held by many of his contemporaries, including Fermat, Hobbes, Barrow and Huygens, and even Descartes and Leibniz, along with a supposition that synthetic geometrical constructions were the ultimate aim of analysis. Newton, however, used his method of first and last ratios to try to develop a synthetic method of fluxions which would extend to problems beyond the reach of any conceivable ancient analysis because they involved infinitesimal quantities.

There is a kind of logic in this from a physical point of view, for all geometrical constructions, as Newton observed, derive from mechanical processes, and all information in physics derives, ultimately, only from observations of changes of spatial coordinates against an assumed progression in time. Isaac Barrow's *Lectiones mathematicae* had also proclaimed that all 'physics' could be reduced to geometry.[70] However, Newton found that many of the gravitational problems he encountered in writing the *Principia*, which could involve multiple sources acting in three spatial dimensions, with perturbations preventing a completely analytic solution, couldn't be handled in this way, and his extremist viewpoint, denying an autonomous existence to the algebraic analysis which could handle them, is incompatible with modern approaches in which the algebraic method has become completely dominant, especially where the systems are highly complex.

It seems that, while recognising that algebra and geometry were dual methods, Newton refused to allow them dual status. This kind of attitude

---

[70] Barrow (1734, 21–27), quoted Guicciardini (2009a, 172).

has become a repeating pattern where Nature presents us with genuine dualities, such as waves and particles, or standard and non-standard analysis. In such cases, one approach has frequently been dominant for a considerable period to the complete exclusion of the other, but there is often no fundamental way in which we can predict which one will lead to the more efficient solution of future problems. Interestingly, the dominance of algebraic methods has not entirely ruled out geometrical approaches, for, in modern theorising about complex systems, the algebraic analyses are often used to provide computerised graphical simulations which give instant overall 'physical' information about the behaviour of the system being examined. This is particularly true in the case of particle accelerators where particle interactions are simulated graphically to give an instant impression of how both the machine and the algorithms are performing. Another modern example occurs with Feynman diagrams where the geometrical lines and vertices code for a mass of algebraic expressions. In these cases the algebraic part does become a 'hidden' algorithm just as Newton's analytical algorithms were in those cases where he was able to produce a synthetic geometrical reconstruction. To some extent he was able to do this with his work on cubic curves, but he had less success with the physical propositions of the *Principia*, where the achievement of a dual analytic/synthetic construction was usually only partial.

Since Newton's views on the relative status of analysis and synthesis — or, as they were also called, the methods of resolution and composition — had their ultimate origin in their relation to physical meaning, it is interesting that he made a point, in both *Principia* and *Opticks*, of extending the same categories to natural philosophy. Natural philosophy, however, posed different problems to mathematics. In mathematics, confidence could be achieved by a synthetic proof from an analytic development, but, in natural philosophy, synthetic developments could only ever approach towards an approximate description of reality, and this was always at least at one further remove from the fundamental laws which had been derived by analysis. It was not, therefore, possible to equate mathematical with physical analysis or mathematical with physical synthesis. Nevertheless, physical analysis had ultimately to be reduced to abstract relations between space and time (and even between absolute space and absolute time) without the interposition of hypotheses based on complex effects, which could only be derived by synthesis. Ironically, this meant that Newton was forced to privilege analysis in natural philosophy by creating a quasi-mathematical abstraction as its basis. While Hobbes and Barrow had seen the dilemma posed by the categories of analysis and synthesis, as introduced respectively for mathematics and natural philosophy by Pappus and Aristotle,[71] it was, characteristically, solved only by Newton, and then only by instinctive practice rather than by philosophical development. In view of its significance for the subsequent development of physics, however, it can be considered his greatest achievement.

2.5. Newton's Mathematical Legacy

The public dispute with Leibniz ended when Newton, from his position as President of the Royal Society, used the 'nuclear weapon'. Clearly, when Newton saw he had the opportunity — 'The Lord has delivered him into my hands' — he despatched it pretty quickly. One thing that is seldom emphasized, however, is that Newton had not begun the dispute and only entered it publicly in 1711, a full 35 years after his first correspondence with Leibniz, and this was only after Keill had pointed out that the statement made by Leibniz in his review of 1705 was effectively an accusation of plagiarism. Newton almost certainly had nothing to do with Fatio's denunciation of 1699, as Leibniz himself seems to have realised, or with George Cheyne's of 1703,[72] of which he and his supporters certainly disapproved, or even, necessarily, of Keill's of 1708, first published in 1710, as he and Keill were probably not then closely acquainted.[73] He only became involved after Keill showed him the passage from Leibniz, from 1705, which he couldn't leave unanswered, especially after Leibniz's further interventions in 1711. At this moment, everything must have seemed to be happening at once. Along with the accusation from 1705, there was a growing attack on Newtonian gravity and dynamics, continuing controversy over the *Opticks* and Newton's religious and philosophical beliefs, rumblings over Whiston's public announcement of his Arianism, and a dispute with Flamsteed over the publication of his astronomical data, as well as anxiety over the second edition of the *Principia* and the correction highlighted by Nicolaus Bernoulli.

Subsequent commentators have inferred on the basis of knowledge not available to Newton that Leibniz had developed his calculus quite independently of Newton's work, although Brown's work now makes this somewhat

---

[71] Guicciardini (2009a, 221–227) discusses the dilemma, citing Shapiro (2004), Ducheyne (2005), and various works by Hobbes and Barrow.
[72] Fatio (1699); Cheyne (1703). Leibniz's published reply to Fatio of May 1700 stated his belief that Newton had not been involved in the denunciation, and reaffirmed Newton's statement of 1687 that only he and Leibniz were the inventors of the calculus (Hall, 2002, 437–438, *MP* VII, 12, n. 4).
[73] Keill (1708).

less certain. Leibniz has been regarded by some as a more sympathetic figure and a more attractive character. He was certainly one of the great men of his age — a philosopher, outstanding mathematician and more than capable physicist — but his actions, though certainly understandable, given his situation, suggest that he brought much of the trouble that followed onto himself by not making a proper acknowledgement in 1684. Newton proved in the end to be the more ferocious and uncompromising warrior when faced with the accusation of 1705, but there seems to be every reason to believe that he wanted to avoid public dispute and tried to hold it off it as long as possible. It is perhaps ironic that Newton's avoidance of printing his mathematical work as a result of the prolonged dispute over his optics, had led to the biggest dispute of them all.[74]

The dispute over the calculus can only be regarded as a tragic episode, damaging to both parties, but it was also used from the early nineteenth century as a moral tale. The legend arose that Newton's methods dominated in Britain but were less effective than those of Leibniz, used on the Continent, the upshot being that mathematics and mathematical physics atrophied in Britain for the best part of a century. In fact, Newton's mathematical successors in Britain — Cotes, Stirling, de Moivre, Taylor, Maclaurin — were ahead of their contemporaries in many areas until about 1740. The late publication of Newton's early mathematical tracts in the early eighteenth century led rather to a revival in the algebraic calculus that Newton had generally avoided in the *Principia*.[75]

There was certainly a decline from the mid-century, but it was due to a whole variety of factors. It was also only a relative decline, and by no means an absolute one. British mathematicians were still making outstanding contributions to both mathematics and mathematical physics in the mid- and late eighteenth century. Names like Thomas Simpson, Thomas Bayes, John Landen, Edward Waring, John Michell, Henry Cavendish, John Robison and Robert Blair spring immediately to mind, and there were a considerable number of others. Some of their contributions are strikingly original and not paralleled by equivalent work on the Continent. In addition to a number of results that are now well known in pure mathematics, statistics, gravitational theory and optics, there were also unique analyses

---

[74] All major works on Newton contain discussions of the dispute, and there are two accounts by Hall (1980, 2002). The succinct review by Blank (2009) is as good a summary as any, but the book it reviews, by Bardi, is described by Blank as 'appalling' (610).
[75] Guicciardini (1989) has an extensive account of Newtonian calculus in the eighteenth century.

of dynamical systems that would now be considered nonlinear or chaotic, originating ultimately in Newton's work on resisted motion. For such difficult physical problems, Newton tended to use geometry to arrive at approximate results but his successors found their answers using the Newtonian fluxional calculus.[76] Powerful though the analytical methods were in many areas of dynamics, they were not well-suited to handling the dynamics of nonlinear oscillators, where the conditions were no longer steady-state and variables were frequently cross-coupled and could not be separated. The fluxionists were able to find solutions in these cases using partial trajectories without needing to establish closed curves or definite integrals. This led to a treatment of dynamics which was more like that of Poincaré at the end of the nineteenth century than that of Lagrange and Laplace a century earlier.[77] However, there was no great central tradition to unify the isolated contributions, and there was no figure of the calibre of Euler, Lagrange or Laplace.

Things started to go wrong with Bishop George Berkeley's ill-timed and religiously-inspired attack on the foundations of the calculus in *The Analyst* (1734), addressed to an 'Infidel Mathematician' (the free-thinking Edmond Halley).[78] Rather than further developing its consequences as the immediate post-Newtonian generation had done, mathematicians put their greatest efforts into re-examining its foundations, and many retreated into less productive work in Euclidean geometry. Maclaurin looked for inspiration from Newton's attempt to make calculus more rigorous using geometrical methods (1742).[79] Newton's quasi-geometrical style very probably had the same potential as the analytic, but perhaps only in the hands of exceptional masters like Newton and Maclaurin. On the Continent, Newton's theorems in the *Principia* were translated into more regular algebraic calculus and led to major new developments in both mathematics and physics.

---

[76] This line of thinking had an early practical consequence in Harrison's work on chronometers, but was continued throughout the century (Hobden, 1981–1982, 2011, Harrison, 1988). The abandonment of the Newtonian calculus in the early nineteenth century led to its subsequent neglect, but a thorough analysis of the sources might be a potentially significant research project.

[77] I am grateful to Mervyn Hobden for drawing my attention to this unique body of work (Hobden, 1981–1982, 2011, Harrison, 1988).

[78] Berkeley was not too happy that Newton had chosen to write on prophecy and theology in the *Observations* of 1733 but he denied that this had motivated his attack on Newton's mathematics (Mandelbrote, 2016, 565).

[79] Maclaurin (1742).

As a result of their isolation, the British mathematicians missed out on the Continental 'revolution' of the 1740s introduced with multivariate analysis and the calculus of variations.[80] It has often been claimed that the Newtonian dot notation was less convenient than that of Leibniz. This is arguable, but the Newtonian notation could still have been used to develop multivariate techniques. Newton himself had worked in similar areas, and John Craig had shown how easy it was to translate between the two systems.[81] Modern calculus, of course, still uses the very elegant notation of Leibniz, but Newton's dot notation (begun early but systematised later in a significantly different way) is also used by physicists who are particularly interested in variations in space and time, and it is especially convenient where x dot becomes a parameter as much as x. It is used especially in the Euler–Lagrange equations and the calculus of variations, which lie at the basis of a great deal of contemporary physics.

Alternative methods are often like alternative versions of technologies. When one becomes dominant in the market-place, it becomes unproductive to follow the other, even if it is technically superior. The mathematical methods of the Continental followers of Leibniz won out in the mathematical market-place and, because they are now more familiar to us, it is tempting to assume that this was because of some innate superiority of notation, or algorithmic procedure, but this is far from established. It is also by no means established that the geometrical methods of Newton's later years were incapable of being extended to the advances that took place in the later eighteenth century. Donald Lynden-Bell, reviewing Chandrasekhar's *Newton's Principia* in an article entitled 'The Wonderful Geometry of Dynamics', spoke of Newton's 'more insightful geometrical methods', and compared the 'relative dullness of the more modern manipulative methods with the greater

---

[80] Guicciardini (1989, 79–81).
[81] Smeenk and Schliesser (2013, 149), point out that the work of both men 'lacked various key concepts, such as that of a function, introduced by Euler' and warned against 'project[ing] back Euler's work onto Leibniz'. It is certainly true that the Leibnizian notation, conceived without any influence from Newton, was in itself a seriously non-negligible achievement and served to emphasize the distinctiveness of the calculus as a new branch of mathematics, a central tenet which distinguishes Leibniz's philosophical from Newton's apparently more utilitarian approach. In effect, Leibniz was more 'modern' than Newton in making calculus an almost automatic consequence of the use of notation, while Newton was more 'modern' than Leibniz in his closer approach to the idea of a function.

insight generated by geometry'.⁸² Roger Penrose's review of the same book says that Chandrasekhar shows that, 'in most cases Newton's geometrical methods are not only more concise and elegant, they reveal deeper principles than would become evident by the use of those formal methods of calculus that nowadays would seem more direct'.⁸³ V. I. Arnol'd makes similar comments 'Comparing the texts of Newton with the comments of his successors, it is striking how Newton's original presentation is more modern, more understandable and richer in ideas than the translation due to commentators of his geometrical ideas into the formal language of the calculus of Leibniz'.⁸⁴

Such comments stand in stark contrast to Whewell's of 1837, which essentially reflected the efforts of the Cambridge Analytical Society to introduce Leibnizian notation and Continental analysis into Britain in the early nineteenth century at the expense of anything connected with the mathematical methods of Newton:

> The ponderous instrument of synthesis, so effective in his [Newton's] hands, has never since been grasped by [any]one who could use it for such purposes; and we gaze at it with admiring curiosity, as on some gigantic implement of war, which stands idle among the memorials of ancient days, and makes us wonder what manner of man he was who could wield as a weapon what we can hardly lift as a burden'.⁸⁵

The legend of the impenetrability of Newton's mathematics, which had begun during his own lifetime, was clearly dominant by the early nineteenth century when it was removed from the Cambridge syllabus, and it certainly remains potent today despite Chandrasekhar's belief that he could expound it for the 'Common Reader'.

Newton gave a clear lead to his successors in both algebraic and geometrical calculus. He anticipated in his own work many of the advances of the later eighteenth century and even some of the nineteenth and twentieth. His work inspired a whole generation of mathematicians in Britain who were more than equal to their Continental contemporaries. It was only with the generation of Euler and Maupertuis in the 1740s that they began to be overtaken. By that time the British mathematicians seem to have become

---

⁸²Lynden-Bell (1996, 253–254).
⁸³Penrose (1995).
⁸⁴Arnol'd (1990).
⁸⁵Whewell (1837, 167), quoted in *MP* VI, 28. Whewell had himself been influential in promoting the Leibnizian notation in his *Elementary Treatise on Mechanics* (1819), two years after it had first appeared in a Cambridge examination.

occupied with the problem of the foundations of the calculus rather than developing it for more practical utility as had happened on the Continent. This development started after Newton's time, and had little to do with the intrinsic utility of either his system or that of Leibniz, but, in the sense that their dispute forced Britain to become isolated from the rest of Europe, the relative failure of British mathematics in the later eighteenth century was ultimately a product of the failure of Newton's contemporaries to understand his work in optics.

Chapter 3

Thermodynamics

3.1. Heat as Motion

The most controversial of Newton's innovations in his own time were those considered the most significant today — the theory of colours, universal gravitation and the calculus — but a mass of equally innovative ideas in other areas passed unnoticed in low key publications or were deliberately held back from publication to avoid controversy. Some would now be considered part of *thermodynamics*, a subject that did not exist until 150 years after Newton's time. Newton's contributions included anonymously published experimental work and theoretical underpinnings in dynamics that caused a controversy, conducted by proxy, but once again involving Leibniz.

Thermodynamics, as we now know it, was created in the mid-nineteenth century in response to technological advances of the late eighteenth and early nineteenth centuries, but some earlier results were significant for its foundations. A well-established tradition held that heat was due to the rapid motion of the *parts* of bodies. A dynamical theory of heat had been held in the Middle Ages by such as Roger Bacon, then, in more recent times, by Francis Bacon, Hooke and Boyle, as well as by Newton. It would later be taken up by Rumford and Davy, among others.[1] Newton's comments in *De natura acidorum*, 1691/92, as taken down by Pitcairne, suggest that there is a degree of randomness in the motion, and that this should be linked with the property of heat in causing rarefaction, especially in the gaseous state: 'Heat is the agitation of particles in every direction'.[2] He certainly believed that all matter was in motion, stating that particles would only be at rest if they were also absolutely cold. According to a draft Query: 'Heat

---

[1] Bacon (1897–1900); Bacon (1620), Book II; Hooke (1665), Observation 7, p. 37; Boyle (1665); Rumford (1798); Davy (1799).
[2] *De natura acidorum*, Corr. III, 210.

consists in the trembling agitation of the smallest parts of bodies all manner of ways; & the parts of all bodies are always in some agitation'.[3]

At various times, Newton stated that heat is the motion of material particles and is produced mechanically; magnetism is affected by heat and is created by mechanical means; light causes heat in bodies by causing a vibrational motion of their particles, and light itself is produced from the vibrating parts of bodies; hot bodies produce light, but light is also produced mechanically, by friction, and chemically, by putrefaction; this also connects with his view that mechanical friction produces electricity, and frictional electricity, in Hauksbee's globe experiment, produces light.

Heat, as the motion of material particles, must be produced mechanically, and can be created or lost, because mechanical motion can. So new motion can be generated, for example, in earthquakes and storms, as well as in chemical processes:

> Now the above-mention'd Motions are so great and violent as to shew that in Fermentations the Particles of Bodies which almost rest, are put into new Motions by a very potent Principle, which acts upon them only when they approach one another, and causes them to meet and clash with great violence, and grow hot with the motion, and dash one another into pieces, and vanish into Air, and Vapour, and Flame.[4]

Clearly, new motion will be created if heat is produced by mixing two cold materials.

Heat can also be generated by the absorption of light, just as light can be generated by heat: 'light & heat have a mutuall dependance on each other & noe generation without heat. Heat is a necessary condition to light & vegetation. Heate exites light & light & light exites heat, heat excites the vegetable principle & that increaseth heat'.[5] In Query 5, which questions whether bodies and light mutually interact, the heat of bodies is assumed to consist of a vibratory motion of their parts: 'Do not Bodies and Light act mutually upon one another; that is to say, Bodies upon Light in emitting, reflecting, refracting and inflecting it, and Light upon Bodies for heating them, and putting their parts into a vibrating motion wherein heat consists?'

The heating process is explained in the *Hypothesis of Light*. Here, also, light causes heat in bodies by causing a vibrational motion of their particles, and light itself is produced from the vibrating parts of bodies.

---

[3] CUL, MS Add. 3970.3, 239^r, NP.
[4] Q 31.
[5] *Of Natures Obvious Laws*, NPA.

For its plaine by the heat wch light produces in bodies, that it is able to put their parts in motion, & much more to heat & put in motion the more tender æther; & its more probable, that it communicates motion to the gross parts of bodies by the mediation of æther then immediately; as, for instance, in the inward parts of Quicksilver, Tin, Silver, & other very Opake bodyes, be generating vibrations that run through them, then by Striking the Outward parts onely without entring the body. The Shock of every Single ray may generate many thousand vibrations, & by sending them all over the body, move all the parts, and that perhaps with more motion then it could move one Single part by an Imediate Stroke: for the vibrations by Shaking each particle backward & forward may every time increase its motion, as a Ringer does a bells by often pulling it, & so at length move the particles to a very great degree of agitation wch neither the Simple Shock of a ray nor any other motion in the æther, besides a vibrating one, could do. Thus in Air shut up in a vessell, the motion of its parts causd by heat, how violent soever, is unable to move the bodyes hung in it, with either a trembling or progressive motion; but if Air be put into a vibrating motion by beating a drum or two, it shakes Glass windowes, the whole body of a man & other massy things, especially those of a congruous tone: Yea I have observed it manifestly shake under my feet a cellar'd free stone floor of a large hall, so as I beleive the immediate Stroke of five hundred Drum Sticks could not have done, unless phaps quickly succeeding one another at equal intervals of time. Æthereal vibrations are therefore the best means by wch such a Subtile Agent as Light can Shake the gross particles of Solid bodyes to heat them.[6]

De vi electrica (c 1712) describes the process in the language Newton had used for many years in connection with optical vibrations:

These vibrations are swifter than light itself and successively pursue and overtake the waves, and in pursuing them they dispose the rays into the alternate fits of being easily reflected or easily transmitted; and when excited by the light of the Sun they excite the particles of bodies and by this excitation heat the bodies; and being reflected at the surfaces of bodies they turn back within the bodies and persist a very long time.[7]

The internal reflections from any surface of a hot body would continue until another surface was brought into contact with it, when conduction would occur.[8]

At the same time, the reverse process occurs when hot bodies produce the emission of light, as they reach a kind of critical point at which they automatically heat themselves by emitting light (a true phenomenon).

---

[6] *Corr.* I, 374.
[7] *Corr.* V, 365–366.
[8] Draft Conclusion to the *Principia*, translated in Cohen and Whitman (1999, 292).

54 Newton — Innovation and Controversy

> Do not great Bodies conserve their heat the longest, their parts heating one another, and may not great dense and fix'd Bodies, when heated beyond a certain degree, emit Light so copiously, as by the Emission and Re-action of its Light, and the Reflexions and Refractions of its Rays within its Pores to grow still hotter, till it comes to a certain period of heat, such as is that of the Sun?[9]

Newton, who had originated the study of spectroscopy, also indicated that spectra probably had an origin in dynamical processes:

> Do not all fix'd Bodies, when heated beyond a certain degree, emit Light and shine; and is not this Emission perform'd by the vibrating motions of their parts? And do not all Bodies which abound with terrestrial parts, and especially with sulphureous ones, emit Light as often as those parts are sufficiently agitated; whether that agitation be made by Heat, or by Friction, or Percussion, or Putrefaction, or by any vital motion, or any other Cause? As for instance; Sea-Water in a raging Storm; Quick-silver agitated in *vacuo*; ...Wood, Flesh and Fish while they putrefy; Vapours arising from putrefy'd Waters, usually called *Ignes Fatui*; ...the vulgar *Phosphorus* agitated by the attrition of any Body, or by the acid Particles of the Air; Amber and some Diamonds by striking, pressing or rubbing them; Scrapings of Steel struck off with a Flint; Iron hammer'd very nimbly till it become so hot as to kindle Sulphur thrown upon it; the Axletrees of Chariots taking fire by the rapid rotation of the Wheels; and some Liquors mix'd with one another whose Particles come together with an Impetus, as Oil of Vitriol distilled from its weight of Nitre, and then mix'd with twice its weight of Oil of Anniseeds.[10]

Internal motions are responsible for the emission of light. 'Accordingly, glowing bodies seem to be distinguished from non-glowing ones solely by the motion of their internal parts. For bodies grow hot by motion and always shine by intense heat. Thus iron grows warm under the hammer, and can be so agitated by hammering that at length it glows'.[11] Queries 9 and 10 ask: 'Is not Fire a Body heated so hot as to emit Light copiously?' and: 'Is not Flame a Vapour, Fume or Exhalation heated red hot; that is, so hot as to shine?'

> For glowing coals and glowing exhalations are fire and flame, and to glow is nothing else than to shine by vehement heat, as one may see in glowing metals and stones. Flame does not differ, however, from glowing coal otherwise than the ignes fatui (which are putrid vapours) do from shining

---
[9] Q 11; Bernal (1972, 234).
[10] Q 8.
[11] Draft *Conclusio*; *Unp.*, 343.

putrid wood. The latter shine from the agitation due to putrefaction, the former from the agitation due to heat.[12]

The strength of the emission is determined by the 'vehemence' of the 'agitation'.

> From the largest particles, then, sensible bodies are formed which allow light to pass through them in all directions, and differ markedly among themselves in density, just as water may be 19 times rarer than gold. The particles of such bodies can very easily be agitated by a vibrating motion and this agitation may last a long time (as the nature of heat requires); through such agitation, if it is slow and continuous, they little by little alter their relative arrangement, and by the force by which they cohere they are more strongly united, as happens in fermentation and the growth of plants... If however that agitation is vehement enough, they will glow through the abundance of the emitted light...[13]

Heat is produced also by electricity, which generates the same kind of vibration in matter of which heat consists, for mechanical friction produces electricity, and frictional electricity, in the globe experiment, produces light: 'That the electric spirit is set in motion by light and this is a vibratory motion, and because of this motion heat exists'.[14] The emission of light leads to heating as a result of radiation reaction. As he knew from his own observations, even magnetism is affected by heat and is created by mechanical means.

3.2. Transfer of Heat

Newton has significant comments to make on all the standard methods of heat transfer. Conduction, for example, is due to vibratory motions being passed along the particles of bodies from end to end and between those of contiguous ones:

> And in the same way that in air that vibratory motion of which sound consists is very quickly propagated by the forces of non-contiguous particles... so a vibratory motion can be very quickly propagated in solids by the forces in non-contiguous particles. For which reason heat soon spreads through the whole of a body, and a hot body quickly communicates its heat to water in which it is immersed.[15]

---

[12]Draft *Praefatio*; *Unp.*, 307.
[13]*ibid.*, 306–307.
[14]Proposition 7, draft for General Scholium (MS C); *Unp.*, 361–364.
[15]Draft *Conclusio*; *Unp.*, 345–346.

> ...in a very long piece of wood a vibratory motion caused by percussion is very rapidly propagated from end to end. And just as that vibratory motion of which sound consists is as well propagated through wood and other long solid bodies by transmission through their contiguous parts as through air by the transmission through the forces of non-contiguous particles; so the vibratory motion of which heat possibly consists can be propagated as well through the forces of non-contiguous parts as through the impulses of contiguous ones.[16]

'... because dense Bodies conserve their Heat a long time, and the densest Bodies conserve their Heat the longest, the Vibrations of their parts are of a lasting nature, and therefore may be propagated along solid Fibres of uniform dense Matter to a great distance...'[17]

The most extensive account of the conduction process is in a letter intended for Flamsteed: 'The Sun will heat the matter within it as certainly as melted Lead would heat an Iron bullet immersed in it. An Iron bullet would heat as fast in a Quart as in an ocean of melted Lead, this difference only excepted, that ye bullet would cool a small quantity of Lead more than great one'. 'If then ye liquid matter swimming on ye Sun be but so thick as not (as to be cooled by ye central parts (as it must be), it will certainly heat ye central parts for it imparts heat to ye contiguous matter as fast as if it were thicker and keeps off all cool environing mediums (the instrument of cooling things) from coming neare ye central parts to cool them'.[18]

He makes a number of incidental references to the transfer of vibrational heat energy from a hot body to a cold body always with the implication that it is a one-way process. In *Opticks*: 'And do not the Vibrations of this Medium in hot Bodies contribute to the intenseness and duration of their Heat? And do not hot Bodies communicate their Heat to contiguous cold ones, by the Vibrations of this Medium propagated from them into the cold ones?'[19] In *De vi electrica* (c 1712): 'But when the hot body is immersed in a cold one they [the vibrations] pass out of the hot body into the cold one and very quickly communicate the heat to it'.[20]

Convection occurs in his explanation of the formation of comets' tails, where convection currents of aether particles are brought about by radiation

---

[16] Draft *Conclusio*; *Unp.*, 346–347.
[17] Q 11.
[18] Newton to Crompton, ? 7 April 1681, *Corr.* II, 358–362, 359 (note 5).
[19] Q 19.
[20] *Corr.* V, 365–366.

pressure.[21] The convection current is, in fact, a very simple principle but it is important in geophysics and oceanography. *Forced* convection is an important aspect of Newton's law of cooling.

Heat radiation occurs in numerous places in his works. In *De systemate mundi*, he say: 'from the experiment with the burning-glass we see that the heat increases with the density of light'.[22] Kepler had earlier found that light intensity followed an inverse square law, which is referred to by Newton in the manuscript treatise *Phaenomena*: 'And since in removing from a lucid body the light thereof decreases in a duplicate proportion of the distance...'[23] In the *Principia*, he applies the same law to heat: 'the heat of the Sun is as the density of its rays, that is, reciprocally as the square of the distance of the places from the Sun',[24] clearly indicating a direct relation between the quantity of light absorbed and the quantity of heat produced. In practical terms, Newton constructed a burning mirror in 1704: 'with a possible maximum radiant intensity of 460 W cm^{-2} or thereabouts, his device was certainly brighter thermally than a thousand Suns ($1{,}000 \times 0.065$ W cm^{-2}) and in achieving that he might not have been surpassed until 1945'.[25] That is, Newton may have produced the greatest intensity of radiation brought about by human agency before the arrival of nuclear weapons in 1945.

One of Newton's discussions of the wave theory produced a notable link between the vibrations of light and those of radiant heat. In his permanent quest to understand the properties of any potential aether, Newton was particularly interested in comparing the respective efficiencies by which radiant heat was transferred in vacuum and air, and Query 18, written in 1706, describes experiments made as early as 1692–1693. The main result, however, came from the experiment performed by Desaguliers using two thermometers to compare rates of cooling in air and in vacuum and intended to show that heat was almost equally transmitted through the two media. Newton designed the experiment and ordered Desaguliers to carry it out in his presence, late in 1716. Newton says in his manuscript account, that 'Within two tall & cylindricall Glasses hollow & closed at the upper

---

[21] *De Systemate Mundi*, § 69 (In what manner the tails may arise from the atmospheres of their heads).
[22] *De Systemate Mundi*, § 17.
[23] *Phaenomena* (MS), *Unp.*, 381.
[24] III, Proposition 41.
[25] Simms and Hinkley (1989). *Brighter than a Thousand Suns* was the title of a book by Robert Jungk (1958) on the production of the first nuclear weapons in 1945.

end, I caused two short Thermometers to be hung from the tops of the glasses...' (Obs. I), implying that it was Newton, as President, who devised the experiment and explained to the Demonstrator how it should be done. According to the Journal Book:

> Mr. Desaguliers Shew'd an Experiment to prove that a Body in a Vacuum of Air was susceptible of heat and cold; by putting two Thermometers of equal Lengths and equally graduated into two tall glass Receivers of equal Contents. The one being exhausted of its Air and the other not, upon bringing the said glasses near the fire, the Spirit rose in the Thermometer that was in vacuo almost as much as the Thermometer that was in pleno; only that in pleno rose something faster for it rose 56 degrees or tenths of an Inch, whilst the Thermometer in the exhausted glass rose only 50. When the glasses were removed from the fire, the Liquor in the Thermometer in the Glass full of Air subsided faster than in the other, much in the same proportion.[26]

When Desaguliers appeared with this demonstration before the Royal Society on 15 November, the minutes of the Society record that,

> The President said that the Experiments must also be try'd where there is heat without Light [so corpuscles could not easily pass through the vacuum and by their impacts warm the fluid in the thermometer] and Desaguliers reported that he had try'd it in a medium so affected namely by holding both the Glasses near a hole where heated Air is convey'd into a Room by means of Iron Cavities round about the fire and the Thermometer in that case rose in vacuo much after the same manner. He further promised to try the Experiment again with a great deal of Care, putting Cement instead of wet Leather under the exhausted Glass because the wet Leather throws a dew upon the Ball of the Thermometer.[27]

The vibrations of heat were thus shown to be transmitted in the same way as light but not to be the same as light, and, significantly, Newton described heat as the motion in bodies caused by this (presumably invisible) vibration, rather than as the substance of the vibration itself. The experiment apparently inspired Newton into incorporating a new version of a universal aether into the edition of *Opticks* he was then preparing for publication. He asked: 'Is not the Heat of the warm Room convey'd through the *Vacuum* by the Vibrations of a much subtiler Medium than Air, which after the Air was drawn out remained in the *Vacuum*?'[28]

---

[26] Guerlac (1967, 49); Simms and Hinkley (1989).
[27] Royal Society Journal Book, XII, 142–143, 15 November 1716; Simms (2004, 75).
[28] Q 18.

In another context, Chandrasekhar, an expert in the area, noticed that *Principia*, I, Proposition 93, Case 2, and Corollary 1 were examples of the use of a 'principle of invariance', 'used extensively in modern treatments of problems in radiative transfer'.[29] One other method of heat transfer known to Newton, though a rather obvious one, is through mass transfer, where moving an object, either hot or cold, from one place to another simultaneously acts as a means of transferring heat.

3.3. Temperature and the Law of Cooling

The dynamic understanding of heat also led to a dynamical definition of temperature and a temperature scale. Newton was one of the first investigators who understood that temperature, not previously distinguished from heat, was a separate concept, and he was more scientific than his predecessors in defining a temperature different from heat energy. He was one of the first investigators to extend thermometry beyond its uses in meteorology and medicine, and to create new types of thermometer, which could investigate phenomena outside the ranges of previous instruments. He was also instrumental in creating the idea of a temperature scale, mostly at the high temperature end, from which more practicable versions would follow soon afterwards. His contribution was expressed both mathematically and experimentally, and can justifiably be called thermodynamics (and even the first work in this field) because the method of temperature measurement he devised is based on dynamical principles. Significant to this development was the idea that a change of state did not require a change of temperature, certainly in the case of a freezing liquid: 'Cold & freezing have divers principles because a thing will freez wthout growing colder & grow colder wthout freezing. also because things freez not proportionably to their fluidity as in water oyle ☿ [mercury] sp[iri]t[us] V[ini] Cold is only rest, freezing is by an agent as fumes of lead coagulate ☿ [mercury]'.[30] This meant that the freezing point of water could be used as a fixed point on the scale.

An outgrowth from his extensive experimental work in alchemy, Newton's observation of the constancy of freezing and boiling points allowed him to put his ideas on temperature measurement into practice and create thermometers with a scale graduated between such limits. Newton may not have been completely certain that changes of state take place at fixed temperatures for some of the changes he observed occurred over small ranges

---

[29] Chandrasekhar (1995, 319).
[30] *Of Natures Obvious Laws*, NPA.

of temperature. His observations were recorded without comment.³¹ He also, for the first time, observed that some materials, including beeswax and various metal alloys, melted and solidified at different temperatures, though again the data were recorded without comment.³² In the 1680s, before the *Principia*, he measured temperatures using his own design of oil-in-glass thermometer, rather than the alcohol-in-glass instrument, then in more general temperature use. His instrument contained linseed oil or olive oil in a bulb 1.5–2 inches in diameter made of thick glass, and a tube which remained open except for a cotton plug which prevented the oil from falling out. He was able to use his thermometer to measure temperatures beyond the range of the alcohol-in-glass thermometer, in particular for an accurate value for the temperature of boiling water, which he referred to in the *Principia*.³³

From his experiments on defining a temperature scale, combined with his earlier theoretical work on resistance to moving bodies, he deduced a famous 'law of cooling', the first significant result in the theory of heat transfer, in which a body loses heat to its surroundings at a rate proportional to the difference in temperature between them: '...the heat which iron gives up to the cold bodies with it in a given time, that is, the heat which iron loses in a certain time, is as the whole heat of the iron, and so if equal times of cooling be taken, the degrees of heat will be in geometrical progression and can therefore easily be determined from a table of logarithms'. A probably more accurate statement of the law concerns the excess temperature ('heat') of a body over its surroundings: '...the excess of the degrees of the heat of the iron, and the particles above the heat of the atmosphere, found by the thermometer, were in geometrical progression when the times are in arithmetical progression...'.³⁴ The distinction which Newton is making may be 'between the total heat emitted (which would depend only on the temperature of the body itself) and the net heat lost (after subtracting

---

[31] Chang (2004, 2007), who quotes Newton's 1701 scale, says that variations in the boiling point of water were 'once widely known', and were extensively investigated by J.-A. De Luc and others in the eighteenth and nineteenth centuries. This variation may have been the reason why Newton did not use the boiling point of water as one of the fixed points on his scale (Simms, 2004, 36).
[32] Simms (2004, 43–44).
[33] *Principia*, III, Proposition 41.
[34] *Scala graduum caloris*, 267–268; Ruffner (1963); Cheng (2009). A logarithmic (or, equivalently, exponential) relationship is characteristic for a system in which the rate of change of a quantity depends on the quantity.

the heat received from surrounding bodies at a rate proportional to their temperature)'.[35]

The law seems to have had its origin in *Principia*, Book II, Propositions 1 and 2 where Newton developed equations for a body moving inertially with resistance proportional to the velocity, and there is a particularly close parallel with Proposition 2.[36] By analogy, a hot body should cool down at a rate proportional to its excess temperature over its surroundings. Mathematically,

$$\frac{dT}{dt} = -T,$$

for a body, with excess temperature T, reaching the temperature of its surroundings in time t. On integration, this yields

$$t = \ln T + \text{constant}.$$

So measuring time t gives us the original temperature excess. The times taken for a body to cool from one known temperature to another, as measured by Newton's linseed oil thermometer, and assuming a uniform coefficient of expansion, could then be used to construct a scale to measure temperature using cooling times.

One way to express the law is to introduce a proportionality constant k and the symbol T_∞ to represent the temperature of the surroundings. Then

$$\frac{dT}{dt} = -k(T - T_\infty)$$

Engineering textbooks rewrite 'Newton's equation' for the solid/fluid interface considered by Newton in terms of the rate of heat transfer dQ/dt, the difference in temperature between the solid 'wall' and ambient fluid $(T_w - T_\infty)$, the surface area of contact A, and a convective heat transfer coefficient h:

$$\frac{dT}{dt} = -hA(T_w - T_\infty)$$

They generally credit Newton with the concept of the coefficient.[37] With Newton's concept of uniform flowing time, the law implies that in any fixed

---

[35] Ruffner (1963).
[36] *ibid.*, The dynamical analogy is a significant result in its own right, leading to such things as the exponential law for the absorption of radiation.
[37] *Scala graduum caloris* (1701).

time interval Δt, the ratio $\Delta T/T$ will be constant, meaning that h is also a constant.[38]

In addition to providing the first theoretical model for heat transfer, the law introduced a completely new idea to temperature measurement, which allowed Newton to access very high temperatures beyond the boiling points of most thermometric liquids, and beyond those achieved by any previous researcher. His linseed oil-in-glass thermometer had already taken him beyond the usual temperature range, and could now be used in conjunction with a block of heated iron to go well beyond the temperature at which even this liquid boiled. In the *Principia* Newton writes that the '...heat of red-hot iron (if my conjecture is right) is about three or four times greater than the heat of boiling water'.[39] (Iron just glowing (depending on the viewing conditions) produces temperatures from about 3.5 to about 5 times that of boiling water.) Probably as early as this, he had worked out the method using the law of cooling and an iron slab at red heat, but at this stage had no thermometer accurate enough to be calibrated at high temperatures. When he finally solved the problems of calibrating his 'slab' thermometer a few years later, Newton was able to establish the assumption he had based on his dynamics as a 'fundamental principle', demonstrated by experiment.[40]

Newton defined 12 degrees of heat as the temperature difference between the freezing point of water (the ice point or temperature of melting snow) and body heat ('the external heat of the body in its natural state'). Assuming the uniform expansion of oil in glass, he calibrated an oil-in-glass thermometer to the point where he could use it with his 'pretty thick piece of iron' at red heat. Newton heated the iron slab until it became red hot and then placed on it various small pieces of different metals, including fusible alloys of tin, lead and bismuth, which immediately melted, and solidified only as the slab cooled down to the melting points of each. The slab seems to have been something like a rectangular bar of dimensions 2.45, 5 and 20 cm, and mass 1.93 kg. To calibrate this iron block he subjected it to the 'degree of heat of live coals in a small kitchen fire, made up of bituminous pit coals, and that burn without using bellows'.[41]

Recording the time when the slab started to cool, and the times at which each of the different metals solidified, and also the time at which

---

[38] Cheng (2009).
[39] III, Proposition 41.
[40] Ruffner (1963, 147).
[41] *Scala graduum caloris*, 267. Cheng (2009) gives the dimensions based partly on Simms (2004).

the block had cooled sufficiently for its temperature to be measured directly using the linseed oil-thermometer, he was able to apply the law of cooling to calculate the temperatures at which the various metals had solidified, and the temperature of the block at the start of the experiment.[42] (The cooling rate for the iron block depends on the background temperature, but this does not affect the actual temperatures measured.) Lead solidified after a little over 15 minutes cooling, a mixture of tin and bismuth in equal proportions set after approximately 30 minutes, while the iron slab cooled down to body temperature after about an hour. The linseed-oil thermometer showed that the tin and bismuth mixture solidified at 48°, and, with body heat at 12°, the temperatures of the other solidifications could be calculated. Also, once the iron calorimeter had been calibrated, the results obtained were then used to establish the validity of the assumption of the uniformity of expansion in the oil thermometer. Newton writes: 'the several degrees of heat thus found had the same ratio among themselves with those found with the thermometer'.[43] During the course of his experiments, he noticed that hot water evaporated off the surface of iron at 35 or 36° but cold water required iron at 37°, the first quantitative recording of this phenomenon.[44]

Newton described his method of measuring high temperatures in the paper he published in 1701, which included a statement of the law of cooling:

> Having discovered these things; in order to investigate the rest, there was heated a pretty thick piece of iron red-hot, which was taken out of the fire with a pair of pincers, which was also red-hot, and laid in a cold place, where the wind blew continually upon it, and putting on it particles of several metals, and other fusible bodies, the time of its cooling was marked, till all the particles were hardened, and the heat of the iron was equal to the heat of the human body; then supposing that the excess of the degrees of the heat of the iron, and the particles above the heat of the atmosphere found by the thermometer, were in geometrical progression when the times are in arithmetical progression, the several degrees of heat were discovered....[45]

Although Newton did not use calculus or fluxions to state the law of cooling in algebraic form, his conclusion that the fact that the excess 'degrees of heat' were 'in geometrical progression' for times 'in arithmetical progression', or equal intervals, which followed from assuming that 'the heat which the iron loses in a certain time is as the whole heat of the iron', can be

---

[42] *Scala graduum caloris*, 267.
[43] *ibid.*, 268.
[44] *ibid.*, 267–268.
[45] *ibid.*, 268.

expressed using modern calculus. Thus, in James A. Ruffner's reconstruction:

$$\frac{dh}{dt} = -ch$$

$$\int_{h_0}^{h_n} \frac{dh}{h} = \int_{t=0}^{t=n} -c\,dt$$

$$\ln h_n - \ln h_0 = -cn.$$

When $n = 1$,

$$-c = \ln \frac{n_1}{n_0} \propto \ln r$$

$$\ln h_n = \ln h_0 + n \ln r$$

$$h_n = h_0 r^n.$$

In Newton's experiment, as specified in his account, heat transfer by conduction would not have been significant, while convection would have played a bigger part than radiation. Convective heat loss does follow Newton's law to a significant extent, but becomes less significant as the temperature increases.[46]

The temperatures derived from such experiments helped to establish a scale against which the temperatures of other high temperature phenomena could be measured. To define his scale Newton used the melting points he had established for various alloys of tin, lead and bismuth. This was the first use by any researcher of the melting or freezing points of metals for a temperature scale. They are still used for this purpose today, notably in ITS 90, the internationally agreed scale of 1990, where they are the exclusive source of information about the high temperatures. The scale of 1701 was, in effect, between about 1/3 and 1/2 of the modern Celsius scale. Melting ice was defined as 0°, though Newton knew that lower temperatures were possible. He even contemplated an absolute zero at which all motion would cease and atoms would be devoid of vacuum, though he made no attempt at a calculation of the value.

On 3 March 1692, a day after Newton had referred to a lost theoretical work on the kinetic theory of matter, and discussed the concept of absolute zero, Pitcairne recorded that 'an incubated egg, or anything else warmed only by the heat of the skin, grows so hot that it raises the oil contained in a thermometer to ten degrees; the heat of blood raises it to twenty degrees,

---

[46] Ruffner (1963).

that of milk to less. But the summer heat of the sun or of the air at that season raises it to 7 1/2, and boiling water to 52'.[47]

A year after this memorandum, in March 1693, Newton recorded 10 phenomena, which appeared in the published paper of 1701, along with a number of others, with temperatures between 0 (the ice point, the temperature of melting snow) and 12° (body heat), with the expansivity of oil over this range being set at 40–39, down from the 41 recorded in 1693.[48] The change here represents a difference of just 6 parts in 10,000 or 0.06%. 33° is the temperature at which water begins to boil (it had been 52° on the scale used in 1692–1693), the boiling becoming vehement at 34°, and not occurring at higher than about 34.5°. Newton had, as we have seen, already established the boiling point of water by the time of the *Principia*, with the heat of red-hot iron set at about three or four times that of boiling water. The melting point of tin (232°C) was 72° on his scale, and may have been the highest temperature for which he used the linseed oil thermometer. Newton writes in his published account: 'first it was found by the thermometer with linseed oil, that if, when it was placed in melted snow, the oil possessed the space of 10,000 parts; then the same oil rarefied with the heat...of a human body possessed the space of 10,256 parts...; therefore the rarefied oil was to the same expanded by the heat of the human body, as 40 to 39'.[49]

3.4. The Dynamics of Cooling

The law of cooling presupposes the second law of thermodynamics — it is a version of this law applied thermodynamically, if not yet a 'law of thermodynamics'. It requires heat always to be transferred from hotter to colder bodies, by necessity if Newton's dynamic model for temperature exchange through momentum transfer is valid, and it is closely linked to the Newtonian idea that some amount of scalar momentum is always lost in an inelastic collision. Versions of this transfer are in Query 18, *De vi electrica* and *Scala graduum*.[50] In addition, radiant heat in vacuum must also pass

---

[47] Pitcairne, 3 March 1692, *Corr.* III, 209.
[48] *Scala graduum caloris*, 267–268.
[49] *ibid.*, 267–268. Newton's mean value for the coefficient of expansivity of linseed oil at the equivalent of about $6.92 \times 10^{-4} K^{-1}$ is 'remarkably good' (Simms, 2004, 41).
[50] The idea of temperature exchange through momentum transfer is found in Newton's work at a relatively early date in an alchemical context. Interpreting Boyle's supposed observation of 'the Incalescence of *Quicksilver* with *Gold*' as an example of 'mediation', he says of the impregnating particles that 'their grossness may enable them to give ye parts of ye gold ye greater shock, & so put ym into a brisker motion than smaller particles

from hotter to cooler environments, because its mode of transfer is assumed to take place through aether particles.

The law of cooling is an arbitrary theoretical device based on logarithmic dynamics. It is true for conduction, convection and radiation; but with different rates for the three processes. The mathematical background begins with *Principia*, Book II, on the motion of bodies subject to resistance. According to Proposition 2: 'If a body is resisted in the ratio of its velocity, and moves, by its *vis insita* [inertia] only, through a similar [homogeneous] medium, and the times be taken as equal, the velocities in the beginning of each of the times are in a geometrical progression, and the spaces described in each of the times are as the velocities'. The kind of mathematical analysis involved is suggested by Proposition 4,[51] where, to solve the differential equation,

$$\frac{dv_t}{dt} = -kv_t,$$

for the motion of a projectile with both horizontal and vertical components, we make

$$v_o/(v_o dt - v dt) = v dt/(v dt - v_2 dt) = \cdots = 1/k.dt$$

from which we derive

$$v_n dt/v_o = (1 - k.dt)_n.$$

In the limit as $n.dt \to 0$ we obtain

$$\ln(v_t/v_o) = -k.t,$$

from which we derive the solution

$$\frac{v_t}{v_0} = e^{-kt}.$$

In principle, a body which is at a higher temperature than its surroundings, will have parts which have a higher degree of scalar momentum, and, by the basic theory of particle collisions, will transfer some to the surroundings. If the temperature difference is higher, the heat or motion will be transferred at a higher rate. It was an operational convenience to assume, in the first instance, that the proportionality was linear, but Newton subsequently

---

could do; much after ye manner that ye saline particles wherewith corrosive liquors are impregnated heate many things wch they are put to dissolve, whilst ye finer parts of common water scarce heat anything dissolved therein be ye dissolution never so quick'. (Newton to Oldenburg, 26 April 1676, *Corr.* II, 1–3, 1.)
[51] Discussed, for example, in Whiteside (1982).

established the validity of the result by his lengthy series of experiments on temperature. The law of cooling subsequently became the foundation for the very important theory of heat transfer presented in Joseph Fourier's *Théorie analytique de la chaleur* (*The Analytic Theory of Heat*) of 1822. Fourier imagined the heat flow between contiguous molecules as proportional to the almost infinitesimal temperature difference between them, and expressed the relation in a partial differential equation, which is still the basic equation for heat diffusion across a thermal conductor. In its one-dimensional form, this becomes

$$\frac{dQ}{dt} = -kA\frac{dT}{dx}.$$

Newton's law of cooling provides a discrete analogue.

In fact, laws in this form, stemming from Newton's work on the dynamics of resisted motion, are so generic that many analogies present themselves. For example, in making $\Delta T/T$ a constant Newton's law could be considered an analogue to Sadi Carnot's theory of the most efficient heat engine of 1824, where the efficiency $\Delta T/T$ is also a constant, and the temperature difference ΔT is similarly the 'driving force' of the process. The significant difference is that Carnot's expression is concerned with producing work from a heat engine while preserving the total quantity of heat, while Newton's is about heat transfer.[52]

Newton's law of cooling also becomes analogous to Kelvin's first absolute temperature scale of 1848 when the time parameter is replaced by temperature.[53] This scale, though it was not the one finally adopted, was an equally valid way of defining absolute temperature, and, in its logarithmic form, had the advantage of suggesting that the absolute zero of temperature would be at $-\infty$, indicating its unattainability. Both Fourier's law of heat conduction and Fick's first law of diffusion could also be considered analogues of Newton's law of viscosity from *Principia*, Book II (see 5.5), another expression arising from resisted motion, the three laws being concerned respectively with the transport of energy, mass and momentum.[54] Later, the Fourier theory of heat also provided an analogy for electrical conduction, while Fourier's use of 'Newton's equation' for defining a 'convective boundary condition' ultimately led to the twentieth century development of eigenvalues and eigenfunctions,[55] which would prove to be of such importance in

---

[52] Carnot (1824); Cheng (2009).
[53] Cheng (2009).
[54] *ibid.*
[55] Herivel (1975).

quantum mechanics. It is remarkable that so many fundamental laws and concepts should have their ultimate origin or analogy in Newton's work on resisted motion in *Principia*, Book II, the widespread application of the basic structure being a direct consequence of its generic nature.

Newton had, in deducing his law of cooling dynamically, considered temperature as an analogue of velocity; this Newtonian analogy must be interpreted to mean that the quantity of heat is to temperature as 'motion' (or momentum: mv) is to velocity (v), so the heating effect must be inversely proportional to the mass of the body. Just as two moving bodies in collision share their momentum according to a certain simple arithmetical rule, so the mixing of hot and cold bodies results in the sharing of heat between them according to an analogous rule. Furthermore, the principle of conservation of motion (momentum) implies a no less fundamental principle: that of the conservation of heat (in the case where no new 'active principle' is brought into operation). Newton got Desaguliers and Brook Taylor to investigate the relation experimentally, and show that the 'actual heat is to the sensible heat as motion [momentum] is to velocity'.[56] On 18 February 1714, we read 'that Mr. Aguiliers be desired to wait on the President to take his directions' for experiments with thermometers for great degrees of heat'.[57] Hot and cold bodies share their heat on mixing by a rule, which is analogous to the sharing of momentum between bodies in collision, and the total heat is conserved in the same way as the total momentum. So Desaguliers and Taylor concluded that the amount of heat needed to warm a quart of water through a given number of degrees would warm two quarts through only half of that number. The experiment contributed significantly to the discovery of specific heat capacity later in the century. It was also the beginning of a distinction between 'heat' and 'temperature' as fundamental quantities.

Newtonian dynamics also led to other thermodynamic ideas. From his understanding of heat as random motion ('the agitation of particles in every direction') Newton proceeded also to the logical conclusion that absolute rest would mean absolute coldness (absolute zero of temperature, in our terms) and that absolute coldness would mean the absence of spaces between particles, that is, zero or minimal volume. Pitcairne recorded him saying in 1692: 'Heat is the agitation of particles in every direction. Nothing is

---

[56] Desaguliers (1763, ii, 296–297); Cardwell (1971).
[57] Guerlac (1967, 56).

absolutely at rest as regards its particles and therefore nothing is absolutely cold except atoms destitute of vacuum'.[58]

There were also insights to be gained from optics. If light was totally absorbed by a body, Newton believed, it would be entirely converted to a heating effect. 'All black things grow hot more quickly and strongly', he had said to Pitcairne in 1692.[59] Query 6 introduced the concept of the black body, whose study famously later led Planck to the quantum theory of radiation, though the putatively particulate nature of the light rays is central here to Newton's argument: 'Do not black Bodies conceive heat more easily from Light than those of other Colours do, by reason that the Light falling on them is not reflected outwards, but enters into the Bodies, and is often reflected and refracted within them, until it be stifled and lost?'

3.5. The Impossibility of Perpetual Motion

All the fundamental concepts of classical thermodynamics are found in Newton's work, at least in embryo form, namely, temperature, conservation of energy, dissipation and absolute zero; these are the quantities that enter into the respective zeroth, first, second and third laws of thermodynamics. Of these, the subtlest concept and the most forward-looking, though at the time it might have seemed to look backward, was the idea of dissipation. In fact, this idea of Newton's seems to have been one of the most misunderstood in modern contexts, leading some authors to think that he had no true understanding of the conservation of energy.[60]

In his youthful *Hypothesis* of 1675 Newton had seen nature as a perpetual worker. By the time of the *Principia* in 1687, when he had studied more thoroughly the dynamical consequences, he saw the cosmos as tending to decay. Physical systems (including the Solar System) could not be maintained in perpetual motion. Friction and inelasticity would induce decay in the motion of any object or system of objects unless this was restored by the application of a new force. And he found himself on a collision course

---

[58] *De natura acidorum*; *Corr.* III, 210.
[59] Pitcairne with Newton at Cambridge, 3 March 1692; *Corr.* III, 212.
[60] Simms (2004, 65–66), for example, completely misunderstands the statement in Q 31 that 'Motion ... is always upon the Decay' as a rejection of the conservation of energy! The Leibniz-Clarke correspondence, among other things, clearly shows that it is not, and that it is, in fact, an ancestor of our present understanding of a thermodynamics based on two fundamental principles rather than one. The first version of Simms's otherwise valuable article was written forty years before publication, that is, before mathematical investigators realised the true significance of the energy conservation theorems in the *Principia*.

with Leibniz, who still viewed the universe as a machine in perpetual motion, as a result of a theory he was developing in which in inelastic collisions the *vis viva* (mv^2), equivalent to our kinetic energy ($1/2\ mv^2$), would be transferred to the internal particles of the colliding bodies.

From his analysis of power in machines and statements that he made at various times on heat and friction, it seems probable that Newton was aware of this and of the role of heat in the process. Roger Cotes, for example, speaking for Newton in the Preface to the second edition of the *Principia*, says bodies meeting resistance in fluids transmit their motion gradually to the fluid, effectively particle by particle, so losing their own motion and being retarded in proportion to the motion lost.[61] Newton had developed his law of convective cooling seemingly based upon this kind of model, in which times and temperatures showed the same mathematical relationship (a geometrical progression) as times and velocities did in the dynamics.

The kinetic theory of heat, based on the dynamics of material particles, and its use in the definition and measurement of temperature, led to the first expression of the concept of the loss of usable energy as a thermodynamic concept. In the Newtonian system, temperature is due to scalar momentum. Momentum transfer equals heat transfer. Scalar momentum cannot be gained without active principles. New scalar momentum is only generated through such principles, allowing work to be done. If we take momentum transfer as due to collisions, as in mechanistic theories, either through material particles or aether particles, and suppose that, for the collision of two bodies with momenta $m_1 u_1$ and $m_2 u_2$, the resulting momenta are $m_1 v_1$ and $m_2 v_2$, and choose symbols such that u_1 is numerically greater than or equal to u_2, with u_1 positive, then Newton's experimental law or Newton's law of restitution states that $u_1 - u_2 = e(v_2 - v_1)$. Identical particles must lose (scalar) motion in collisions unless $e = 1$, but, in fact, $1 > e > 0$ in all cases. It is experimentally observed that the coefficient of restitution, e, is never truly 1, because there is always some resistance; in all real systems, $e < 1$. It is also never exact (due to perturbation) when (as always) there are more than two 'bodies'.[62]

Since $u_1 > u_2$, the second body must gain scalar momentum at the expense of the first. The tendency of momentum transfer is also from the system with greater scalar momentum to the system with less; from the higher 'temperature' to the lower. The transfer of 'heat' is all

---

[61] Cotes (1713, xxxi).
[62] *Principia*, Axioms, or Laws of Motion, Scholium.

one way. In addition, the 'heat' is an intensive quantity, depending on the mass; for higher mass particles, or for a greater number of particles, the scalar momentum transfer is greater. An exponential relationship occurs in cooling because it depends on the number of (material or aether) particles involved in the collision process. We may compare the cooling process with the exponential relation in radioactive decay; if it is a many-body process, it will be effectively random.

One consequence of this process is the fact that perpetual motion cannot be true. Early in life, in the *Quaestiones*, Newton had toyed with perpetual motion devices, wheels and windmills turned by gravity and magnetism. He asked, for instance, whether they could be accomplished by 'gravitational radiation', 'that endless flux of aethereal matter causing gravity', and also causing the 'matter in motion' aspect of a Cartesian universe (or a similar magnetic one): 'Whither ye rays of gravity may be stopped by reflecting or refracting y^m, if so a perpetuall motion may bee made one of these two ways'.[63] A drawing shows 'rays of gravity' being reflected from a gravitational shield. Newton imagined a wheel suspended horizontally, so that half of it would be exposed to the rays of gravity, while the other half was protected by the shield; the exposed section would absorb more rays and so become heavier, causing the wheel to turn without requiring a supply of power. A second drawing shows a device based on gravitational 'refraction': this was a horizontal windmill with its vanes turned by a stream of gravitational rays striking them. Other devices were powered by a stream of 'magneticall rays', rather than gravitational ones, with a loadstone turning 'a red hot iron fashioned like wind-mill sailes'. Perhaps the rays could be reflected by cold iron, or 'blow a candle move a red hot copper or iron needle, or passé through a red hot plate of copper or iron'.[64] He also asked whether a magnetic pendulum was perpendicular to the horizon, or whether magnetised iron was heaviest when the north pole or south pole was uppermost. (Interestingly, the magnetic pendulum now provides a classic example of chaos.)

Perpetual motion has a long history, and seems to be perpetual in itself. As early as the eighth century, in Bavaria, there was a 'magic wheel', turning on an axle and driven by magnetic lodestones. The French architect and mason Villard de Honnecourt has a drawing of one from about 1230. Leonardo da Vinci thought of using overbalanced wheels, though elsewhere

---

[63] QQP, Of Attomes', f. 121^v, NP. Westfall (1971, 330–331) (for quotation).
[64] QQP, Of Attomes', f. 121^v, 'Atraction Magneticall', f. 101^v, NP. Newton retained an interest in gravitational refraction in later work applying it to the deflection of light rays.

he shows a degree of scepticism about the idea. Other devices, based on fluids, were proposed by Newton's contemporaries Robert Boyle and Johann Bernoulli.[65]

As he developed his dynamics, Newton came to realise that such devices were not conformable to the laws of nature. Much later, as President of the Royal Society, he rejected them out of hand. In 1725, for example, he refused to entertain at his house one Johann van Hatzfeld who tried to demonstrate a perpetual motion machine to him.[66] He also wrote in his 'Remarks upon the Observations made upon a Chronological Index of Sir Isaac Newton... publish'd at Paris', the last contribution he ever made to the Royal Society's *Philosophical Transactions*: 'But I hope that these Things [i.e. disputes about his optical experiments], and the perpetual Motion, will be the last Efforts of this Kind'.[67] On 7 August 1721, 'sGravesande had written to him supporting perpetual motion, but Newton's reply has not survived.[68] However, later in the same year (9 November), when he was in the chair, Desaguliers made some 'Remarks on some attempts towards a perpetual motion' to the Society, which used Newtonian arguments to denounce the idea as a fallacy.[69] On 24 June 1724, the Society rejected as 'a designed affront to the Society' a book by 'Mr. Francis', which claimed success with a perpetual motion machine at the same time as mounting a concerted attack on Newton's natural philosophy.[70] Rejection of perpetual motion gradually became the normal response. By 1775, the French Académie in Paris would officially refuse to consider any more perpetual motion devices, because they wasted time and resources on

---

[65] Verance and Dircks (1916); Dircks (1968); Ord-Hume (1977).
[66] *Corr.* VII, 253, 1725.
[67] 'Remarks upon the Observations' (1725, 321). Simms (2004, 65–66), seems to have missed these references, in saying that there was no *direct* evidence that Newton had rejected perpetual motion in his mature years, though it was assumed that he had. On 22 March 1715, in a less clear-cut example, Newton had written to John French: 'I have received your Letter (dated yesterday) by the hands of Mr. John Vat & can acquaint you that his Project for the Longitude is as impracticable as to make a perpetual motion like that of the heart but much more uniform...' (*Corr.* VI, 211–212).
[68] *Corr.* VII, 143–146.
[69] Desaguliers (1721, 237); Simms (2004, 66).
[70] *Journal Book of the Royal Society*, XII, 486–487, 24 June 1724; Simms (2004, 66). The later presentation by Arbuthnot of the claims of M. Bourgeois on 22 December 1726 (Simms, 66) is irrelevant, as Martin Folkes, not Newton, was in the chair, and Newton was now very near the end of his life and often absent from the Society.

projects that could never be made workable.[71] That did not, of course, stop individual inventors of perpetual motion machines, and there has been a steady stream of them up to the present day.

3.6. The Decay of Motion

Newton's thermodynamic reasoning is actually a much subtler concept than the mere absence of perpetual motion, and, as usual, his most casual-seeming statements, frequently stemming from a relentless search for new angles on basic truths, often contain hints of more significant things to come. Newton seems to have been the first to express the concept of the loss of usable energy in a thermodynamic context. He created a definition of temperature, based on dynamics, which, more than heat, is fundamentally thermodynamic, and his law of cooling and its dynamic background presuppose what we now consider a version of the second law of thermodynamics. The Newtonian background to the second law is important for a number of reasons: (1) it helps to explain Newton's ideas, especially in regard to his controversies with Leibniz; (2) it influenced the nineteenth century understanding of the second law, especially that of William Thomson (later Lord Kelvin); (3) it allows us to see how and why nonscientific ideas may influence the content of science and become part of it; (4) it may lead to a better understanding of this most enigmatic of laws itself.

In stating that new motion cannot be created without active principles, Newton had observed the decay of motion in collisions and the *necessity for conserving* it. Of course, there is a metaphysical or theological aspect to this, with Newton wanting to specify the continual action of *God* in preserving the universe as a counter to materialism and atheism, where the eternal aspect could be specified for material things. Whereas, in Cartesian theory, the material objects themselves, preserved eternal motions, in Newtonian theory, the forces between particles of matter implied God's continual intervention in the universe.

But Newton always seems to seek to specify that natural *appearances* will be maintained. Inertia (or *vis inertiae*) is merely passive; it cannot generate motion to replace that which is seemingly being continually lost in every system involving the motions of material bodies.

---

[71] *Encyclopedia Britannica*, 18th edition, 1986, 'Perpetual motion'.

> The *Vis inertiae* is a passive Principle by which Bodies persist in their Motion or Rest, receive Motion in proportion to the Force impressing it, and resist as much as they are resisted. By this Principle alone there never could have been any Motion in the World. Some other Principle was necessary for putting Bodies into Motion; and now they are in Motion, some other Principle is necessary for conserving the Motion. For from the various Composition of two Motions, 'tis very certain that there is not always the same quantity of Motion in the World.[72]

Newton based his idea that motion was always lost in collisions on his researches on inelastic impact, that is on the impact of all real bodies. Absolutely hard bodies would not rebound on impact.

> For Bodies which are either absolutely hard, or so soft as to be void of Elasticity, will not rebound from one another. Impenetrability makes them only stop. If two equal Bodies meet directly *in vacuo*, they will by the Laws of Motion stop where they meet, and lose all their Motion and remain at rest, unless they be elastick, and receive new Motion from their Spring. If they have so much Elasticity as suffices to make them re-bound with a quarter, or half, or three quarters of the Force with which they come together, they will lose three quarters, or half, or a quarter of their Motion. And this may be try'd, by letting two equal Pendulums fall against one another from equal heights. If the Pendulums be of Lead or soft Clay, they will lose all or almost all their Motions: If of elastick Bodies they will lose all but what they recover from their Elasticity. If it be said, that they can lose no Motion but what they communicate to other Bodies, the consequence is, that *in vacuo* they can lose no Motion, but when they meet they must go on and penetrate one another's Dimensions.[73]

The vector addition of momenta shows that the scalar value of momentum (the quantity of motion as imagined by Descartes) is 'always upon the Decay', because of friction, the resistance of fluids, and the inelasticity of collisions between material bodies. This quantity of motion could not be universally constant:

> For if two Globes joined by a slender Rod, revolve about their common Center of Gravity with an uniform Motion, while that Center moves on uniformly in a right Line drawn in the Plane of their circular Motion; the Sum of the Motions of the two Globes, as often as the Globes are in the right Line described by their common Center of Gravity, will be bigger than the Sum of their Motions, when they are in a Line perpendicular to that right Line. By this Instance it appears that Motion may be got or lost. But by reason of the Tenacity of Fluids, and Attrition of their Parts, and the

---

[72] Q 31.
[73] *ibid.*

Weakness of Elasticity in Solids, Motion is more apt to be lost than got, and is always upon the Decay.[74]

The idea behind such statements is slightly more anthropomorphic than most of Newton's theological positions, and can be traced to a view of a universe in decay widely held in his day. However, as always with Newton, there is a truly analytical element, and his observation of motions through resistive media also led to his logarithmic law of loss of excess heat.

Vortical motion provided even further evidence for the inelasticity of the parts of material substances. Referring obliquely to Descartes' description of vortical motion in the heavens and his own analysis of such motion in the *Principia*, Newton stated that viscosity quickly destroys the vortical motion produced by stirring real liquids, such as pitch, oil and even water; only in an ideal fluid 'void of all Tenacity and Attrition of Parts, and Communication of Parts', which, of course, cannot exist, would the vortices remain for more than a very short time.[75] If three identical round vessels are filled, respectively, with water, oil, and molten pitch, and then stirred to produce a vortical motion, the pitch loses its motion first, then the oil, and finally the water. But even the water will lose its motion in a short time. To continue indefinitely, a vortex would have to be created in a nonexistent kind of matter free of all viscosity and friction in its parts.[76]

Though motion is lost in every impact and collision, in every interaction which causes dynamic change, the motion is restored. God, who had created the original hard particles involved in these collisions, was the ultimate cause also of restoring the motion. However, the physical causes which allowed the restoration were what Newton called 'active Principles', though these were in some sense manifestations of the direct power of God. In stating that new motion cannot be created without active principles, Newton observed both the decay of motion in collisions and the *necessity for conserving* it.

Newton's opponents argued that his view that the world was imperfect, and inclined to run down without the direct action of God, meant that God had to regularly wind up the clock he had originally set in motion. The whole issue led to a debate with Leibniz, largely carried out by proxy, with Samuel Clarke standing for the Newtonian position, though, it is believed, with the active cooperation of Newton himself. Though Leibniz, in particular, assumed that God's action in restoring the motion meant

---

[74] *ibid.*
[75] *ibid.*
[76] *ibid.*

regular supernatural intervention, it seems that Newton believed God acted continually and that the 'active Principles' were part of the natural order that God had created.[77]

> Seeing, therefore, the variety of Motion which we find in the World is always decreasing, there is a necessity of conserving it and recruiting it by active Principles, such as are the Cause of Gravity, by which Planets and Comets keep their Motions in their Orbs, and Bodies acquire great Motion in falling; and the cause of Fermentation, by which the Heart and Blood of Animals are kept in perpetual Motion and Heat; the inward Parts of the Earth are constantly warm'd, and in some places grow very hot; Bodies burn and shine, Mountains take fire, the Caverns of the Earth are blown up, and the Sun continues violently hot and lucid, and warms all things by his Light. For we meet with very little Motion in the World, besides what is owing to these active Principles. And if it were not for these Principles, the Bodies of the Earth, Planets, Comets, Sun, and all things in them, would grow cold and freeze, and become inactive Masses; and all Putrefaction, Generation, Vegetation and Life would cease, and the Planets and Comets would not remain in their Orbs.[78]

Two active principles which he describes, 'gravity and fermentation, regarded as sources of energy, would now be called potential energy and energy released by chemical change'.[79] While scalar momentum is lost through the collisions of inelastic hard particles (the ultimate components of matter, which cannot themselves wear out), the active principles serve to restore it. Though God is the ultimate, and even direct, source of such principles, periodically keeping his clock in regular order to use the metaphor of the time (though it is not Newton's), it would seem that this comes about through natural rather than supernatural manifestations, a continual miracle, not an extraordinary one.

The implication, elsewhere, however, is that even these active principles cannot prevent the ultimate breaking down of the system, with the complexity of multibody processes inevitably leading to a progressive increase in disorder. Newton emphasizes the effect of attrition and dissipative forces in organized systems, especially in the Solar System, explicitly denying the perpetual motion which Leibniz thought would result from the conservation of kinetic energy in mechanical collisions. Here, Leibniz writes:

> I have affirmed that active forces are preserved in the world. The author (Clarke) objects, that two soft or un-elastic bodies meeting together, lose

---

[77] ibid.
[78] ibid.
[79] Alexander (1956), xviii, n. 1.

some of their force. I answer, no. 'Tis true, their wholes lose it with respect to their total motion; but their parts receive, being shaken by the force of the concourse. And therefore that loss of force, is only in appearance. The forces are not destroyed, but scattered among the small parts. The bodies do not lose their forces; but the case here is the same, as when men change great money into small.[80]

Clarke replies for Newton:

In order to show that the *active forces* in the world (meaning the quantity of motion or impulsive force given to bodies) do not naturally diminish; this learned writer urges, that two soft un-elastic bodies meeting together with equal and contrary forces, do for this only reason lose each of them the motion of their whole, because it is communicated and dispersed into a motion of their small parts. But the question is; when two perfectly HARD un-elastic bodies lose their whole motion by meeting together, what then becomes of the motion or active impulsive force? It cannot be dispersed among the parts, because the parts are capable of no tremulous motion for want of elasticity. And if it be denied, that the bodies would lose the motion of their wholes; I answer: then it would follow, that elastic hard bodies would reflect with a double force; viz. the force arising from the elasticity, and moreover all (or at least part of) the original direct force: which is contrary to experience.[81]

In the next paragraph he says:

That active force, in the sense above defined, does naturally diminish in the material universe; hath been shown in the last paragraph. That this is no defect, is evident; because 'tis only a consequence of matter being lifeless, void of motivity, unactive and inert. For the inertia of matter causeth...that solid and perfectly hard bodies, void of elasticity, meeting together with equal and contrary forces, lose their whole motion and active force, (as has been shown above) and must depend on some other cause for new motion.[82]

By 'active force', of course, Clarke is concerned with the loss of the total scalar value of linear momentum, not with Leibniz's *vis viva*. Now, it is possible that inelastic collisions can conserve the vector momentum without conserving the scalar value of momentum or the *vis viva*, and this certainly happens for a system defined as a particular set of bodies; scattering the *vis viva* among the small parts of such bodies would still not keep the bodies in

---

[80] Alexander (1956, 87–88). In fact, even Leibniz's last example of money changing is denied by the kind of everyday experience which tells us that us that it can be unproductive in terms of time and effort to collect small change.
[81] Alexander (1956, 111).
[82] Alexander (1956, 112).

motion. Also, even Leibniz's process cannot take place through percussion alone because matter itself is inert, and the ultimate particles of matter are incapable of percussion. According to Newton, the ultimate particles of bodies are not intrinsically elastic, because elasticity is a result of the internal forces (or active principles) which create the hierarchical substructure of matter; without the substructure, there is no elasticity.

The impact of hard bodies, of course, would require an instantaneous collision and, therefore, an infinite force. Newton declared in the *Principia* that 'no such bodies are to be found in Nature'.[83] If the ultimate particles of matter were 'absolutely hard', they were never found separated from physical bodies. (To satisfy all the conditions, in fact, the ultimate particles would have to be points, subject to infinite forces at zero displacement, exactly like modern ultimate particles.) Newton's (or Clarke's) rejection of Leibniz's arguments seems to contain two components: one is that the ultimate particles themselves have no elasticity — this has to be supplied by an active principle; the other is that there is no such thing as true percussion in any case because the processes of 'collision' involve active principles, and so the process described as percussion cannot be said to conserve the total kinetic energy of the particles.

Leibniz's notion that the living force of a body does not disappear even in inelastic collisions but is distributed between the parts of the body was an important insight, but it takes no account of the work done in compression. Newton's understanding of the process is perhaps more original and more subtle. Something is lost in the process, something is lost in *any* process, and the presence of an active principle is evidence of it. Whatever happens in the internal parts of bodies, the bodies themselves, however defined, lose part of their kinetic energy through friction or inelasticity unless an active principle is brought in to restore it, or, as we would say, unless work is done. But doing work brings a change in the system. In our terms, kinetic energy lost through dissipation in the smaller parts of bodies cannot be restored to a form which can give back motion to the body itself.

Newton, of course, had only a partial grasp of this very important idea, but he certainly had a very powerful sense of the dissipative nature of friction and percussion and saw natural systems as showing a tendency to long-term decay, which is its effective equivalent. His ultimate particles, being so completely hard as to be inelastic, were therefore unable to preserve motion by percussion itself. New motion was generated only by the active principles

---

[83] Scholium to the Laws of Motion.

at work between them; hence, a degree of dissipation always resulted from any actual percussion.

Fontenelle's *Elogium* of Newton for the French Academy of Sciences (1728) provided a useful summary of the theory and its cosmological implications:

> ... Hardness of bodies is the mutual attraction of their parts which closes them together and if they ... are capable of being everywhere joined, without leaving any void spaces, the bodies are then perfectly hard. Of this kind there are only certain small bodies, which are primitive and unalterable, and which are the elements of all other bodies... If a certain degree of motion that is once given to anything by the hand of God, did afterwards only distribute itself according to the laws of Percussion, it appears that it would continually decrease in motion by contrary Percussions, without ever being able to recover itself, and the Universe would soon fall into such a state of rest, as would prove the destruction of the whole. It may likewise happen ... that the System of the Universe may be disordered, and require, according to Sir Isaac's expression, a *hand to repair it*.[84]

We know that Newton believed that even the Solar System would be subject to long-term instability because of the dissipative effects of friction or perturbative forces. From the tendency for motion to decay, he concluded that eventually the Universe must 'run down', and 'all things ... grow cold and freeze, and become inactive Masses'.[85] Leibniz, by contrast, believed that the Solar System not only exhibited signs of perfection, but would remain in motion *in perpetuo* because 'the same force and vigour remain always in the world and only passes from one part of matter to another, agreeably to the laws of nature, and the beautiful pre-established order'.[86] His Solar System was a perpetual motion machine of the second kind, one which preserved the first law of thermodynamics but violated the second. Newton, with his first-hand knowledge of many-body systems and perhaps also his theological belief that only God was a true measure of perfection, instinctively rejected it.

3.7. The Laws of Thermodynamics

The first and second laws of thermodynamics were established in the mid-nineteenth century as the fundamental principles of a new science whose

---

[84] Fontenelle, *Elogium*, 1728, 20–21, Cohen, *Papers and Letters* 462–463; quoted Scott (1970, 5). Maclaurin (1748) also emphasizes the significance of denying the perpetual motion of Descartes and Leibniz on the basis of Newton's analysis of inelastic collisions.
[85] Q 31.
[86] Alexander (1956, 12).

original purpose had been to explain heat engines, the working devices (power sources or prime movers) of the Industrial Revolution. Like Newton's first two laws, one cannot be defined without the other; the second law of thermodynamics — the one-way flow of heat, the increase in entropy or disorder in the universe after each physical event — also requires the existence of the first — the conservation of energy — before it can be defined. And the first law, though a statement of the conservation of energy, is more than this, for it introduces the concept of heat, without which the second law has no meaning. In even more fundamental terms, the two laws of thermodynamics could probably be replaced by the conservation of mass (or mass-energy) and the continuity, or irreversibility of time.

William Thomson first used the terms first and second law of thermodynamics in his paper of 1851, showing a degree of generosity in attributing them respectively to Joule and Clausius. Earlier ideas date from the winter of 1850. Thomson put in more explicit form what Clausius had implied in his paper of February 1850. The contribution of a third figure, W. J. McQuorn Rankine, is somewhat more obscure. Rankine made a complete statement of thermodynamics including the entropy function, contemporary with Clausius's work, in February 1850, but based on a specific model of the nature of matter.[87] In a way typical of what Clifford Truesdell has described as the 'tragicomic history of thermodynamics', the same concepts were rediscovered again and again.[88] Essentially, the basic principles are relatively simple, but they are difficult in application because heat is not really classical in origin but is made to appear as if it is, with the consequent generation of such arbitrary macroscopic concepts as the 'perfect gas'.

The second law is, in some ways, a law without an author: it has so many alternative statements that it is difficult to decide which is definitive. Both Thomson and Clausius depended on the work of Sadi Carnot of 1824, previously referred to in connection with Newton's law of cooling.[89] Carnot's published work is based on the caloric theory, a material theory which implies the conservation of heat, with no interconversion between heat and work, as would be established by Joule, Mayer, Helmholtz and others in the 1840s. The first law of thermodynamics, in which this interconversion was established, was the result of work by Joule and others in the 1840s.[90] Carnot used a model cycle, with efficiency $T_1 - T_2/T_1$, of which the air

---

[87] Thomson (1851); Clausius (1850); Rankine (1850, 1851).
[88] Truesdell (1980).
[89] Carnot (1824).
[90] Joule (1847); Mayer (1842); Helmholtz (1847).

engine patented by Robert Stirling in 1816 was the nearest technological (or empirical) realisation.[91] H. L. Callendar and G. Falk have claimed that caloric, especially as used by Black in the eighteenth century, is the same as entropy.[92] Perhaps we could say they were isomorphic, with the same effects and properties, though differently conceived; one could entirely replace the other in fundamental laws. A half-conservation law applies in the sense that heat (caloric) can be created but not destroyed.

The second law of thermodynamics has many unusual characteristics. In terms of macroscopic observation, it is statistical, as was proved by Maxwell and Boltzmann in the nineteenth century.[93] It is also the only law of physics which gives a direction to time. It has no single fixed statement, and it is not clear whether any particular one can be considered 'primary' with respect to the others. It is the classic case of a law that is generic in a fundamental way, which seems to suggest that it has an abstract form which is prior to any physical realisation. Aside from the purely mathematical formulations, there are many alternatives of varying precision and accuracy, and many related statements — some of which seem, at first sight, close to everyday experience. So, entropy always increases in irreversible interactions. No heat engine is more efficient than a Carnot engine (the version named after Kelvin and Planck). It is impossible for heat to pass from a colder to a hotter body without doing work (the version named after Clausius). A system left to itself breaks down. There is no such thing as reversibility. There is a tendency to disorder in all natural processes. Information is always lost in any exchange. Perpetual motion is impossible even when energy is not lost from a system. Clearly, there is a one-to-one correspondence between these statements and the fact that time is irreversible. This in turn has a correspondence with the fact that the velocity of light cannot be exceeded, which also means that the principle of causality cannot be violated. Another version might propose, more simply, that time is an absolute continuum.[94] It is the realisation that all the versions produce equivalent physical effects that constitutes the second law as a fundamental principle.

As a uniquely time-asymmetric principle, the second law cannot be derived from other laws of physics which are time-symmetric, whether classical, relativistic or quantum. In particular, it cannot be derived from the laws of dynamics. One branch of the subject, however, statistical mechanics,

---

[91] Stirling (1816).
[92] Callendar (1911); Falk (1985).
[93] Maxwell (1860); Boltzmann (1871a–c, 1872, 1877a).
[94] Rowlands (2007, 235–236).

does attempt to *relate* the laws of thermodynamics to those of particle mechanics. It still doesn't solve the problem of the connection with the irreversibility of time. Very likely, the main problem has been defining the direction of time from the second law, rather than defining the second law from the direction of time. In this connection, Newton has an interesting passage, which appears to refute the ergodic hypothesis, that given an infinite period of time, any given state of the universe will occur at some point: 'And hee may tell you further that ye Author of mankind was destitute of wisdome & designe because there are no final causes & that matter is space & therefore necessarily existing & having always the same quantity of motion, would in infinite time run through all variety of forms one of wch is that of a man'.[95]

In addition, though the law is fundamental, it appears to be expressed in something other than fundamental terms. It has all the appearance of a truly classical law, like Newton's laws of motion or Maxwell's equations for the electromagnetic field, but it nevertheless uses terms which do not relate well to other areas of physics. Unlike other fundamental physical laws, it uses macroscopic parameters, like temperature and entropy — that is, ones that appear to depend on a specific model for the macroscopic organisation of matter — not purely abstract ones like space, time, mass and charge. Though the law establishes a great fundamental truth, the normal way of expressing it depends on empirical concepts. It is a basic law expressed in a non-basic 'concrete', rather than fundamentally simple, language.

The problem is really that, though classical in appearance, the law is not really classical. In the usual formulations, it is based on the concept of heat, which, though generally treated in a semi-classical way, is not classical in origin at all. Heat is really a quantum phenomenon. The fact is that thermodynamics is a case in which fundamental and abstract concepts can lead to inevitable results concerning matter in a collective state which can then be treated as simple and fundamental because they are a direct reflection of the fundamental conditions. Historically, the treatment in collective terms came first, and has continued to define the concepts by which we tackle the fundamental problems.

The question now is: where do Newton's ideas on thermodynamic issues fit into this overall picture, especially in relation to the second law, and did they contribute to the later, nineteenth century, development? And it would

---

[95] Westfall (1971, 419–420). Schliesser (2012) interprets this passage as an attack on Spinoza or a Spinozist position.

seem that the answer is that Newton's technical treatment of the concepts which were later given formal status in the second law of thermodynamics, while having antecedents in a very ancient prescientific way of thinking, were very considerably in advance of those of any contemporary, and in fact of any other scientist before the nineteenth century, and they had a very perceptible influence on the nineteenth century interpretation, with a causal link that can be established from the historical record.

Newton was certainly influenced by ancient ideas of a world in decay, as with the Renaissance view of a Golden Age in an earlier period, and in similar views by more ancient writers, like Hesiod. It certainly fitted in with his belief in *prisca sapientia*, a lost wisdom of the ancients which needed to be recovered.[96] There was also a background in Stoic thinking, and in his emphasis on the analogy of nature. The truth of this analogy is, of course, one of the reasons for the apparent success of earlier ideas. The second law of thermodynamics and some other fundamental laws of physics are true for collective phenomena as well as for individual particles on the micro-scale. Because it represents the irreversible direction of time, it simply builds up relentlessly from the small scale to the large, like the parallel phenomenon of gravitational attraction. Life and death are the results of thermodynamics, and, in particular, the second law. If applied by analogy from the animate to the inanimate world, a correct analogy has been made. Stoics and other ancient philosophers made this analogy. Newton made use of it. Newton had the particular mental capacity to select correct analytical ideas from whatever source, in a way that cannot be analysed by a linear process, but then to transform them to be compatible with his more advanced scientific thinking.

This is what he did with the concept of a world in decay. In realising that motion — in effect, scalar momentum — is always on the decay, and that new motion must be generated by active principles which cause further degradation of the system, Newton's analysis ends up looking more modern than that of his contemporaries. Leibniz, for example, as a convinced Cartesian, believed in the reality of perpetual motion and the maintenance of perfect order, but Newton, though believing that motion lost was restored by active principles, instinctively realised that the use of such principles implied a degree of permanent change in a system that would eventually lead to its instability. In Leibniz's theory, the universe would be in perpetual motion, but Newton was unable to believe that any system could be maintained in

---

[96]See McGuire and Rattansi (1966).

perpetual motion, and his view of the dynamic evolution of the Solar System takes us to the edge of chaos.[97]

Although Newton did not specifically propose in his discussion of the loss of motions in Query 31 that such motion would be converted into heat, various statements that he made on power in machines as well as on heat and friction, in addition to Cotes's statement in the Preface to the second edition of the *Principia* about the gradual retardation of bodies produced by communicating their motion to the particles of an ambient fluid, with the retardation proportional to the motion communicated, suggest that Newton was not unaware of the transfer of motion to the internal particles of bodies in inelastic collisions, as advocated by Leibniz.[98] As we have seen, Newton's law of convective cooling seems to have followed from this kind of model, showing the same kind of relationship (a geometrical progression) between times and velocities as between times and excess temperatures. Newton, also, certainly believed that friction and inelasticity were responsible for the decay of the motion of any particular object or system of objects not subjected to a new force.

Newton's emphasis on dissipation was lost on his successors, as Laplace began to show that many of the long term instabilities in the Solar System postulated by Newton were connected with periodic variations, while the progressive changes were largely cancelled out, though Laplace may have overemphasized the stability of the Solar System. The issue remains unresolved even today, as to the degree of stability, but, in principle, there is no doubt that such a complicated large-scale system cannot remain stable over an indefinitely long period of time. The contemporary geologist, John Playfair saw the perpetual order of Laplace's universe as enhancing the claims of a uniformitarian geology. In a passage of the *Illustrations of the Huttonian Theory* (1802), subsequently quoted by William Thomson, he wrote: 'The Author of nature has not given laws to the universe, which like the institutions of men, carry in themselves the elements of their own destruction. He has not permitted in His works any symptoms of infancy, or of old age, or any sign by which we may estimate either their future or their past duration'.[99]

---

[97] The modern description of 'chaos', of course, implies specific features that were only identified in relatively recent times, but a fundamental requirement is a system whose behaviour, while following deterministic laws, is highly sensitive to small changes in the initial conditions.
[98] Cotes (1713).
[99] Playfair (1802, 119–121); Smith and Wise (1989, 90).

William Thomson's father and teacher, Dr. James Thomson, in his introductory lecture at Glasgow College, on 6 November 1832, referred to the mathematicians and astronomers of the Laplacean school when he wrote:

> They have also arrived at one of the most interesting conclusions to which science has ever led: they have found that contrary to some appearances, the eternal stability of the Solar System is fully provided for in its own mechanism: that it contains no seed of its own dissolution, no principle leading to old age or decay; but that every phenomenon appearing to give such indication is periodical, and is regulated by a corrective power arising from the principle of universal gravitation, which prevents it from exceeding a fixed amount.[100]

However, the Newtonian argument, which was available in readily-accessible form in both *Opticks* and the *Principia*, was never completely forgotten, and it was always remembered that Laplace's argument was specifically directed at showing that the Solar System could be regarded as an exception which, by a fortunate combination of circumstances, would avoid this fate. Then, in 1833, a year after Dr. Thomson's lecture, observations on Encke's comet seemed to imply that its motion was retarded by the action of a resisting medium, exactly as Newton had said would happen to comets in such circumstances.[101] The behaviour of the comet led many to suggest that a resistive medium in space would lead to a decay in the motion of the Solar System, in exactly the way that Newton had suggested. Laplace's argument gave way to Newton's. The Solar System was ultimately unstable after all.

Through William Whewell (who invoked Newton's authority), this became part of the background to William Thomson's doctrine of the dissipation of energy resulting from the second law of thermodynamics. Whewell wrote in his Bridgewater Treatise, *Astronomy and general physics considered with reference to natural theology*, published in the year of the comet (1833), that: 'Since there is such a retarding force perpetually acting, however slight it be, it must in the end destroy all the celestial motions.... The smallness of the resistance... does not allow us to escape this certainty'.[102] Whewell discussed his ideas with the natural theologian, Thomas Chalmers, at the British Association meeting in 1833, and influenced him into 'seeing inevitable decay as more relevant to natural theology than eternal order',[103] while Dr. Thomson added a footnote to his lecture: 'It may

---

[100] Quoted Smith and Wise (1989, 91).
[101] Smith and Wise (1989, 91).
[102] Whewell (1833, 191–209); Smith and Wise (1989, 91–92).
[103] Smith and Wise (1989, 94).

be proper to mention that a supposed change in the movements of one of the comets had led some to believe that there is a highly rarefied medium which offers some slight resistance to its motion from the lightness of its parts. This however requires confirmation and nothing of the kind has been observed in regard to the planets'.[104]

William Thomson was, as we have seen, Dr. Thomson's son and pupil, and soon after he had laid down his version of the foundational principles of thermodynamics, Thomson saw that his version of the second law led to a doctrine of the dissipation of energy in all natural processes. The idea was subsequently formalised mathematically in Clausius's doctrine of entropy and a related idea in the work of Rankine. Entropy was defined by Clausius as the integral of the heat exchanged over the absolute temperature. Boltzmann later supplied a statistical interpretation in which entropy could be measured as a function of the number of available microstates in a system, which increased for the universe with every physical interaction.[105] The apparently irreversible increase in entropy in every known process led to predictions of a 'heat death' of the universe in which everything would be at a uniform temperature, making further physical processes impossible. This concept has never gone away, although it is not yet firmly established whether it is likely to happen. There is no doubt that part of the background to this way of thinking, as instituted by Thomson, was the work of Whewell, who had in turn invoked Newton's own authority.

---

[104] Smith and Wise (1989, 92).
[105] Thomson (1852); Clausius (1865); Boltzmann (1877b).

Chapter 4

The Microstructure of Matter

4.1. Atoms and Fundamental Particles

The controversy over the structure of matter is one which never happened, though the implied challenge of Newton's ideas to the Cartesian plenum would certainly have led to major trouble had he openly promoted them. The theory of matter is an intractable subject that has required several centuries of systematic and sustained attempts using increasingly more sophisticated and expensive machinery just to begin to understand. For Newton, before the era of high-energy physics, it was even more intractable. It is nonetheless, fascinating to see how his extraordinary grasp of scientific method was brought to bear on an enormously complex problem where he had no hope of easy solutions or insights based on his mathematics. His writing on the subject is full of brilliant intuitions and astonishing insights, and collectively the sum is greater than the parts. In the most uncompromising and single-minded fashion he applies his relentless logic to make nature 'simple and conformable to herself'. Never satisfied with explaining some phenomenon by 'feigning' some hypothesis which simply multiplies the starting assumptions, he strives endlessly to fit everything into a unifying and semi-abstract pattern of particles of matter acted on by attractive and repulsive forces, though never sure that he has successfully achieved his object.

No single work was devoted to this subject — the main published sources were queries, cryptic hints and casual asides in the *Principia* and *Opticks*. Newton many times thought of producing a more systematic treatment and wrote numerous drafts in the form of prefaces, conclusions and additions to his two main works. But every time he suppressed them at the last moment. Though he always knew the level of certainty to be associated with any particular idea, and that science could only advance through the proper use of ideas at different levels of certainty, including some that might be considered risky, he also knew that his contemporaries did not, and that one false step would brand him as a speculator, and reduce the certainties which he had accomplished in the *Principia* and *Opticks* to the level of hypotheses.

Again, some of the main insights came relatively late in Newton's career and have been seen more as an extension of his earlier work in optics than as a separate investigation. Newton's career has been fitted to the familiar pattern of youthful achievement, consolidation in middle life, and decline in old age. It has been said that Newton's true originality and creativeness ended in 1666, or 1687, or 1696. However, although he ceased to do his own experiments after 1696, as President of the Royal Society from 1703 to 1727, he supervised, suggested or interpreted many of those of Hauksbee, Taylor and Desaguliers. These experiments led to some quite profound suggestions on the electrical nature of chemical operations and the electrical vibrations responsible for light and the forces of electricity and magnetism, and some searching investigations of the nature and strength of capillary action. Through the work of his disciples they influenced men such as Bošković and Priestley later in the century and ultimately led to the creation by Faraday and Maxwell of the concept of the electromagnetic field. Newton had not ceased creativity with the end of his work in optics and mechanics; he had rather, with a typical instinct for choosing the most significant problems, transferred his attention to a new frontier of research. Newton's main immediate influence, however, was on eighteenth century chemistry, and, because chemistry ultimately moved in a different direction, historians of chemistry have seen little to interest them in the speculations of Newton and the eighteenth century Newtonians.

Newton's ideas on the structure of matter filtered through only in unconnected fragments. This was a disadvantage for the development of science in his own time, but it is helpful to us, because it makes it possible to build up a picture of the way he thought. In 1672 he had sent to the Royal Society a brilliant account of his discovery of the analysis of white light, presented as a chain of deductive inferences.[1] Though we know that science does not really happen this way, presentations of this kind are important rhetorical strategies, necessary to convince others of the validity of the work. For the theory of matter, however, there was no such formal presentation. Even the Queries, which were the main published source, have the appearance of a collection of random notes, and the main manuscript sources are drafts, letters and even records of conversation. The

---

[1] 'A New Theory', *Corr.* I, 92–102.

very unfinished state of this type of material brings us closer to the actual processes by which the original insights came about.

The ultimate origin of Newton's line of thinking on matter lay, as with many of his contemporaries, in classical philosophy. Of all the books which were rescued from antiquity during the Renaissance, one was especially important. This was Lucretius's poem *De Rerum Natura*, discovered in the fifteenth century in a single copy. In the seventeenth century this led to a major revival of atomism by Pierre Gassendi.[2] Gassendi's follower, Walter Charleton, became a significant early influence on Newton.[3] Newton was a convinced atomist from his student days, but atoms were not necessarily for him the ultimate particles. In his very early *Quaestiones quaedam philosophicae*, he considered four possibilities for the 'first matter' or ultimate material particles: 'Whither it be mathematicall points: or Mathematicall points & parts: or a simple entity before division indistinct: or individualls i.e. Attomes'.[4] He immediately opted for atoms and he never deviated from this choice.

> Not of mathematical points since what wants dimensions cannot constitute a body in their conjunction...Not of parts and mathematical points: for such a point is either something or nothing...Not of simple entity before division indistinct for this must be an union of the parts into which a body is divisible...It remains therefore that the first matter must be atoms. And that matter may be so small as to be indiscerptible...[5]

His original notion of atom was influenced by the *minima naturalia* of Henry More, which were bodies almost infinitesimally small but also indivisible, bodies with solidity but minimum extension, and his subsequent conception of ultimate particles was not significantly different.

> Suppose then a number of mathematical points were indued with such power as that they could not touch nor be in one place...Then add these as close in a line as they can stand together every point added must make some extension to the length because it cannot sink into the former's place, or touch it...The distance then twixt each point is the least that can be and so little may an atom be and no less...[6]

Newton was, of course, aware of the classical origins, stating, in a draft addition to the *Principia*: 'That all matter consists of atoms was a very

---

[2] Gassendi (1649).
[3] Charleton (1654).
[4] QQP, 'Off the first mater', f. 88ʳ, NP.
[5] QQP, 'Off the first mater', f. 88ʳ, f. 89ʳ, NP.
[6] QQP, 'Of Attomes', f. 89ʳ, NP.

ancient opinion'.[7] The concept was connected with both solidity and gravity. In a draft conclusion for *Opticks*, he asserted that 'All solid bodies are (composed of?) Atoms'.[8] And in a draft Corollary to Proposition 6 of the *Principia*, aimed against the followers of Descartes: 'Without atoms gravity cannot be explained mechanically'.[9] If atoms exist, then there must also be vacuum. 'If all the solid particles of all bodies are of the same density, nor can be rarefied without pores, a void, space, or vacuum must be granted'.[10]

Newton, however, distinguished atoms and fundamental particles. It is notable that he seldom used the word 'atom' in later publications, except in historical contexts, apparently preferring terms such as 'particle', which were more noncommittal and allowed scope for a hierarchical theory of matter with component bodies at different levels of organisation, in line with an alchemical tradition also represented in Boyle, but abstracted in Newton to a hierarchical theory of interactions. Even in the early *Quaestione*, there was the possibility of something like molecules as well as atoms, or structured atoms as well as indestructible basic particles: 'The first matter must be homogeneous...and of the same constitution will all the parts be into which it is divisible...'[11] 'But it may be that the particles of compound matter were created bigger than those which serve for other offices'.[12]

His ultimate particles are fundamental particles in our terms.

> All these things being consider'd, it seems probable to me, that God in the Beginning form'd matter in solid, massy, hard, impenetrable, moveable Particles, of such Sizes and Figures, and with such other Properties, and in such Proportion to Space, as most conduced to the End for which he form'd them; and that these primitive Particles being Solids, are incomparably harder than any porous Bodies compounded of them; even so very hard, as never to wear or break in pieces; no ordinary Power being able to divide what God himself made one in the first Creation. While the Particles continue entire, they may compose Bodies of one and the same Nature and Texture in all Ages: But should they wear away, or break in pieces, the Nature of Things depending on them, would be changed. Water and Earth, composed of old worn Particles and Fragments of Particles, would not be of the same Nature and Texture now, with Water and Earth composed of entire Particles in the Beginning. And therefore, that Nature may be

---

[7] Draft addition to *Principia*, III, MS related to Classical Scholia, 1690s; McGuire and Rattansi (1966, 115).
[8] Draft conclusion for *Opticks*, Book IV, c 1690, Hypothesis IV; Cohen (1966).
[9] Draft Corollary to *Principia*, III, Proposition 6; McGuire (1966, 210).
[10] *Principia*, III, Proposition 6, Corollary 4.
[11] QQP, 'Off ye first mater', f. 88^v (cancelled), NP (diplomatic text).
[12] QQP, 'Conjunction of bodys', f. 90^v, NP.

lasting, the Changes of corporeal Things are to be placed only in the various Separations and new Associations and Motions of these permanent Particles; compound Bodies being apt to break, not in the midst of solid Particles, but where those Particles are laid together, and only touch in a few Points.[13]

In a draft version, he says:

While these particles continue entire corporeal nature may continue the same, & produce the same sorts of fluids & solids in all ages but should these be broken into less particles, the nature of things would be altered. For the broken particles would scarce {move} & convene & stick together any more in the same manner & form as they do at present, unless reunited by a divine power.[14]

We now call the ultimate particles 'fermions'. Though they come in different varieties, we can regard them as versions of a single type; and they have the same permanence. Their number never changes; it is conserved.[15] Exactly twelve fermions are known: electron, muon and tau, with their respective neutrinos, and up, down, charm, strange, top and bottom quarks; and there are also twelve corresponding antifermions. They are the only truly fundamental particles known, bosons appearing only as the consequences of fermion interactions.

For Newton, the corpuscles of matter had inner structure and inner motions. He was certainly no believer in a universe composed of billiard-ball atoms; his hard impenetrable particles were the ultimate components of matter, not the composite atoms which took part in chemical operations and reflected light. He says in a draft Query: 'I reccon these particles to be solid, least compound bodies should be porous in infinitum [I reccon them the least in Nature least matter should be actually divided in infinitum &]'.[16] Interactions between particles took place without physical contact. Matter was an extraordinarily intricate web of interacting particles in motion behaving as centres of various forces; the true inner structure of material particles had yet to be determined.

Newton had, of course, no conception of antimatter, but his discussion of holes in space, with properties identical to those of particles, in *De*

---

[13] Q 31.
[14] Draft Q 31.
[15] That is the *fermion number*, which counts every fermion as 1 and every antifermion as -1. Fermions can only be created or destroyed at the same time as the same number of antifermions.
[16] Draft Q for *Opticks*, UL Cambridge, Add. 3970.3, 243^r, NP.

gravitatione et aequipondio fluidorum, is of great interest in showing that the category of thought which would allow a description of antimatter in those terms was already present in his work, and that its origin was purely abstract, deriving more from metaphysics and theology than from induction from experiment.[17] This abstract definition of particle properties is continued in *Principia*, with an appeal to self-similarity:

> We no other way know the extension of bodies than by our senses, nor do these reach it in all bodies; but because we perceive extension in all that are sensible, therefore we ascribe it universally to all others also. That abundance of bodies are hard, we learn by experience; and because the hardness of the whole arises from the hardness of the parts, we therefore justly infer the hardness of the undivided particles not only of the bodies we feel but of all others. That all bodies are impenetrable, we gather not from reason, but from sensation. The bodies which we handle we find impenetrable, and thence conclude impenetrability to be an universal property of all bodies whatsoever. That all bodies are moveable, and endowed with certain powers (which we call the *vires inertiæ*) of persevering in their motion, or in their rest we only infer from the like properties observed in the bodies which we have seen. The extension, hardness, impenetrability, mobility, and *vis inertiæ* of the whole, result from the extension, hardness, impenetrability, mobility, and *vires inertiæ* of the parts; and thence we conclude the least particles of all bodies to be also all extended, and hard, impenetrable, and moveable, and endowed with their proper *vires inertiæ*. And this is the foundation of all philosophy. Moreover, that the divided but contiguous particles of bodies may be separated from one another, is matter of observation; and, in the particles that remain undivided, our minds are able to distinguish yet lesser parts, as is mathematically demonstrated. But whether the parts so distinguished, and not yet divided, may, by the powers of Nature, be actually divided and separated from one another, we cannot certainly determine. Yet, had we the proof of but one experiment that any undivided particle, in breaking a hard and solid body, offered a division, we might by virtue of this rule conclude that the undivided as well as the divided particles may be divided and actually separated to infinity.[18]

4.2. Internal Forces and Matter

Newton felt that he had successfully explained the large-scale motions in the universe on the basis of gravitational forces. He wrote in a draft preface to the *Principia*: 'How the great bodies of ye Earth Sun Moon & Planets gravitate towards one another what are ye laws and quantities of their gravitating

---

[17] *De gravitatione*, Unp., 139, quoted in *Newton and the Great World System*, 4.1.
[18] III, Rule III, 1713.

forces at all distance from them & how all ye motions of these bodies are regulated by their gravities I shewed in my Mathematical Principles of Philosophy to the satisfaction of my readers'.[19] He was, of course, well aware that other forces existed and actively pursued them, in particular, the electric, magnetic and optical forces, which he eventually conjectured had a common origin.

After the *Principia*, he felt that the next stage in the elucidation of nature would be to explain the forces that determined the small-scale structure of matter:

> I wish we could derive the rest of the phænomena of nature by the same kind of reasoning from mechanical principles; for I am induced by many reasons to suspect that they may all depend upon certain forces by which the particles of bodies, by some causes hitherto unknown, are either mutually impelled towards one each other, and cohere in regular figures, or are repelled and recede from one each other; which forces being unknown, philosophers have hitherto attempted the search of nature in vain...[20]

He had come to realise that gravity was very different from the interparticulate forces involved in the generation of light and heat. In early drafts of the General Scholium he emphasized that the laws to which these small-scale forces are subject 'are completely different from the laws of gravity', and, from the time of the *Principia*, he laid stress on the fact that gravity, unlike all other forces, depended only on the quantity of matter and could not be changed by other agencies. Nevertheless, the analogy of nature suggested that the large-scale and small-scale forces should have some similarities.

Some of the small-scale forces, such as electricity and magnetism, were known, but there may be other forces, as yet unknown.

> There are however innumerable other local motions which, on account of the minuteness of the moving particles cannot be detected, such as the motions of the particles in hot bodies, in fermenting bodies, in putrescent bodies, in growing bodies, in the organs of sensation and so forth. If any one shall have the good fortune to discover all these, I might almost say that he will have laid bare the whole nature of bodies so far as the mechanical causes of things are concerned. I have least of all undertaken the improvement of this part of philosophy. I may say briefly, however, that nature is exceedingly simple and conformable to herself. Whatever reasoning holds for greater motions, should also hold for lesser ones as well. The former depend upon the greater attractive forces of larger bodies, and I suspect that the latter

---

[19] Draft conclusion for *Opticks*, Book IV, c 1690; Westfall (1971, 374) and (1980, 521).
[20] *Principia*, Preface.

depend upon the lesser forces, as yet unobserved, of insensible particles. For, from the forces of gravity, of magnetism and of electricity it is manifest that there are various kinds of natural forces, and that there may be still more kinds is not to be rashly denied. It is very well known that greater bodies act mutually upon each other by those forces, and I do not see why lesser ones should not act on one another by similar forces.[21]

He was especially aware that the electric force, and others so far undetected, might well extend to short distances beyond the range of observation: 'The Attractions of Gravity, Magnetism, and Electricity, reach to very sensible distances, and so have been observed by vulgar Eyes, and there may be others which reach to so small distances as hitherto escape Observation; and perhaps electrical Attraction may reach to such small distances, even without being excited by Friction'.[22] (The statement about 'electrical Attraction' at small distances was added in 1717 to a Query originally published in 1706, so ensuring that the significance of the electrical force was not completely excluded from the *Opticks*.)

We should, therefore, set up inquiries to investigate by experiment whether such forces exist within the structure of matter and to specify their properties.

I therefore propose the inquiry whether or not there be many forces of this kind, never yet perceived, by which the particles of bodies agitate one another and coalesce into various structures. For if nature be simple and pretty conformable to herself, causes will operate in the same kind of way in all phenomena, so that the motions of smaller bodies depend upon certain smaller forces just as the motions of larger bodies are ruled by the greater force of gravity. It remains therefore that we inquire by means of fitting experiments whether there are forces of this kind in nature, then what are their properties, quantities and effects. For if all natural motions of great or small bodies can be explained through such forces, nothing more will remain than to inquire the causes of gravity, magnetic attraction and the other forces.[23]

The idea that other forces existed apart from gravity, electricity and magnetism, could be seen as a notable intuition, for it was made without direct evidence, and it combined effectively with his belief, with its alchemical overtones, that they must be seated deep within matter. There had to be other forces besides the more easily observed gravity, electricity and

---

[21] Draft *Conclusio*; Unp., 333.
[22] Q 31.
[23] Draft *Praefatio*; Unp., 307.

magnetism, if nature was 'very consonant and conformable to her self'.[24] The last point was a statement he repeated on many occasions; nature was 'conformable to herself' because simple and basic principles hold universally.

Writing of the forces that prevailed in the microstructure of matter, Newton had a very definite grasp of the idea that smaller distance involved stronger attractions, so there must be a hierarchy of attractions — exactly the modern reasoning. In this hierarchical theory of matter, interparticulate forces were stronger at shorter distances. He was able to point to the immense strengths of these interactions as several calculations showed.[25] A structural hierarchy allowed Newton to accommodate a force, which his experiments on electricity suggested was mostly hidden or screened, exactly as in modern theories of matter. Postulated in a draft of the 1690s, such hidden or screened forces would emerge, as he knew, from interactions dependent on inverse powers of the distance greater than 2. Newton had experience of forces based on these powers in special cases of the actions of gravity and magnetism, but he did not know, as we do now, that such forces often result from the presence of equal numbers of sources of opposite sign or effect, and that the cancellation of their effects on the large scale can usually be attributed to the reduction in range due to dipolar and other such combinations of the sources.

In a draft of the 1690s for the conclusion to a suppressed fourth Book of the *Opticks*, Newton invokes an inverse power of distance greater than 4 as the critical case for the screening effect in intermolecular attractions. He also postulates spheres of activity within which the particles of bodies attract or shun one another, foreshadowing the eighteenth century concepts of point-centres of force, with extension representing spheres of influence, alternately attractive and repulsive. He says how he had explained the interactions of large bodies, in the *Principia*, as resulting from gravity, but that he had avoided discussing the smaller-scale interactions between the particles of matter in order to avoid prejudicing readers against the book's main conclusions, '& yet I hinted [at them] both in the Preface & in ye Book it self where I speak of the [refraction] of light & of ye elastick power of ye Air: but [now] the design of y^t book being secured by the approbation of Mathematicians, [I have] not scrupled to propose this Principle in plane words'. The 'plane words' in manuscript, however, did not result in plain words in print.

---

[24] Q 31.
[25] Q 31.

Hypoth. 1. The particles of bodies have certain spheres of activity wth in wch they attract or shun one another. For ye attractive vertue of the whole magnet is composed of ye attractive vertues of all its particles & the like is to be understood of the attractive vertues of electrical & gravitating bodies. And besides these the particles of bodies may be endued with others not yet known to us. For if they have any forces decreasing in any ratio greater than the quadruplicate one of the distance from their centers, they may attract one another very strongly & yet great bodies composed of such particles shall not attract one another sensibly. And if Nature be most simple & fully consonant to her self she observes the same method in regulating the motions of smaller bodies wch she doth in regulating those of the greater.

Hypoth. 2. As all the great motions in the world depend upon a certain kind of force (wch in this earth we call gravity) whereby great bodies attract one another at great distances: so all the little motions in ye world depend upon certain kinds of forces whereby minute bodies attract or dispel one another at little distances . . . .

Hypoth. 3. The spheres of activities of bodies do not resist or promote the motion of other bodies through them any otherwise then by causing them to be attracted or repelled to or from the centers of those spheres. For the cause of gravity retards not the motions of the heavenly bodies, nor does the cause of magnetical. Nor is light resisted in its passage through uniform mediums.[26]

The truth of this Hypothesis I assert not because I cannot prove it is, but I think it is very probable because a great part of the phaenomena of nature do easily flow from it wch seem otherways inexplicable: such as are chymical solutions precipitations philtrations, detonizations, volatizations, fixations, rarefactions, condensations, unions, separations, fermentations: the cohesion texture firmness fluidity & porosity of bodies, the rarity & elasticity of air, the reflexion & refraction of light, the rarity of air in glass pipes & ascention of water therein, the permiscibility of some bodies and impermiscibility of others, the conception & lastingnesse of heat, the emission & extinction of light, the generation & destruction of air, the nature of fire & flame, ye springinesse or elasticity of hard bodies.[27]

The draft does not indicate how Newton calculated the minimum inverse fourth-power law for cohesion and the screening effect in intermolecular attractions, but in Proposition 86 of *Principia*, Book I, citing examples 2 and 3 of Proposition 81, he had shown that an inverse cube or an inverse fourth power attraction on a corpuscle by the particles of a sphere would tend to

---

[26] Draft conclusion for *Opticks*, Book IV, c 1690; Cohen (1966); Bechler (1974, 197).
[27] *ibid.*, Westfall (1971, 380) and (1980, 521–522).

produce an infinite degree of attraction at the sphere's surface, which rapidly falls away further out.[28] If the force is proportional to, say, $1/r^n$, where r is the typical interatomic or intermolecular distance, then the potential energy varies as $-1/r^{n-1}$. If we take out the effects on molecules due to their nearest neighbours, and assume the number of the more distant molecules is proportional to the volume, we find that the total potential energy is proportional to the integral of $4\pi r^2 dr\, r^{n-1}$, taken over the dimension of the whole body. For n less than or equal to 4, the integral increases with the size of the body; but for n greater than 4 it decreases, tending to 0 as the size of the body becomes infinite. So, for bodies subject to intermolecular forces varying with inverse powers of distance greater than 4, the total potential energy depends only on the interactions between nearest neighbours, and so the effect of the forces is only short-range, as Newton had proposed.

In Propositions 85 and 86, Newton showed that, if the attraction of the body attracted is vastly stronger when it is separated 'by a very small interval; the forces of the particles of the attracting body decrease', in a ratio more than the square of the distance of these particles,[29] and that, if the forces of the particles which compose the attracting body decrease as the cube of the distances, 'the attraction will be vastly stronger in the point of contact' than when the attracting body and the body attracted are separated by only a small interval; he was then able to extend these results to bodies of any shape.[30] The cohesion of bodies in direct contact is now explained by a van der Waals force which usually varies as something like the inverse-seventh power of the distance but this relationship ultimately emerges from component dipole-dipole forces which vary with the inverse-fourth power, and which are, of course, dual to inverse-seventh power forces.

Principia, I, Proposition 87 (with Corollary 1), gives the scaling law of Newton, for the force exerted at an external point by 'an arbitrary distribution of mass specified by a density function' when the law of centripetal attraction, by the elements of the mass of the distribution, is 'inversely as the nth power of the distance'.[31] 'Therefore', he says in Corollary 1, 'if, as the distances of the corpuscles attracted increase, the

---

[28] Newton writes that 'the attraction is infinitely increased when the attracted corpuscle comes to touch an attracting sphere of this kind'.
[29] I, Proposition 85. This proposition begins the Section (XIII) on attractions to nonspherical bodies (projected in a Scholim at the end of the preceding section) which continues until Proposition 93.
[30] I, Proposition 86.
[31] See Chandrasekhar (1995, 306).

attractive forces of the particles decrease in the ratio of any power of the distances, the accelerative attractions towards the whole bodies will be as the bodies directly, and those powers of the distances inversely'. 'Hence, on the other hand', according to Corollary 2, 'from the forces with which like bodies attract corpuscles similarly situated, may be found the ratio of the decrease of the attractive forces of the particles as the attracted corpuscle recedes from them, in which case 'that decrease is directly or inversely in any ratio of the distances'. Proposition 88 gives the case where the attractive forces of the particles are 'as the distance of the places from the particles'. The force is the same as that of 'a globe, consisting of similar and equal matter, and having its centre in the centre of gravity'. This is the only example in which we can find the attraction by an arbitrary mass distribution. The next theorem, Proposition 89, provides a generalisation in which the point-masses of Proposition 88 are replaced by bodies of finite size with density distributions arranged to be spherically symmetric, and composed of particles attracting particles external to them with a force proportional to the distances between them. In both cases, the result is elliptical motion of the attracted body round the centre of gravity of the mass distribution.[32]

4.3. A Hierarchical Structure

Newton saw that, if matter was constructed on the basis of internal forces, then it would be mostly composed of empty space, though this in itself would not give us the true inner structure of material particles. He suggested one or two ways in which matter could be organised to be highly porous, mainly to calculate as simply as possible the relative portions of vacuum to matter, though he did not commit himself to these or to any other kind of hypothetical model, saying that the true inner structure of material particles was still unknown. In reference to this theory, Joseph Priestley wrote, in 1777, that: 'It has been asserted... that... all the solid matter in the solar system might be contained within a nut-shell'.[33]

Observing the structure called 'The Net' in alchemical experiments, may have led Newton to use nets as an analogy for the microstructure of matter.[34] Here was a way in which seemingly solid matter could be replaced

---

[32] Corollaries to Propositions 88 and 89.
[33] Priestley (1777, 14).
[34] Dobbs (1975, 161–164). KCC, Keynes MS 19, f. 4^v; MS Add. 3975, f. 43^r; *Sendivogius Explained*, Keynes MS 55, 1680s, ff. 73^v–8^r, 13^v.

by nearly point-like particles, bound together in regular 3-dimensional arrays in structures arranged like nets, with forces acting between them across a nearly empty space.

> One must have recourse to a certain wonderful and exceedingly artificial texture of the particles of bodies by which all bodies, like networks, allow magnetic effluvia and rays of light to pass through them in all directions and offer them a very free passage: and by such an hypothesis the rarity of bodies may be increased at will. Salts form regular figures in congelation and certain of these such as nitre and sal ammoniac are always transformed into branches. Why should it not be that the first seeds of all things com(e) together geometrically in net-like figures by the force of nature...?[35]

Gravity proportional to mass demanded that particles of matter must be structured in more subtle ways than a compactified mass.

> They are mistaken therefore who join the least particles of bodies together in a compact mass like grains of sand or a heap of stones. If any particles were pressed together so densely, the gravitating cause would act less towards the interior ones than towards the exterior ones and thus gravity would cease to be proportional to the matter. Other textures must be devised by which their interstices are rendered more ample.[36]

The same was required to explain why magnetism could act through 'solid' gold.

> Truly, so great is the abundance of pores in gold that the action of a magnet on iron is propagated through intervening gold very freely and without sensible diminution. Gold indeed must be believed to consist of larger particles and these of smaller ones; and the aforesaid liquids must be supposed to permeate the interstices between the larger ones, and magnetic effluvia to pass through the interstices between the smaller ones.[37]

'Gold has particles which are mutually in contact: their sums are to be called sums of the first composition and their sums of sums, of the second composition, and so on'.[38]

S. I. Vavilov, in the Tercentenary volume, compared Newton's ideas on matter of the first and second compositions to Rutherford's theory of the

---

[35] Proposed corollary to *Principia*, III, Proposition 6, Corollaries 4 and 5, c 1693; *Unp.*, 317. Westfall (1980, 509), gives the following version of the beginning of this passage: 'We must have recourse to some wonderful and very skillfully contrived texture of particles whereby all bodies, on the pattern of nets, lie open and offer unrestricted passage in all directions both to magnetic effluvia and to rays of light'.
[36] Draft addition for *Principia*, III, Prop. 6, Coroll. 4 and 5, 1690s; *Unp.*, 315.
[37] *ibid.*, 316–317.
[38] *De natura acidorum*, 1691/92; *Corr.* III, 211.

atom,[39] and so it is interesting that Newton, on one occasion, suggested that material particles might be composed of a hard nucleus surrounded by a repulsive sphere of a more 'tenuous matter'.[40] Though this very probably had an alchemical origin, it might also be thought of as a natural outcome of a hierarchical theory of matter in which the smaller units were more impenetrable than the larger, and matter was mostly empty space, though giving the impression of bulk.

Newton, as we have said, suggested some ways in which the porous structure might be accomplished, though only by way of illustration, and not as a precise hypothesis.

> Now if we conceive these Particles of Bodies to be so disposed amongst themselves, that the Intervals or empty Spaces between them may be equal in magnitude to them all; and that these Particles may be composed of other Particles much smaller, which have as much empty Space between them as equals all the Magnitudes of these smaller Particles: And that in like manner these smaller Particles are again composed of others much smaller, all of which together are equal to all the Pores or empty Spaces between them; and so on perpetually till you come to solid Particles such as have no Pores or empty Spaces within them: And if in any gross Body there be, for instance, three such degrees of Particles, the least of which are solid; this Body will have seven times more Pores than solid Parts. But if there be four such degrees of Particles, the least of which are solid, the Body will have fifteen times more Pores than solid Parts. If there be five degrees, the Body will have one and thirty times more Pores than solid Parts. If six degrees, the Body will have sixty and three times more Pores than solid Parts. And so on perpetually. And there are other ways of conceiving how Bodies may be exceeding porous. But what is really their inward Frame is not yet known to us.[41]

Whatever precise structure was assumed, it was clear that materials such as water and gold must include a considerable portion of vacuum between the particles.

> If the proportion of the particles of both kinds to their respective interstices is assumed to be as that between sand and the vacant spaces between them, which is as 7 to 6, the vacuum in gold will be to its matter as 5 to 2 roughly and the vacuum in water to its matter as 65 to 1. And yet the parts of water are so arranged among themselves that they cannot be

---

[39] Vavilov (1947, 48).
[40] *De aere et aethere*, Unp., 223, quoted in 6.2.
[41] *Opticks*, 1706/17, addendum to Book II, Part 3, Proposition 8. A draft Query 16 takes the calculation to 30 degrees and 10^9 more pores than parts before saying '& so on perpetually' (CUL Add MS 3970.3, f. 234).

condensed by compression. Moreover, since air in our regions is 900 times lighter than water, its solid matter will scarcely fill the 50,000th part of the space through which it is spread. In higher regions it will be more rare still ad infinitum . . . . . . if anyone shall devise an Hypothesis by which the body of water can be so rare that its solid parts scarcely fill the sixtieth part of its own bulk, and yet does not yield at all to compression, then he may easily understand how gold can be as rare as we proved water to be, and how the rarity of water and air may be increased in the same proportion; but the rarity of gold and of all bodies he will increase at will.[42]

The porous structure of bodies allows light to pass through transparent ones.

> The Magnet acts upon Iron through all dense Bodies not magnetick nor red hot, without any diminution of its Virtue; as for instance, through Gold, Silver, Lead, Glass, Water. The gravitating Power of the Sun is transmitted through the vast Bodies of the Planets without any diminution, so as to act upon all their parts to their very centers with the same Force and according to the same Laws, as if the part upon which it acts were not surrounded with the Body of the Planet. The Rays of Light, whether they be very small Bodies projected, or only Motion or Force propagated, are moved in right Lines; and whenever a Ray of Light is by any Obstacle turned out of its rectilinear way, it will never return into the same rectilinear way, unless perhaps by very great accident. And yet Light is transmitted through pellucid Bodies in right Lines to very great distances. How Bodies can have a sufficient quantity of Pores for producing these Effects is very difficult to conceive, but perhaps not altogether impossible.[43]

Many statements in the *Opticks* concerned the relationship between the density of substances like water and the passage of light resulting from their transparency and porosity, but, in this passage, we have a good example of Newton's almost automatic method of synthesizing the disparate aspects of his knowledge into ever more powerful unifications. Here the synthesis involves the structure of matter allowing the passage of light and magnetic influences, while simultaneously explaining the seemingly separate, and far from obvious, fact that gravity cannot be shielded.

Bodies must be composed of hard particles, but how such particles hold together with so few points of contact was hard to understand.

> The Parts of all homogeneal hard Bodies which fully touch one another, stick together very strongly. And for explaining how this may be some have invented hooked Atoms . . . ; and others tell us that Bodies are glued together

---

[42] Draft addition for *Principia*, III, Prop. 6, Coroll, 4 and 5, 1690s; *Unp.*, 317.
[43] *Opticks*, 1706/17, addendum to Book II, Part 3, Proposition 8.

by rest...; and others, that they stick together by conspiring Motions... I had rather infer from their Cohesion, that their Particles attract one another by some Force, which in immediate Contact is exceeding strong, at small distances performs the chymical Operations above-mention'd, and reaches not far from the Particles with any sensible Effect. All Bodies seem to be composed of hard Particles: For otherwise Fluids would not congeal... Even the Rays of Light seem to be hard Bodies... And therefore Hardness may be reckon'd the Property of all uncompounded Matter. At least, this seems to be as evident as the universal Impenetrability of Matter. For all Bodies, so far as Experience reaches, are either hard, or may be harden'd; and we have no other Evidence of universal Impenetrability, besides a large Experience without an experimental Exception. Now if compound Bodies are so very hard as we find some of them to be, and yet are very porous, and consist of Parts which are only laid together; the simple Particles, which are void of Pores, and were never yet divided, must be much harder. For such hard Particles, being heaped up together, can scarce touch one another in more than a few Points, and therefore must be separable by much less Force than is requisite to break a solid Particle, whose Parts touch in all the Space between them, without any Pores or Interstices to weaken their Cohesion. And how such very hard particles which are only laid together and touch only in a few Points, can stick together, and that so firmly as they do, without the assistance of something which causes them to be attracted or press'd towards one another, is very difficult to conceive.[44]

Newton saw the colours produced in thin films as a source of information on the microstructure of matter. According to H. W. Turnbull, he proceeded 'from the assumption that the structure of matter is analogous to that of a developed layer in a Lippmann photograph'.[45] In this kind of early colour photograph, the development process creates submicroscopic grains of silver in a series of parallel layers, whose spacing is that of the wavelengths corresponding to the colours to be reproduced. He writes: '...the several parts or corpuscles of which a shining body consists...must be supposed of various figures sizes and motions...'[46] 'The bigness of the component parts of natural Bodies may be conjectured by their Colours'.[47] 'Between the parts of opake and colour'd Bodies are many Spaces, either empty, or replenish'd with Mediums of other Densities; as...for the most part Spaces void of both Air and Water, but yet perhaps not wholly void of all Substance, between the parts of hard Bodies'. 'For...there are many Reflexions made by the internal parts of Bodies, which...would not happen if the parts of

---

[44] Q 31.
[45] H. W. Turnbull, *Corr.* I, 391.
[46] Newton to Oldenburg, 3 April 1673 (in reply to Huygens); *Corr.* I, 264.
[47] *Opticks*, Book II, Part 3, Proposition 7.

those Bodies were continued without any such Interstices between them; because Reflexions are caused only in Superficies, which intercede Mediums of a differing density....'[48]

The smallest parts of bodies must be transparent, and opacity due to multiple internal reflections. 'The Parts of Bodies on which their Colours depend, are denser than the Medium which pervades their Interstices'.[49] 'The least parts of almost all natural Bodies are in some measure transparent: And the Opacity of those Bodies ariseth from the multitude of Reflexions caused in their internal Parts. That this is so has been observed by others, and will easily be granted by them that have been conversant with Microscopes'.[50] 'The Parts of Bodies and their Interstices must not be less than of some definite bigness, to render them opake and colour'd. For the opakest Bodies, if their parts be subtilly divided, (as Metals, by being dissolved in acid Menstruums, & c.) become perfectly transparent'.[51] They must behave in the same way as transparent plates in being subjected to fits of easy reflection and easy refraction, and their thicknesses must be less than the intervals of such fits. 'The transparent parts of Bodies, according to their several sizes, reflect Rays of one Colour, and transmit those of another, on the same grounds that thin Plates or Bubbles do reflect or transmit those Rays'.[52]

Newton considered it possible that the colours of various bodies could be used to conjecture the sizes of the component parts which produced them, by analogy with the production of colours by thin films or plates.[53] While this has generally been considered one of his less successful speculations, as the mechanism of colour production is actually quite complicated and only a few natural objects, such as the feathers of peacock's tails, directly involve colouring by thin film effects (or, in some cases, by Bragg reflection off a liquid crystalline structure), it actually involves a profound principle. There is a sense in which all the processes of colour production depend, *at the quantum level*, on length scales for units of matter (that is, for groups of fundamental particles) determining the wavelengths of light which they can emit, absorb or reflect. Though we now know that this is a much more involved process than the simple one generated by the Newton's rings

---

[48] *ibid.*, Proposition 3.
[49] *ibid.*, Proposition 6.
[50] *ibid.*, Proposition 2.
[51] *ibid.*, Proposition 4.
[52] *ibid.*, Proposition 5.
[53] *ibid.*, Proposition 7.

experiment, there is nevertheless a fundamental relationship, at all levels where quantum effects (including emission, absorption and reflection) are important, between the 'size' of a matter unit, for example a nucleus and electron combination, and the wavelength of radiation with which it will interact. In all such cases, a material length scale defines a distinction between wavelengths allowed and wavelengths rejected. This principle would later become the basis of the application of quantum constraints to the atom in the successive theories of Bohr, de Broglie and Schrödinger.

In his attempt at deriving the sizes of material particles by the colours of material bodies, Newton made much of the possibility of identifying the *order* of the colouring with relation to the orders in the Newton's rings pattern. Metals, which were, in his view, the materials closest to the most primitive form of matter at the base of everything, and were presumed from their densities to have relatively small component particles with minimal spacings, seemed to reflect white of the *first* order, that of the central spot in the ring pattern, where the spacing between lens and glass plate is as close as it can be, and where no characteristic distance determines the transmission or reflection of a single colour. In the modern theory of metals, the reflection is due to free electrons, the smallest conceivable material units within ordinary matter, which reflect light of all wavelengths without the restrictions imposed upon the electrons within atomic orbitals. For most metals, this is the dominant effect, as the electrons in orbitals produce emissions outside the visible region. However, a few such as copper and gold also have visible emissions from the confined electrons at visible wavelengths. Newton saw this as a result of their having bigger component particles or structural units, especially as he thought their colours were of the second or third order. As is often the case with Newton's more conjectural ideas, the principle survives the failure of the specific details, and, as in other cases, this is because of the intrinsic validity of his analytical method working *inwards* from general principles rather than *outwards* towards them.[54] In the modern theory, as an *atomic*, rather than electronic, effect, the additional colouring observed in metals like copper and gold is due to larger structures within the material. In the case of gold, the large proton number intensifies the electrostatic attraction and brings the s and d orbitals closer together with a significant relativistic effect on the orbital velocities, thus allowing a transition in the visible region.

---

[54] In fact, it might be considered a specially significant case in respect of the very definite failure, on this occasion, of the details.

In establishing the periodicity of light as a fundamental aspect of its relationship to its colour-producing property, Newton correctly used the analogy of nature to infer that it would be significant at the level of the microscopic component particles of matter as well as at that of the macroscopic objects with which he was able to experiment, and his inference was based on a deep synthesis created from a range of ideas and observations, in his usual manner, not on an unsupported conjecture. His work on thick polished plates had shown that periodicity was a *coherent* phenomenon over distances thousands of times larger than the wavelengths of the vibrations, and that colour production could occur at the micro-level while being observed through the coherent ordering as a macro-process. He believed, however, on the evidence of his own experiments, that the colours of natural bodies were produced only by the largest particles in a hierarchy determined by a corresponding hierarchy of forces. In terms of the modern quantum theory, length scales involve corresponding levels of force or energy, and, while colours are easily changed in chemical reactions and through natural biological processes, they do not affect the innermost parts of matter which he suspected must be considerably smaller than the characteristic length scales involved in optical phenomena. In modern terms, visible light, like chemistry, is a manifestation of changes within electron orbitals and does not involve the atomic nucleus. It occurs at the atomic or molecular level of structural organisation within matter, rather than within its central core.

Optical microscopes, Newton believed, could eventually be improved to lead to the discovery of the structural units on which colours ultimately depend, but they would not be able to see at the level where particles became transparent.

> ...it is not impossible but that Microscopes may at length be improved to the discovery of the Particles of Bodies on which their Colours depend, if they are not already in some measure arrived to that degree of perfection. For if those Instruments are or can be so far improved as with sufficient distinctness to represent Objects five or six hundred times bigger than at a Foot distance they appear to our naked Eyes, I should hope that we might be able to discover some of the greatest of these Corpuscles. And by one that would magnify three or four thousand times perhaps they might all be discover'd, but those which produce blackness. ...it will add much to our Satisfaction, if those Corpuscles can be discover'd with Microscopes; which if we shall at length attain to, I fear it will be the utmost improvement of this Sense. For it seems impossible to see the more secret and noble Works of Nature within the Corpuscles by reason of their transparency.[55]

---

[55] *Opticks*, Book II, Part 3, Proposition 7. Newton had proposed in 'A New Theory' (1672) that a reflecting miscroscope might produce improvements in microscopy even greater than

Newton clearly did not believe that the sizes of 'atoms' or ultimate particles could be found from the size of the equivalent thin film reflecting light of a particular colour. It was rather the size of the minimum unit of matter causing that reflection which would be revealed in this way; and he was perfectly aware that colour or opacity was not a property of the smallest parts of bodies. Vavilov pointed out that, though Newton was about two orders of magnitude out with regard to the visibility of molecules under microscopes, the idea was right in principle, and, for the first time in the history of science, presented a testable conclusion based on the theory of atoms.[56]

Newton saw the problem in making 'wavelengths' greater than particle sizes:

> if the thickness of the Body be much less than the Interval of the Fits of easy Reflexion and Transmission of the Rays, the Body loseth its reflecting power. For if the Rays, which at their entering into the Body are put into Fits of easy Transmission, arrive at the farthest Surface of the Body before they be out of those Fits, they must be transmitted. And this is the reason why Bubbles of Water lose their reflecting power when they grow very thin; and why all opake Bodies, when reduced into very small parts, become transparent.[57]

As the reference to the reflecting power of bubbles might suggest, Newton, like Thomas Young after him, also compared 'molecular' separations, revealed by surface tension and capillarity, with optical 'wavelengths'. The black spot at the contact between glass and lens in the Newton's rings experiment was produced, at its perimeter, by a separation which could be no more than a fifth or sixth of the interval of fits of the adjacent red. The thickness of the plate separation in Hauksbee's capillary experiment with oil of oranges, 'where it appears very black, is three-eighths of the ten hundred thousandth part of an Inch', and the thickness of 'a single Particle of the Oil' had to be even smaller.[58] In unpublished calculations, the thickness of the oil drop was taken as 10^{-7} inches or 2.54 nanometres.[59] Young was certainly familiar with this earlier work of Newton — his originality lay in his *fixing* of the molecular size/optical wavelength ratio, which his predecessor had left undetermined.[60]

---

those of reflecting telescopes in telescopy, because only one reflecting surface was needed. He sketched out a design but never built the device.
[56] Vavilov (1947).
[57] *Opticks*, 1704, Book II, Part 3, Proposition 13.
[58] Q 31.
[59] *De vi electrica*; *Corr.* V, 368.
[60] Young (1807), 1: 459, 464; Young (1805); Young (1816).

Newton's early experimental work in optics had also led him to see the processes of reflection and refraction in terms of mechanisms which we would now describe as *field*-induced effects, and this again affected the way he saw the structure of matter:

> ...the Reflexion of a Ray is effected, not by a single point of the reflecting Body, but by some power of the Body which is evenly diffused all over its Surface, and by which it acts upon the Ray without immediate Contact... Now, if Light be reflected, not by impinging on the solid parts of Bodies but by some other principle; it's probable that as many of its Rays as impinge on the solid parts of Bodies are not reflected but stifled and lost in the Bodies... And hence we may understand that Bodies are much more rare and porous than is commonly believed. Water is nineteen times lighter, and by consequence nineteen times rarer, than Gold; and Gold is so rare as very readily and without the least opposition to transmit the magnetick Effluvia, and easily to admit Quicksilver into its Pores, and to let Water pass through it. For a concave Sphere of Gold filled with Water, and solder'd up, has, upon pressing the Sphere with great force, let the Water squeeze through it, and stand all over its outside in multitudes of small Drops, like Dew, without bursting or cracking the Body of the Gold, as I have been inform'd by an Eye witness. From all of which we may conclude, that Gold has more Pores than solid parts, and by consequence that Water has forty times more Pores than Parts. And he that shall find out an Hypothesis, by which Water may be so rare, and yet not be capable of compression by force, may doubtless by the same Hypothesis make Gold, and Water, and all other Bodies, as much rarer as he pleases; so that Light may find a ready passage through transparent Substances.[61]

The same was true also of refraction.

Once Newton had postulated a hierarchical, porous matter, composed of small or even point-like particles subject to internal forces, he had a general mechanism for explaining all the bulk physical properties of matter, and the transformations effected in nature. The vital agent acting to produce these transformations may have had an alchemical or Neoplatonic origin, but in later work it took on the characteristics of an 'electric spirit' and the body of light. He recognised that some things could not be changed by ordinary physical and chemical processes, but his hierarchical theory of matter enabled him to explain how some transformations could only take place at a more fundamental level of composition. This meant that matter could be all made of the same material, avoiding a multiplication of categories, but not subject to the same changes. There could be changes

---

[61] *Opticks*, 1706/17, addendum to Book II, Part 3, Proposition 8.

at different levels as well as matter at different levels. In this concept of the ultimate unity of matter, the traditions of Platonic philosophy, alchemy, theology and mechanistic philosophy converged in such a way as to suggest that they represented some more fundamental abstract truth of a simpler kind than could be derived from any of the individual traditions. This was the spirit in which he searched for the ultimate truths which he thought lay hidden behind the shifting figurative language of the alchemical authors.

Also resulting from this work was Newton's reversion to the mass model of optical dispersion, which he had originally entered in his student notebook alongside the velocity model. This model, in which the colours were dispersed according to the masses of their characteristic particles, was first made public in Query 22/29 of the Latin edition of the *Opticks* of 1706. The model, however, was never given a full mathematical treatment, according to Bechler, because it would have meant the introduction of new types of forces and fields — ones which involved nonlinear functions of mass. And it was because of this that Newton decided 'that the fundamental and universal force of the micro-realm was electricity, a decision which he could not publish without eschewing his former methodology', which assumed that the small-scale processes going on within matter were of the same kind as the large-scale processes due to gravitation.[62]

A conclusion of this kind would certainly have coincided with his realisation, from the experiments of Hauksbee in 1705–1706 and his own long-standing aethereal speculations, that the electric force was responsible for the emission, refraction, reflection and diffraction of light. The mass model of dispersion had been originally introduced with the intention of establishing light as the active principle in nature, and, by a process which is no coincidence, it had already led Newton to the conclusion that matter is interconvertible with light, and that light is the active principle, or, in more modern terms, the source of potential energy, within matter. At the same time, it also implied that there must be an inverse relation between wavelength and acceleration, of a similar kind to the modern conception of an inverse relation between wavelength and momentum. (For an impulse force, similar to a discrete quantum transition with action h, the Newtonian relation would be between wavelength and change of momentum.) These conclusions and several others, both quantitative and qualitative, were incorporated into manuscript drafts which Newton wrote for the *Queries* which

---

[62] Bechler (1974), quoting 189.

he added in the later editions of the *Opticks*, and which reveal the full story of Newton's struggle with the implications of the particle concept of light in a way that could not be at guessed if we only had the published sources.

Newton may have thought that the micro-forces looked 'very strange', but they could be viewed also as 'regular forces freshly reorganised'.[63] If the force on a light corpuscle was inversely proportional to mass,[64] perhaps more bulk would imply more cancellation of electric effects. Newton's light corpuscle, the modern photon, now has a place in the particle hierarchy. It is not a foundational particle, however, like the fermion, but a boson, which can be created or destroyed, absorbed or emitted. It is a form of matter, but, having no rest mass, is not matter in the ordinary sense. In addition, although it is the carrier of the electric interaction, electrical forces do not act on the photon (unless it can be split into fermion and antifermion). As Newton found, the rays of light are not changed by refraction.[65]

Newton's progress in trying to comprehend the nature of light and its interaction with matter, however, has many remarkable features, and it had significant importance for his theory of the structure of matter, and the forces that hold it together. One example we have seen is his idea that optical reflection and refraction results from action over the surface as a whole, which may be taken as the beginning of the idea of field theory in the case of the electric force.[66] Of course, the field theory of electricity would eventually lead to relativistic or 4-vector principles and to forces which are nonlinear with rest mass, and consequently also to the conversion of mass and energy. It would be this interconversion, together with the quantum condition relating energy and frequency, which would make it possible to reconcile the conflicting principles of least time and least action which are valid, respectively, for the actions of waves and particles, and so to reconcile the predictions of refraction for the wave and particle theories of light. And, ancestral to the successful application of least action was Newton's mass-dispersion model, with its essentially valid approach to refraction in terms of *momentum*.[67]

---

[63] Bechler (1974, 191–192). The fact that photons showed significant differences from particles of ordinary matter would eventually lead to the separation of fundamental particles into bosons and fermions.

[64] See *Newton and Modern Physics*, 4.4.

[65] The *interconvertibility*, but not equivalence, of light and ordinary matter in Newtonian theory becomes in a more modern perspective the interconvertibility, but not equivalence, of bosons and fermions in the weak interaction, though this also requires antifermions.

[66] *Opticks*, 1706/17, addendum to Book II, Part 3, Proposition 8.

[67] Bechler (1973, 1974).

4.4. Capillarity and the Force of Attraction

The true theory of capillarity as a force of cohesion came relatively late in Newton's work and is one of the few instances where he clearly rejected an erroneous earlier theory on the basis of subsequent experimental evidence. His original explanation had assumed that capillarity was a result of the unsociability of air with glass, and was based on the earlier ideas of Hooke and Boyle.[68] 'This also may be the principall cause of the cohæsion of the parts of Solids & Fluids, of the Springines of Glass & other bodyes whose parts Slide not one upon another in bending, and of the Standing of the Mercury in the Torricellian Experiment, sometimes to the top of the Glass, though a much greater height then 29 inches'.[69]

> ... and it may be demonstrated in several ways. As first that water ascends within a very narrow pipe whose lower end is immersed in stagnating water higher than the external level, and ascends the higher in proportion to the narrowness of the pipe, so that it will rise several inches in the narrowest pipes... And next, that when a glass jar filled with compressed ashes stands with its mouth immersed in stagnating water, it imbibes much of the water although no air escapes. And again, that water rises spontaneously in that paper or sheet of which filters are made. A rope, even when stretched by a heavy weight, is so swollen by the force of the absorption of water that the weight is lifted up... For this reason standing water creeps little by little up the sides of vessels...; and water commonly clings to the surfaces of all substances, or as we ordinarily say, wets everything.[70]

After Hauksbee had shown that the effect also took place in a vacuum, Newton developed the now accepted explanation that it was due to a force of adhesion between the liquid and the walls of the containing vessel. The revision of Newton's theory of capillarity did not actually require major changes in his work; descriptions of capillarity from such early works as *De aere et aethere* remained valid but were effectively reinterpreted in terms of the explanation presented in later works such as Query 31.

> ... if a large Pipe of Glass be filled with sifted Ashes well pressed together in the Glass, and one end of the Pipe be dipped into stagnating Water, the Water will rise up slowly in the Ashes, so as in the space of a Week or Fortnight to reach up within the Glass, to the height of 30 or 40 Inches above the stagnating Water. And the Water rises up to this height by the Action only of those Particles of the ashes which are upon the surface of the elevated Water; the Particles which are within the Water, attracting

---
[68] Boyle (1660); Hooke (1661).
[69] *Hyp., Corr.* I, 367.
[70] *De aere et aethere, Unp.*, 221.

or repelling it as much downwards as upwards. And therefore the Action of the Particles is very strong. But the Particles of the Ashes being not so dense and close together as those of Glass, their Action is not so strong as that of Glass, which keeps Quick-silver suspended to the height of 60 or 70 Inches, and therefore acts with a Force which would keep Water suspended to the height of above 60 Feet.

By the same Principle, a Sponge sucks in Water, and the Glands in the Bodies of Animals, according to their several Natures and Dispositions, suck in various Juices from the Blood.

In a draft *Conclusio* to the *Principia* (c 1687), Newton had written: 'It is probable that the particles of homogeneous bodies, whether solid or fluid, cohere by a mutual attraction, and thence it is that whenever the particles of quicksilver in a Torricellian tube are everywhere contiguous to each other and to the tube, the quicksilver can be suspended and sustained to a vertical height of 40, 50, 60 inches and more'.[71] Now, in the later editions of the *Opticks*, this becomes:

> The Atmosphere by its weight presses the Quick-Silver into the Glass, to the height of 29 or 30 Inches. And some other Agent raises it higher, not by pressing it into the Glass, but by making its Parts stick to the Glass, and to one another. For upon any discontinuation of Parts, made either by Bubbles or by shaking the Glass, the whole Mercury falls down to the height of 29 or 30 Inches.[72]

It was important now to find out where these forces originated. 'There are therefore Agents in Nature able to make the Particles of Bodies stick together by very strong Attractions. And it is the business of experimental Philosophy to find them out'.[73]

> ... what we have thus far said about these forces will appear less contrary to reason if one considers that the parts of bodies certainly do cohere, and that distant particles can be impelled towards one another by the same causes by which they cohere, and that I do not define the manner of attraction, but speaking in ordinary terms call all forces attractive by which bodies are impelled towards each other, come together and cohere, whatever the causes be.[74]

'But so great a force is abundantly sufficient to provide the cohesion of bodies'.[75]

---

[71] Draft *Conclusio*, *Unp.*, 336.
[72] Q 31.
[73] *ibid.*
[74] Draft *Praefatio*, *Unp.*, 345.
[75] *De vi electrica*, *Corr.* V, 368.

The case of capillarity was unusual for Newton in that the change in his views took place at such a late period. The use of the ideas of such predecessors and contemporaries as Descartes, Hooke and Boyle normally occurred only at an early stage in his thought. Though he often began with the standard hypotheses of his day, he usually proceeded to an independent analysis based on his own abstract categories. But such an analysis could only be based on whatever experimental information was then available, and the standard theory on this occasion fitted closely into his own mode of thinking on attractive and repulsive forces. The episode certainly provides strong backing for Newton's claim that, for him, experimental evidence was the starting point for all scientific investigation. Newton calculated the force of capillarity at 10^{-7} inches because this was the order of magnitude which he had derived from his optical studies for the sizes of material corpuscles. The strength but short range of this force probably led him to imagine that there were perhaps even greater attractive forces at an even shorter range within these atoms and that, to be short range, as he had already explained in the *Principia* and elsewhere, the forces must be proportional to inverse powers of the distance greater than 2.

The experiments of Hauksbee on capillary action were extensions of earlier ones in which he had long been interested. 'That the particles of bodies attract each other mutually at short distances is also manifest from this experiment'.[76] 'If two polished plane pieces of glass, with their parallel surfaces as contiguous as possible, are immersed in still water, the water ascends between the pieces of glass above the surface of the water, and the height of ascent will be inversely as the distance between the glasses'.[77] 'If the plates are slanted towards one another to meet along a line perpendicular to the horizon, the water will ascend between them and with its upper surface trace a hyperbolic line, one of whose asymptotes is the line of junction of the glass plates while the other asymptote is the line parallel to the horizon made by the surface of the standing water'.[78]

Now, the new experiments had shown that it was nothing to do with atmospheric pressure. 'This experiment succeeds in the Boylian vacuum and so does not depend on the weight of the incumbent atmosphere. The parts of the glass near to the rising water attract the water close to and below themselves, and make it rise'.[79] 'The particles of glass at the upper surface

---

[76] *ibid.*, 367.
[77] Draft for General Scholium (MS A), *Unp.*, 354.
[78] *De vi electrica*, *Corr.* V, 367.
[79] Draft for General Scholium (MS A), *ibid.*, 354

of the rising water attract the particles of the water nearest to themselves and below them and by attracting them draw them up, and these particles attract others and these yet others down to the standing water'.[80] 'The attraction is the same when the pieces of glass are at various distances apart and draws up the same weight of water, so that it causes the water to ascend higher, the closer the glasses are together. And for a similar reason water ascends in narrow glass tubes, and the higher when the tubes are narrower; and all liquids ascend in spongy substances'.[81]

The capillarity which causes the rise of a column of liquid in a narrow tube with a force inversely proportional to the diameter of the tube is due to the upward surface tension force on the exposed liquid molecules. Since the energy depends on the surface area, the force is inversely proportional to the length of a line on the surface. The surface tension T is measured in Nm^{-1}, and the force around the circumference where the liquid surface touches the tube is $2\pi r T = \rho g \pi r^2 h$, where ρ is the density of the liquid and g the acceleration of free fall. For a tube of radius r, the liquid will rise to a height $h = 2T/\rho r g$, explaining why the height is inversely proportional to the radius or diameter, and why the liquid assumes a hyperbolic surface at the boundary with the glass.

Newton's account of capillarity continues with a description of the experiment made by Hauksbee with oil of oranges. In a draft for the General Scholium of the second edition of the *Principia* (1713), he writes:

> Two polished plane pieces of glass were prepared, 20 (or 25) inches long and (3 or 4) inches wide. One lay parallel to the horizon, and upon one end of it was a drop of oil of oranges. The other was so placed upon the first that they touched one another at the other end; and at that end where the drop lay they were about 1/16th of an inch apart, and the upper glass touched the drop. As soon as this happened the drop began to move towards the meeting-point of the glasses. And the closer it was to the meeting-point, the faster it moved. This experiment also succeeds in vacuo. And the origin of this motion lies in the attraction of the glasses.[82] (The figures, which are not in the original draft, are in the published Query 31 in the *Opticks*.)

> If the glasses are raised slightly at their meeting-point so that the lower glass is inclined to the horizontal, the drop will ascend, and the upper glass will maintain its position with respect to the lower one. The ascending drop moves more slowly than at first and the greater the inclination of the glass,

---

[80] *De vi electrica*, *Corr.* V, 367.
[81] Draft for General Scholium (MS A), *Unp.*, 354.
[82] *ibid.*

the slower the motion of the drop, until it will come to rest, its weight being equal to the attraction of the glass. Thus from the inclination of the lower glass the weight of the drop is given, and from the weight of the drop the attraction of the glasses is given.[83]

In the *Opticks*, he says:

...the thickness of the Plate where it appears very black, is three eighths of the ten hundred thousandth part of an Inch. And where the Oil of Oranges between the Glasses is of this thickness, the Attraction collected by the foregoing Rule, seems to be so strong as, within a Circle of an Inch in diameter, to suffice to hold up a weight equal to that of a cylinder of Water an Inch in diameter, and two or three Furlongs in length. And where it is of a less thickness the Attraction may be proportionately greater, and continue to increase, until the thickness do not exceed that of a single Particle of the Oil.[84]

And in *De vi electrica*:

...furthermore the force of attraction is equal to the weight of a cylinder of the oil whose base is the same as the base of the drop and whose height is to 33.44/80 inches as 1/80 is to the thickness of the drop. Let the thickness of the drop be 10^{-7} inches and the height of the cylinder will be $33.40 \times 105/64$ or 52250 inches, which is 871 paces; and the weight of this will equal the force of attraction.[85]

According to Hauksbee's own account in *Physico-Mechanical Experiments* (1709): '...the attractive forces here do decrease in a greater proportion than that by which the squares of the distances do increase'.[86] The comment probably stems from Newton.

Since the intermolecular distance was derived directly from the study of optical interference phenomena, the force of capillary action at this distance could be compared directly with the optical force of refraction. The revised draft of Newton's refraction calculation on f. 621^r of Cambridge University Library MS Add 3970, based on the centripetal force per unit mass on a light corpuscle, is immediately followed by a statement about the capillary forces involved in the Torricellian experiment which was surely intended to link the two forces on account of their short range and great strength: 'The Atmosphere by its weight presses the Quicksilver into the glass, to

---

[83] *ibid.*, 354–355.
[84] Q 31. In a draft version, the thickness is supposed to be of 10^{-7} inches and the cylinder of water it supports is 'above a mile in length' (CUL Add. 3970.3, f. 622^r, NP).
[85] *De vi electrica*, *Corr.* V, 368.
[86] Hauksbee (1709/1719), 201. Guerlac (1967).

the height of 29 or 30 Inches. And some other Agent raises it higher, not by pressing it into the glass, but by making its parts stick to the glass, & to one another'.[87] Experiments by Hauksbee and others, going back to an original demonstration by Jean Picard in 1675, showed that shaking the mercury in a Torricellian tube caused emission of light at the same time as overcoming the force of cohesion, suggesting a possible link between optical and capillary effects.[88] In 1717, this was incorporated directly into Query 31 almost immediately preceding the description of Hauksbee's experiments.

4.5. Newton's Influence on the Explanation of Material Structure

There was a genuine Newtonian tradition linking the corpuscular theory of light with the structure of matter and with chemistry. Newton himself believed that an electro-optical force could be responsible for chemical reactions and had seen the emission of light particles in such processes as analogous to the expulsion of material particles in 'fermentations' as 'vapours' and 'exhalations'. Light, he believed, was not only interconvertible with matter, but also, in principle, one of its components. In the hierarchical structure of matter which he devised, the composite particles responsible for chemical reactions were the same as those producing colours. Though Newton was more explicit about such things in his manuscripts than he was in his published works, he allowed some fairly strong hints to appear in his optical Queries, while others could be found in the *Introductio ad philosophiam Newtoniam* which his first Continental follower, Wilhelm 'sGravesande, produced in 1720, some years after visiting London as secretary to an embassy from the Dutch government.[89] 'sGravesande here presented a theory in which light corpuscles interacted and combined with material particles, and in which the vibratory motions of aethereal 'atmospheres' within materials like glass were responsible for the effects of frictional electricity. The work was translated into English by Desaguliers, and, sometimes in conjunction with a non-Newtonian material theory of heat, proposed by Homberg in 1705, 'informed the theory of chemistry up to, and including, Lavoisier'.[90] Lavoisier, notably, included light and caloric at the top of his table of chemical elements.

---

[87] Draft Queries, CUL, MS Add. 3970.3, f. 621ʳ, NP (published version in Q 31, 390).
[88] Hauksbee (1704–1705).
[89] 'sGravesande (1720).
[90] Homberg (1705, 88); Thackray (1970).

Following Newton, there was an eighteenth century Newtonian tradition of matter theory in which what would now be chemistry became high-energy physics. Important among these followers were Robert Green (although he opposed Newton in some respects), John Rowning, Bošković, Michell and Priestley.[91] Green excluded the material aspect entirely, as did Bošković, whose point-centres of force and alternate spheres of attraction and repulsion were influential on later physicists. Ideas on spheres of attraction and repulsion were part of a developing tradition beginning with Newton himself, with later writers like Priestley being influenced by earlier ones like Rowning as well as, subsequently, by Bošković. Newton also considered the idea of point-centres but did not commit himself to accepting or denying it.[92] The idea was a logical development from his 'nutshell' theory of matter[93] and was quickly taken up by various followers. The ultimate result of this eighteenth century Newtonian tradition was the development of the concept of the electromagnetic field in the work of Michael Faraday. Faraday is generally thought to have been strongly influenced by the eighteenth century point-atom tradition but whether through Bošković, Priestley or some other source is still uncertain.

Newton's work in this area and on forces of affinity did not lead to the systematic chemistry of Lavoisier, and John Herschel claimed in 1833 that 'premature generalisation[s]', such as Newton's, have a way of 'repressing curiosity, by rendering further inquiry apparently superfluous, and turning attention into unproductive channels'.[94] This may be a genuine hazard in scientific research, but there is no compelling evidence that Newton's views really had such an effect, and Herschel's views may be regarded as a product of their particular historical moment before the true microstructure of matter was revealed.

In fact, Newton did exercise a strong influence on the more conventional aspects of atomic theory, and in interestingly diverse ways. For Arnold Thackray,[95] Lavoisier's *concept* of the elements as the simplest components of matter discoverable by purely *chemical* analysis was a direct product of the Newtonian 'nutshell' view of matter and its separation of the ultimate physical particles from the chemical structural units created from them. Thackray sees both the hierarchical distinction between the physical and

---

[91] Green (1727); Rowning (1735–1743); Bošković (1746, 1748, 1749); Priestley (1772, 1777).
[92] *Principia*, I, Definition 8.
[93] As designated by Priestley (1777, 14), see 4.3.
[94] Herschel (1833, 401).
[95] Thackray (1968, 1970).

chemical units and the concept of chemical element as the last point in chemical reduction in a succession of eighteenth century authors, such as G. E. Stahl, P. J. Macquer, Guyton de Morveau and Richard Watson. To a definition of element which had become relatively conventional, Lavoisier then added the systematic separation of elements and compounds, and the creation of a list of elements, looking very recognisable today, which incorporated carbon, sulphur, phosphorus, and the newly-discovered gases oxygen, nitrogen and hydrogen alongside the ancient metals which Newton himself had believed could not be decomposed by chemical means.

For Thackray, the nutshell theory had systematically removed the materiality of atoms, culminating in the work of Bošković and Priestley, and, in the case of the latter's *Disquisitions Relating to Matter and Spirit*, implying that the matter-spirit duality was redundant as well. It may not have been a coincidence, then, that a theory with such dangerous theological implications should have been countered most strongly by two members of a by-then conservative Christian sect, the Quakers. While Thomas Young saw his adoption of the Huygenian wave theory as a means of avoiding the extreme porosity in matter required by Newton's light corpuscles, John Dalton's direct revival of material atoms led to the next major piece of work in chemical theory after Lavoisier, a new system of chemical *philosophy* in which the elements were no longer merely the minimal products of such chemical reductions as had been accomplished to date but fundamental structural elements of matter each composed of a specific type of irreducible atom.

Interestingly, Dalton's atomic theory itself resulted directly from a misreading of Newton's Book II, Proposition 23, along with a double misreading of Newton's views about atoms! Dalton was a meteorologist with no apparent interest in chemistry. He collected data about rainfall throughout his entire life, and his last recorded act was to write down the weather for that day in a shaky hand. The big question for him was, if air was composed of several gases of different densities, why didn't it separate out into layers? He hit upon Newton's model of gases being composed of particles repelling each other mutually, and then decided that each type of gas only repelled molecules of its own type. He saw atmospheric gases as solvents for each other. This theory of mixed gases went down badly. But his friend William Henry had shown that gases dissolved in inverse proportion to their density. So Dalton decided to use Henry's law to shore up his theory.[96]

---

[96] Henry (1802).

But he needed data on the relative masses of the gas particles. To interpret the chemical data to justify his theory of gases he had to assume that elementary chemical substances were composed of unbreakable atoms, each having characteristic weights.[97]

His justification for these assumptions were two (misinterpreted) paragraphs from Newton's *Opticks*. The first was Newton's statement about ultimate particles, which Dalton took to be his atoms.[98] Dalton also creatively misread Newton's statement that: 'And since Space is divisible *in infinitum*, and Matter is not necessarily in all places, it may also be allow'd that God is able to create Particles of Matter of several Sizes and Figures, and in several Proportions to Space, and perhaps of different Densities and Forces, and thereby to vary the Laws of Nature, and make Worlds of several sorts in several Parts of the Universe'.[99] He took it to mean that 'atoms' of different densities and forces existed in *our* part of the universe. So, though there were many seventeenth century scientists who were atomists of a sort, Newton, according to W. L. Scott, was 'the man responsible for transferring atomism from philosophy into the great currents of physics and chemistry', and the one 'from whom Dalton received the theory directly', for the decisive break with the past came only with Newton's definitive statement on ultimate particles in the 1706 edition of the *Opticks*.[100]

Dalton's theory, like Young's, was opposed by those scientists, like Humphry Davy, who supported the Newtonian 'nutshell' theory of matter and who wanted to connect optical science with atoms and chemistry. Ultimately, however, it prevailed, and, from that moment, the historiography of chemistry would be structured in terms which excluded the kind of hierarchical atomic theory which had its origins in Newton.[101] This seems to have been the philosophy behind Herschel's statement on 'premature generalisation', written when he and his colleagues were trying to obtain proper recognition for Dalton. The hierarchical theory, of course, would be revived, beginning at the end of the nineteenth century, on the basis of field theory and the analysis of spectra, but, by then, it was no longer considered to be anything to do with chemistry which had long since generated its own narrative concerning its foundations, one which emphasized its distinctness from physics. Newton, however, though largely excluded from this narrative,

---

[97] Dalton (1805, 1808).
[98] Q 31.
[99] *ibid*.
[100] Scott (1970, 5).
[101] Thackray (1968, 1970).

had contributed significantly to all three streams which led to the theory of matter which is prevalent today: the distinction between physical and chemical interactions leading to Lavoisier's concept of an element, the Daltonian atomic theory, and the theory of fields and spectra which led to the modern version of the hierarchical theory.[102]

There are many valid approaches to the work of a thinker as varied and prolific as Newton. Newton's influence on the development of science was probably greater than that of any other individual. If, as many claim, the historical importance of any scientist depends mainly on such influence, then Newton's work on the structure of matter does not bear comparison with his work in mathematics, mechanics, astronomy or optics. Modern ideas on the structure of matter, however similar to those of Newton, probably owe little direct inspiration to him, except in so far as he influenced the work of Dalton on atomic theory. An alternative approach — and one now highly favoured — would look to use Newton's ideas on the structure of matter to seek out the sources of inspiration for his more immediately successful work in other areas. Newton is now revealed to us as an extremely complex personality with extensive interests in such nonscientific areas as alchemy, theology, prophecy and chronology; his scientific work is seen only as a fragment of an attempted synthesis at a world-view involving all of these studies. Such studies, rather than merely revealing Newton as a 'man of his times', serve to reveal the profundity of his scientific imagination. Individual concepts may be related to sources in theology, alchemy or the mechanistic philosophy but they are all utterly transformed into ideas which we immediately recognise as scientific.

But there is another way of approaching Newton's views on the structure of matter. Regardless of their actual influence in their own time, regardless of the extent to which they reveal the various nonscientific patterns of thought which lay behind the great successes of the *Principia* and *Opticks*, they reveal, without encumbrance of mathematics or experimental detail, the very categories of thought in which he operated. It is extraordinary how valid many of them seem to be, even in twenty-first century terms. It was his unique achievement to create a programme for the whole future development of physics. No lasting success has ever been achieved outside

---

[102]Though the modern hierarchical theory of matter has distinct origins in the late nineteenth century, postdating direct interest in Newton's optical work (which had then long been out of print), this does not preclude a 'subliminal' influence of the kind which can be seen in those ideas which, though not being actively explored, are striking enough to remain in the collective consciousness of the scientific community.

this programme, though the attempt has been made many times. There seems to be something in the structure of his thought which makes it the perfect exemplar of scientific method even today, and this must be intimately bound up with his extraordinary powers of abstraction and simplification. In seeking for a unified theory which will explain everything in physics, we are confronted with a multiplicity of approaches, many of fearful complexity and 'mechanistic' in principle, being built on particular models of the universe. The Newtonian paradigm suggests that the final result will be not be complex, but of an extraordinary simplicity, and, unlike all other physical theories, be expressible in terms of complete abstraction, without any 'mechanistic' element whatsoever.

Chapter 5

The States of Matter

5.1. Solids

Contemporaries of Newton saw physical theories as being dependent on the specific structures within matter on which they were based. Newton's strategy was to avoid hypotheses which posited particular material structures. There was one area, however, where he felt he had no option but to create material models, even though they were framed in an abstract and generic, rather than hypothetical, mode. In order to counteract Descartes' all-pervading theory of celestial vortices he had to make specific assumptions about fluids and fluid resistance, although without assuming that they had to be physically true. As he would certainly have expected, controversy and criticism was to follow. Right up to the present day this has been the most criticised portion of Newton's work. Yet, it departed in no significant way from the methodology he always adopted and close attention to the details shows that it was in many ways just as successful as the most recognised part of his work.

The modern theory of matter sees material structures as being composed of molecules connected by varying degrees of cohesive force and with varying degrees of random motion, vibrational, rotational and translational, in the three states of matter, solid, liquid and gaseous, the transitions between them being effected by increases in temperature as solids become liquid and liquids become gaseous. Newton saw the states as determined at the highest level of structural composition within matter, in which the component units are not themselves structurally changed by the transformations between them. His statements concerning the basic states of matter and their transformations may not require his deepest level of thinking, but are essential to a complete understanding of his ideas about the structure of matter.

Many passages discuss transformations between the states of matter in general terms. 'And it may be further argued from the great capacity of (one & the same matter) by corruption & generation to put off & put on all manner of forms, & be changed into all sorts of shapes'.[1] 'And the... Question may be put concerning all, or almost all the gross Bodies in Nature. For all the Parts of Animals and Vegetables are composed of Substances volatile and fix'd, fluid and solid, as appears by their Analysis; and so are Salts and Minerals, so far as Chymists have been hitherto able to examine their Composition'.[2]

The changes may be attributed to the actions of a 'spirit' which, in the later writings, becomes the 'electric spirit'.

> Hence also, in proportion as the particles of bodies the more easily, or with more difficulty glide among themselves, the bodies by the agitation of this spirit either become soft and ductile or else become fluid and are converted into liquids. And some bodies become fluid and put on liquid form by very small agitations of this spirit, as do water, oil, spirit of wine, quicksilver; others by greater agitations, such as wax, cebum, resin, bismuth, pewter: others by very great agitation, as clay, stones, copper, silver, gold. And in proportion as the particles of the bodies the more firmly cohere or are the more easily separated from each other, some of them by the agitation of this spirit are quickly converted into vapour and smoke, as are water, spirit of wine, spirit of turpentine, spirit of urine: others with more difficulty, as oil, spirit of vitriol, sulphur, quicksilver; others with very great difficulty, as lead, copper, iron, stones, while others even with the strongest fire remain stable, as gold and diamond.[3]

The physical properties of the materials can be attributed to the configurations of their component particles. 'Depending on the force and manner of the coming together and cohering of the particles, they form bodies which are hard, soft, fluid, elastic, malleable, dense, rare, volatile, fixed; emitting, refracting, reflecting or stopping light'.[4]

> If the Body is compact, and bends or yields inward to Pression without any sliding of its Parts, it is hard and elastick, returning to its Figure with a Force rising from the mutual Attraction of its Parts. If the Parts slide upon one another, the Body is malleable or soft. If they slip easily, and are of a fit Size to be agitated by Heat, and the Heat is big enough to keep them in Agitation, the Body is fluid; and if it be apt to stick to things, it is humid;

---

[1] Draft Q for *Opticks*; McGuire (1968, 158).
[2] Q 31.
[3] CUL, Add MS 6970, f. 604ʳ; McGuire (1968, 177).
[4] Draft *Praefatio*, *Unp.*, 306.

and the Drops of every fluid affect a round Figure by the mutual Attraction of their Parts, as the Globe of the Earth and Sea affects a round Figure by the mutual Attraction of its Parts by Gravity.[5]

Changes of state occur as heat or cold changes the configuration of the particles, though some of the changes are what we now call chemical and were identified as such in other passages:

> Water, which is a very fluid tasteless Salt, she changes by Heat into Vapour, which is a sort of Air, and by Cold into Ice, which is a hard, pellucid, brittle, fusible Stone; and this Stone returns into Water by Heat, and Vapour returns into Water by Cold.... Mercury appears sometimes in the form of a fluid Metal, sometimes in the form of a hard brittle Metal, sometimes in the form of a tasteless, pellucid, volatile white Earth, call'd *Mercurius Dulcis*; or in that of a red opake volatile Earth, call'd Cinnaber; or in that of a red or white Precipitate, or in that of a fluid Salt; and in Distillation it turns into Vapour, and being agitated *in Vacuo*, it shines like Fire. And after all these Changes it returns again into its first form of Mercury.[6]

'It is agreeable to reason... that bodies, like their cohering particles (which according to the variation in their magnitudes, shapes and forces move either by gliding between one another more or less easily, or by receding from each other in some way) should be fluid or solid, soft or hard, ductile or elastic and more or less apt to be liquefied by fire'.[7]

The most compactified form of matter is the solid state. According to a draft Query, 'if air was so far comprest as to make its particles touch one another it might become a stone'.[8] 'So gold, which is the most fixed of all bodies, seems to consist of compound particles, not all of which, on account of their massiveness, can be carried up by the agitation of heat, and whose component parts cohere with one another too strongly to be separated by that agitation alone'.[9] 'For particles of one and the same nature draw one another more strongly then particles of different natures do...'[10] 'And when many particles of the same kind are drawn together out of the nourishment they will be apt to coalesce in such textures as the particles which drew them did before because they are of the same nature as we see in the particles

---

[5] Q 31.
[6] Q 30.
[7] Draft *Conclusio*, *Unp.*, 336.
[8] 'Various Conjectures', CUL MS Add. 3970.3, 239^r, NP.
[9] Draft *Conclusio*, *Unp.*, 335.
[10] Manuscript connected with *Opticks*, 1717; Westfall (1971, 414).

of salts which if they be of the same kind always crystallise in the same figures'.[11]

The configurations of particles are not random, as we see from the structures of crystals, such as snow and salt:

> For through the forces by which contiguous bodies cohere, and distant ones seek to separate from one another, these particles will not collect together in the composition of natural bodies like a heap of stones, but they coalesce into the form of highly regular structures almost like those made by art, as happens in the formation of snow and salts. Undoubtedly, following the laws of geometry they can be formed into very long and elastic rods, and by the connexion of the rods into retiform particles, and by the composition of these into greater particles, and so at length into perceptible bodies. And such bodies suffer light to pass through them every way, and there can be great differences of density between them. Thus gold is nineteen times denser than water yet it is not destitute of pores through which mercury can penetrate, and light (if the gold be dissolved in aqua regia and reduced into crystals) very freely be transmitted. Bodies of this kind easily receive the motion of heat by means of the free vibrations of the elastic particles and they will conserve (heat) a long time through that motion if it is slow and long-lasting. Their particles coalesce in new ways and by means of the attractive forces of contiguous ones they come together more densely...[12]

Notes which Newton had made at a very early stage from Hooke's *Micrographia* concerned the geometry of crystals: 'All the natural regular figures of bodies may be imitated by compositions of equilateral triangles. And perhaps caused by globules which do naturally lie in that form perhaps some may make a 4 square posture (or they may lie in a six square which in effect varies not from a 3 square) and solid are made by conveniently laying one such(?) surface on another'.[13]

The regularity of such structures is due to the nature of the forces acting on the component particles. '...such forces account for the fact that the particles of bodies do not collect together like a heap of stones, but like snow and salts coalesce into regular figures. From the very smallest particles bigger ones are formed, and from these the largest ones, all in a lattice structure'.[14]

> When any saline Liquor is evaporated to a Cuticle and let cool, the Salt concretes in regular Figures; which argues, that the Particles of the Salt before they concreted, floated in the Liquor at equal distances in rank and

---

[11] Manuscript connected with *Opticks*, 1717; Westfall (1971, 414).
[12] Draft *Conclusio*, Unp., 341.
[13] *Out of Mr. Hooks Micrography*, 1660s; Unp., 405.
[14] Draft *Conclusio*, Unp., 306.

file, and by consequence that they acted upon one another by some Power
which at equal distances is equal, at unequal distances unequal. For by such
a Power they will range themselves uniformly, and without it they will float
irregularly, and come together as irregularly.[15]

Slow and steady heating causes rearrangements of the particles: 'Furthermore through the slow and continued motion of heat the particles can change their arrangements and coalesce in new ways and by the attractive forces of contiguous particles (which are stronger than expulsive ones) come together more densely'.[16]

5.2. Liquids

The next level of compactification, in which the components units are more separated from each other, freer to move, and less interactive, forms the liquid or fluid state. According to Newton, the particles of materials in the fluid state are subject to the same forces as those in the solid state, but with a reduction in strength as the particles become more separated from each other: 'what ever cause makes the parts of ice & other hard bodies stick together the same cause will make them stick together when the bodies are melted, tho perhaps not so firmly. For in fusion the parts of bodies are in a motion of sliding amongst themselves'.[17]

Clifford Truesdell made some famously negative comments about Newton's main treatment of the fluid state, in Book II of the *Principia*, which also included that of 'elastic fluids' or gases.[18] However, Julián Simón Calero has written:

> In spite of Truesdell's opinion, we must consider that the *Principia* marks the definitive point of the birth of fluid mechanics as a science, as it provides

---

[15] Q 31.
[16] Draft *Conclusio*, *Unp.*, 346.
[17] Draft Q 25/31, CUL, Add. 3970.3, f. 245^r, NP.
[18] Writing at a relatively early stage in the study of this subject, Truesdell (1960, 7), said that Book II 'is almost entirely original... and much of it is false'. He also said: 'However successful was Book I, Book II was a failure as an essay toward a unified, mathematical mechanics. With its bewildering alternation of mathematical proof, brilliant hypotheses, pure guessing, bluff, and plain error, Book II has long been praised, in whole or in part, and praised justly, as affording the greatest signs of Newton's genius. To the geometers of the day, it offered an immediate challenge: to correct the errors, to replace the guesswork by clear hypotheses, to embed the hypotheses at their just stations in a rational mechanics, to brush away the bluff by mathematical proof, to create new concepts so as to succeed where Newton had failed' (1968b, 149). He even claimed that Book II was 'the part of the *Principia* that historians and philosophers, apparently, tear out of their personal copies' (*ibid.*). Later, however (1976), he said that: 'For ingenuity and insight, and, above all, for sense of problem', it was 'the most brilliant fertile work ever written in mechanics'.

a new and innovative attempt to submit fluids to a theoretical treatment explaining their behaviour as a function of their constitution, the conditions in which they move, and some general laws. It is quite true that Newton only does this for air and not for liquids, but this underlines precisely the coherence of his hypotheses, as these cannot be applied indiscriminately, but only to well-defined constituting models. What is more, he does not even apply them strictly to air, which he identifies as an elastic fluid, but to what he calls a 'rare medium' (*medium rarum*) which is a kind of imaginary fluid, about which he conjectures that its behavior, when faced with a body that moves in it, comes closer to that of real air.[19]

In fact, it seems to have been something like what we now call an *ideal gas*, a concept used by later physicists to define thermodynamic properties.[20]

Newton defines a fluid in *De gravitatione et aequipondio fluidorum*: 'A fluid body is one whose parts yield to an overwhelming pressure'.[21] In *Principia*, Book II, this becomes: 'A fluid is any body whose parts yield to any force impressed upon it, and, by yielding, are easily moved among themselves'.[22] In *De gravitatione* he says that: 'All the parts of a non-gravitating fluid, compressed with the same intension in all directions, press each other equally or with equal intension. And compression does not cause a relative motion of the parts'.[23] In *Principia*, Book II, this is incorporated into Proposition 19: 'All the parts of a homogeneous and unmoved fluid included in any unmoved vessel, and compressed on every side (setting aside the consideration of condensation, gravity, and all centripetal forces), will be equally pressed on every side, and remain in their places without any motion arising from that pressure'.[24] In *The Elements of Mechanicks*, which postdates the *Principia*:

---

[19] Calero (2008, 91).
[20] A different concept of ideal gas is used in Newton's derivation of Boyle's law. See Section 5.3.
[21] Definition 18; *Unp.*, 150–151. The date of *De gravitatione* has been placed at different times within the period 1668 and 1684, but Shapiro noted that Newton made a 'sound application of his newly wrought concept of hydrostatic pressure' in refuting Descartes' theory of the coronas observed in distorting the shape of the eye in *Optica*, II, Lecture 14, 133, *OP*, I, 582–585, dating from c 1672. Henry (2011) argues for an early date against Dobbs (1991), a possibility also considered by Ruffner (2012). Iliffe (2016), 488, dates it to the 1670s.
[22] II, Section V, The Definition of a Fluid. This hydrostatics section includes Propositions 19–23.
[23] Propositions 1, 2; *Unp.*, 152.
[24] Newton clearly considered this property important; his discussion of fluids for the Proposition provides an extensive qualitative discussion outlining seven separate Cases and concluding with a Corollary restating the main conclusion.

The parts of stagnating water are pressed and press all things in proportion to their depth from the surface of the water besides the pressure which arises from the weight of the incumbent atmosphere. And the parts of all fluids press and are pressed in proportion to the incumbent weight. Whence water ascends in pumps and mercury in the barometer by the weight of the atmosphere...[25]

Pressure acts equally in all directions. 'Gravity tends downwards, but the Pressure of Water arising from Gravity tends every way with equal Force, and is propagated as readily, and with as much force sideways as downwards, and through crooked passages as through strait ones'.[26] 'The internal parts of a fluid press each other with the same intension as that by which the fluid is pressed on its external surface. If the intension of the pressure is not everywhere the same, the fluid does not remain in equilibrium'.[27]

Principia, Book II, Proposition 20, with its nine Corollaries, pursues the subject of hydrostatics, demonstrating all the essential results. A spherical vessel with a flat base will 'sustain the weight of a cylinder' of fluid, of the same area as the base and of the same height as the fluid in the vessel. Only the fluid that is directly above the base will be supported by it, 'the rest of the weight being sustained... by the spherical figure of the fluid' (Corollary 1); the 'quantity of the pressure is the same always' in any direction 'at equal distances from the centre' (Corollary 2); by Proposition 19, the weight of a heavy fluid produces no motion among its parts 'except that motion which arises from condensation' (Corollary 3); a body of the same specific gravity immersed in this fluid 'will neither descend nor ascend nor change its figure' (Corollary 4). A body that is denser than the fluid will sink, one that is less dense will rise, for 'that excess or defect is the same thing as an impulse' (Corollary 5). Bodies placed in fluids of any kind, including air, have two types of gravity: the true and absolute gravity, and the relative and common, or apparent, gravity, which is measured as the difference between the true gravity and that of the ambient fluid (Corollary 6).

Archimedes' Principle is therefore a consequence of the vector nature of forces; so the same result would occur if any other centripetal force were acting in place of gravity (Corollary 7). Newton has effectively succeeded in incorporating hydrostatics into dynamics. For a force acting on the body which is more powerful than the force on the medium, the motive (resultant)

---

[25] *The Elements of Mechanicks*, *Unp.*, 166.
[26] Q 28.
[27] *De gravitatione*, Corollaries 1, 2; *Unp.*, 155.

force on the body will be centripetal; but for a force on the medium more powerful than that on the body, the motive force on the body will be centrifugal (Corollary 8). The parts of bodies immersed within a fluid 'will be agitated with the same motions as if they were placed in a vacuum' and 'will not change the situation of their internal parts in relation to one another', unless the bodies are 'condensed by the compression' or the fluid resists their motions (Corollary 9).

Elastic and non-elastic bodies are distinguished in *De gravitatione*: 'An elastic body is one that can be condensed by pressure or compressed within narrower limits; and a non-elastic body is one that cannot be condensed by that force'.[28] 'Water cannot be compressed because its parts are contiguous, while air can be since its parts do not yet touch each other...'[29] The expansion of liquids by heat is attributed to the vibrations of the particles: '...liquors are expanded by heat and by consequence by the vibrating agitations of this spirit. If the agitations be of short continuance they expand the liquors without heating them for want of time to do it. If lasting...they heat the body by degrees...'[30] 'It is not necessary, however, that this vibration (of which I suppose heat to consist) should always be in a straight line, for the particles may revolve in curves. Nor need the rarefaction through heat arise only from the vibratory motion...'[31]

5.3. Gases

The least compactified state, in which the components units are freest and the interactions between the units are least, is, of course, the gaseous state. For Julián Simón Calero: 'The claim that it was Newton who established the basis of fluid mechanics is justified by his introduction of three decisive hypotheses: one on the constitution of air, the second on the way in which air and a body interact, and the last on the laws regulating this'.[32] Newton, like several of his contemporaries, was interested in gases or 'airs' and their production, as these represented the ultimate units of matter in their least restricted state, existing '...in a rare medium, consisting of equal particles freely disposed at equal distances from each other...'[33] Chemical reactions ('fermentations') produced different types of 'air'. 'Dense Bodies by

---

[28] Definition 16; *Unp.*, 150.
[29] Pitcairne with Newton at Cambridge, 2 March, 1692; *Corr.* III, 211.
[30] Draft Query 22 for *Opticks*; Home, 198–200.
[31] *De aere et aethere*, c 1679; *Unp.*, 224. Revolving in curves presumably signifies rotation.
[32] Calero (2008).
[33] *Principia*, II, Proposition 34.

The States of Matter 129

Fermentation rarify into several sorts of Air, and this Air by Fermentation, and sometimes without it, returns into dense Bodies'.[34] '... in Fermentations the Particles of Bodies which almost rest, are put into new Motions by a very potent Principle, which acts upon them only when they approach one another, and causes them to meet and clash with great violence, and grow hot with the motion, and dash one another into pieces, and vanish into Air, and Vapour, and Flame'.[35]

Newton provides a possible model for the process in his letter to Boyle of 1679:

> In ye solution of metals, when a particle is loosing from ye body, so soon as it gets to that distance from it where ye principle of receding... begins to overcome ye principle of acceding described...: the receding of ye particle will be thereby accelerated, so yt ye particle shall as were wth violence leap from ye body, & putting ye liquor into a brisk agitation, beget & promote yt heat we often find to be caused in solutions of Metals. And if any particle happen to leap of thus from ye body before it be surrounded with water, or to leap of wth that smartness as to get loos from ye water: the water... will be kept of from ye particle & stand round about it like a spherically hollow arch, not being able to come to a full contact with it any more. And severall of these particles afterwards gathering into a cluster, so as by ye same principle to stand at a distance from one another without any water between them, will compose a buble. Whence I suppose it is yt in brisk solutions there usually happens an ebullition.
>
> This is one way of transmitting gross compact substances into aereal ones. Another way is by heat.[36]

Most authors of the time are vague about the motions of particles in gases. Newton continues:

> For as fast as ye motion of heat can shake off ye particles of water from ye surface of it: those particles by ye said principle will flote up & down in ye air at a distance both from one another & from ye particles of air, & make yt substance we call vapor. Thus I suppose it is when ye particles of a body are very small (as I suppose those of water are) so yt ye action of heat alone may be sufficient to shake them asunder. But if ye particles be much larger, they then require the greater force of dissolving Menstruums to separate them, unless by any means the particles can be first broken into smaller ones.[37]

---

[34] Q 30.
[35] Q 31.
[36] Letter to Boyle, 28 February 1679: *Corr.* II, 293.
[37] *Corr.* II, 293.

In the more mature presentation in Query 31 of the *Opticks*, Newton proposes that:

> The Particles of Fluids which do not cohere too strongly, and are of such a Smallness as renders them most susceptible of those agitations which keep Liquors in a Fluor, are most easily separated and rarified into Vapour, and in the language of the Chymists, they are volatile, rarifying with an easy Heat, and condensing with Cold. But those which are grosser, and so less susceptible of Agitation, or cohere by a stronger Attraction, are not separated without a stronger Heat, or perhaps not without Fermentation. And these last are the Bodies which Chymists call fix'd, and being rarified by Fermentation, become true permanent Air; those Particles receding from one another with the greatest Force, and being most difficultly brought together, which upon Contact cohere most strongly.

But he had said much the same in his letter to Boyle, with a specifically aethereal interpretation: '...the small particles of vapours easily come together & are reduced back into water unless ye heat wch keeps them in agitation be so great as to dissipate them as they come together: but ye grosser particles of exhalations raised by fermentation keep their aerial form more obstinately, because the æther within them is rarer'.[38]

Newton's explanation of the expansion of air with heat was an increased amplitude of vibration of the material units. 'Among the properties of air its great rarefaction and condensation are remarkable. Of these there are three chief causes: expansion, compression, heat and the proximity of bodies. The former accounts for the rest and the whole nature of air...'[39] The spacing of the component particles is proportional to their increased momentum. 'Moreover we need not wonder that air is expanded by heat if we consider that its parts when agitated by heat must vibrate, and, by vibrating, propel hither and thither the neighbouring parts...all are scattered through a wider space proportionate to the quantity of motion'.[40]

The particles of airs and vapours, being effectively free of each other, have a tendency to 'recede from one another' and to strive to escape from any container.

> Also yt ye particles of vapors exhalations & air do stand at a distance from one another, & endeavour to recede as far from one another as ye pressure of ye incumbent atmosphere will let them: for I conceive ye confused mass

---

[38] Letter to Boyle, 28 February 1679; *Corr.* II, 294.
[39] *De aere et aethere*; *Unp.*, 221.
[40] *De aere et aethere*; *Unp.*, 224.

of vapors air & exhalations wch we call ye Atmosphere to be nothing els but ye particles of all sorts of bodies of wch ye earth consists, separated from one another & kept at a distance by ye said principle.[41]

It seems also to follow from the Production of Air and Vapour. The Particles when they are shaken off from Bodies by Heat or Fermentation, so soon as they are beyond the reach of the Attraction of the Body, receding from it, and also from one another with great Strength, and keeping at a distance, so as sometimes to take up above a Million of Times more space than they did before in the form of a dense Body.[42]

'The aire too being continually shaken & moved in its smallest parts by vaporous particles every where tossed up & downe in it as appeares by its heate, it must needs strive to get out of all such cavity which doe hinder its agitation...'[43]

In his calculations of atmospheric refraction, Newton considered that he must take into account 'the rarefaction and condensation of ye Air by heat and cold'[44] as well as the pressure-volume relationship of Boyle. He had a value for the expansion coefficient of air, though it was not a very accurate one by today's standards. He calculated that air lost a fifth of its volume in reducing from body temperature to the freezing point of water, an interval of about 37 K.[45] The expansion and contraction of air with heat and cold is also used by Newton to explain that the velocity of sound, which is proportional to the square root of the density, will be greater in summer and less in winter.[46]

He was aware that pressure would change the boiling point:

For if Water be made warm in any pellucid Vessel emptied of Air, that Water in the *Vacuum* will bubble and boil as vehemently as it would in the open Air in a Vessel set upon the Fire till it conceives a much greater heat. For the weight of the incumbent Atmosphere keeps down the Vapours, and hinders the Water from boiling, until it grow much hotter than is requisite to make it boil *in vacuo*.[47]

Vehement 'action' would increase the separation of particles from the liquid and so increase the generation of gas or vapour.

---

[41] Letter to Boyle, 28 February 1679; *Corr.* II, 291.
[42] Q 31.
[43] QQP, 'Attraction Electricall & Filtration', f. 103^v, NP, Westfall (1971, 334).
[44] Newton to Flamsteed, February 1695, *Corr.* IV, 86.
[45] ULC Add MS 3975, 45–46, n. 6; Brewster (1855), II, 366–367, n. 2; Simms (2004, 41).
[46] *Principia*, II, Proposition 50, Scholium.
[47] Q 11.

> Lastly, from these principles the generation of air is easily learned. For this nothing else is required save a certain action or motion which tears apart the small parts of bodies; since when separated they mutually flee from one another, like the other particles of air. And thus it is that every vehement agitation (like friction, fermentation, ignition and great heat) generates the aerial substance which reveals itself in liquids by ebullition; and the more vehement the action the more copiously that substance is generated. So filings of lead, brass, or iron dissolved in aqua fortis produce a great ebullition; these, however, dissolve without ebullition in vinegar or in the same aqua fortis sufficiently weakened by mixing with water.[48]

'For ye most fixed bodies, even Gold it self, some have said will become volatile only by breaking their parts smaller'.[49]

Newton discovered the phenomenon of relative humidity, or the fact that humid air is less dense than dry air, writing in Query 31: 'And because the Particles of permanent Air are grosser, and arise from denser Substances than those of Vapours, thence it is true that Air is more ponderous than Vapour, and that a moist Atmosphere is lighter than a dry one, quantity for quantity'. In modern terms, Avogadro's law says that a given volume of any gas will contain the same number of molecules at the same temperature and pressure. Molecules of water vapour (with mass 18 u) are less massive than those of nitrogen (28 u) and oxygen (32 u), so if water vapour molecules displace those of nitrogen or oxygen in a given volume of gas under the same conditions of temperature and pressure, then the overall density of the gas will be reduced.

Different kinds of air would be the subject of a great deal of eighteenth century chemistry.

> Moreover, aerial substances are very different according to the nature of the bodies from which they were generated. Metals by corrosion [i.e. dissolution in acids] give true permanent air [essentially hydrogen, observed by Boyle]; vegetable and animal substances by corrosion, fermentation or burning give an air of short duration like an exhalation [carbon dioxide]; and volatile substances rarefied by heat give an air least lasting of all, which we call a vapour. Among these are great differences, the vapour of water condenses more quickly than the vapour of spirit of wine and the vapour of certain saline spirits...[50]

---

[48] *De aere et aethere*; *Unp.*, 226.
[49] Letter to Boyle, 28 February 1679; *Corr.* II, 293.
[50] *De aere et aethere*; *Unp.*, 226.

Different airs have different densities and the atmosphere is composed of many kinds of air.

> There is a diversity in the weight of air, for air generated by the force of fire from vegetable and animal spirits is so much lighter than the rest of the surrounding air that not only does it not descend itself but it also carries away the associated thicker fumes with it. So vapours of every kind seek to rise above the rest of the air. And there is no doubt but that vapours and exhalations have various degrees of gravity; for there appear to be as many kinds of air as there are of substances on the Earth from which its origin comes. So the atmosphere is composed of many kinds of air.[51]

These airs have quite different properties. '...I remember I once read in ye P[hilosophical] Transactions how M. Hugens at Paris found that ye air made by dissolving salt of Tartar would in two or three days time condense & fall down again, but ye air made by dissolving a metal continued wthout condensing or relenting in ye least'.[52] '...this latter air serves (as the almost indestructible nature of metals demands) neither for the preservation of fire nor for the use of animals in breathing, as do serve some of the exhalations arising from the softer substances of vegetable matter or salts'.[53] 'The air also is ye most gross unactive part of ye Atmosphere affording living things no nourishment, if deprived of ye more tender exhalations & spirits that flote in it...'[54]

The explosion of gunpowder is accompanied by the production of a large amount of air, greatly enhancing its explosive power. 'Gunpowder fired in vacuo produces common air. The first explosion seems to produce more; but after two, or more days there is about 1/5 or 1/6 of the powders weight of air that is permanent'.[55]

> So nitre, melted and ignited by charcoal thrown upon it, emits much of the aerial substance, with much fixed salt remaining in the bottom; but if the action of the fire is hastened by the due admixture of sulphur and charcoal as is used in making gunpowder, almost all the substance of the mixt is changed by vehement agitation into an aerial form, the huge force of this powder arising from its sudden expansion as is the nature of air.[56]

---

[51] *De aere et aethere*; *Unp.*, 226–227.
[52] Letter to Boyle, 28 February 1679; *Corr.* II, 294.
[53] *De aere et aethere*, *Unp.*, 227.
[54] Letter to Boyle, 28 February, 1679; *Corr.* II, 294.
[55] Memorandum by D. Gregory of comments by Newton, 21 December 1694.
[56] *De aere et aethere*; *Unp.*, 226.

Anyone setting out to describe the properties of gases had to accommodate Boyle's law, the experimental result, discovered in the early 1660s, which said that the pressure of a gas at constant temperature was proportional to its density, or inversely proportional to its volume. As expressed in *De aere et aethere*, which interestingly credits the relationship to Hooke: 'In just the same remarkable manner (air) rarefies and is condensed according to the degree of pressure'.[57]

> The whole weight of the incumbent atmosphere by which the air here close to the Earth is compressed is known to philosophers from the Torricellian experiment, and Hooke proved by experiment that the double or treble weight compresses air into the half or third of its space, and conversely that under a half or third or even thousandth part of that weight is expanded to double or treble or even a hundred or a thousand times its normal space, which would hardly seem to be possible if the particles of air were in mutual contact; but if by some principle acting at a distance (the particles) tend to recede mutually from each other, reason persuades us that when the distance between their centres is doubled the force of recession will be halved, when trebled the force is reduced to a third and so on, and thus by an easy computation it is discovered that the expansion of the air is reciprocal to the compressive force.[58]

By the time of the *Principia*, Newton was ready to subject this law to a more precise mathematical analysis, but it was not an analysis based on a physical hypothesis. Newton saw that the atmosphere which the Earth retained through gravitation would necessarily become gradually less dense with height, and, assuming Boyle's law to be true, he examined a number of hypothetical mathematical models for the centripetal law of attraction between the particles of a uniform medium arranged in concentric spherical layers. In Book II, Proposition 21, he investigates an attraction varying with the inverse first-power of the distance from the centre of the sphere, which will give a constant decrease in density with height. By contrast, an attraction varying inversely with the square of the distance from the centre, in Proposition 22, produces densities varying in geometrical progression for distances taken in harmonic progression. This becomes significant in Newton's theory of atmospheric refraction,[59] for, if the density gradient $d\rho/dr$ at distance r from the centre of the sphere is proportional to ρr^{-2}, then, if R is the Earth's radius, the density is proportional to $e^{1/r - 1/R}$,

---

[57] *De aere et aethere*; *Unp.*, 223.
[58] *ibid.*, 223–224.
[59] Whiteside (1980); Lehn (2008). Corollaries show how to use the relationships to find actual densities. For atmospheric refraction, see *Newton and the Great World System*.

and the density gradient or 'refractive power' becomes proportional to $r^{-2}e^{1/r-1/R}$.

A Scholium looks at other force laws for the gravitational attraction, including inverse cube, inverse fourth power, constant force, direct proportionality to distance, and so on. For a constant force, with distances taken in arithmetic progression, the densities will be in geometric progression, 'as the distinguished gentleman Edmond Halley has found'. It would also be possible to take other laws relating pressure and density, for example, $P^3 \propto \rho^4$, $P^4 \propto \rho^3$, $P^3 \propto \rho^5$, $P \propto \rho^2$. The Scholium concludes, however, with the statement: 'But as to our own air, this is certain from experiment, that its density is either accurately, or very nearly at least, as the compressing force; and therefore the density of the air in the atmosphere of the earth is as the weight of the whole incumbent air, that is, as the height of the mercury in the barometer'.[60]

In *The System of the World*, §68, Newton uses Proposition 22 to calculate the rarity of the air at different heights. If the pressure of air at the Earth's surface is equivalent to that of a column of water 33 feet height (with the density 1200 times less than that of water), then at 5 miles height it will be reduced to not much more than half of this (17.8515 feet). The pressure of the atmosphere then falls dramatically with height. The numbers in the table for the compression are so small and for the expansion so large that Newton has to invent a symbolism to represent powers of 10, so that o.xvii 1224 (the compression at 400 miles height) becomes our 0.1224×10^{-17}, while 26,956 xv (the expansion) is our $26{,}956 \times 10^{15}$. He says that

> a sphere of that air which is nearest to the earth, of but one inch in diameter, if dilated with that rarefaction which it would have at the height of one semidiameter of the earth, would fill all the planetary regions as far as the sphere of Saturn, and a great way beyond, and at the height of ten semidiameters of the earth would fill up more space than is contained in the whole heaven on this side the fixed stars, according to the preceding computation of their distance.[61]

The derivation of Boyle's law for gases, in *Principia*, Book II, Proposition 23, was important as the first mathematical prediction to emerge from Newton's theory of matter; but the theory of the composition of gases on which it seemingly depends does not correspond exactly to ours, in which gases are composed of molecules in completely random motion,

---

[60] *Principia*, II, Proposition 22, Scholium; 'the distinguished gentleman Edmond Halley' is 'Dr. Halley' in Motte's version (1729).
[61] *The System of the World*, §68. The distance to the nearest stars is calculated in §57.

causing collisions with each other and with the walls of the container. In fact, however, Newton's derivation doesn't depend on any physical structure, and doesn't present one. The structure it presents is purely mathematical, a scaling argument based on observable parameters rather than microstructure. The analysis is, in fact, a good example of the way in which Newton preferred to develop theories in terms of general abstract concepts rather than using specific concrete or mechanistic models. Like his first mv^2/r derivation, it is an example of a mathematical model, which is not committed to being *physically* true, because it expresses a fundamental symmetry of some kind, not the effects of the model. The later development of Lagrangian dynamics, using generalised coordinates, would be a development of this kind, with the abstract truth more fundamental than the apparent model, and the same would be true of quantum mechanics, which took inspiration from the Lagrangian method.

Newton's analysis derives the proportionality in a gas between pressure and density from an assumption that there is a force between adjacent gas molecules inversely proportional to the distance between them, because this is what is effectively observed, but he makes it clear that the result is not based on any belief in the validity of a *physical* model of a gas composed of stationary molecules subject to mutually repulsive forces. He states the basic assumption of his mathematical model in Proposition 23 of *Principia*, Book II:

> If a fluid be composed of particles mutually flying each other, and the density be as the compression, the centrifugal forces of the particles will be reciprocally proportional to the distances of their centres. And, *vice versa*, particles flying each other, with forces that are reciprocally proportional to the distances of their centres, compose an elastic fluid, whose density is as the compression.

In fact, the derivation of Boyle's law from the modern kinetic theory of gas molecules in random motion is essentially the same as that of Newton from the static model, for the kinetic behaviour of the molecules is not a factor in deriving the potential energy term with which Boyle's law is really concerned. In the modern theory, we assume in effect that the gas molecules are stationary overall — that is, that they occupy the same positions on a time average — and calculate the outward force exerted by each molecule to be inversely proportional to the length of the container in which it is placed or to any fixed portion of this length; this is exactly the same thing as making the outward force exerted by any molecule on its nearest neighbour inversely proportional to the separation of the two molecules. It is not, of course, necessary to assume that this is due to a physical *interaction* between the molecules. (The actual force is entropic not intermolecular.)

We know, from our measurement of pressure, that gas molecules exert on average over a fixed distance (say the length of a container) a force inversely proportional to this distance outward in all directions. Because average molecular spacings are fixed for a fixed volume, this is in principle equivalent to each molecule at any instant exerting an outward force inversely proportional to the molecular spacing on each of its nearest neighbours. Now, the molecules of a gas in any confined space may move individually with random rectilinear motion but overall they are static. It is, therefore, perfectly possible to assume for mathematical convenience, as Newton did *and said he did*, that the molecules of a gas act as if they were static, though vibrating, particles, exerting repulsive forces on each other, and Boyle's law is then derived correctly because the aspects of gas behaviour on which it depends require no more than this simple model.

It is then a purely formal matter whether we describe the gas in terms of the average kinetic energy of the individual molecules or an equivalent averaged-out potential energy of the gas as a whole. It is only by applying the virial theorem to the potential energy term PV, derived from Boyle's law, that we are able to bring in the kinetic nature of the gas and to explain its temperature behaviour. The fact that gas molecules are not actually static in our modern theory, based on new observations, is immaterial because the normal derivation of Boyle's law assumes that they are static overall.

That Newton's approach to the problem is essentially abstract is apparent from the fact that he derives, for the general case of forces inversely proportional to the power n of the distance, the corresponding proportionality of pressure to the power $(n+2)/3$ of the density.[62] This is the same kind of argument that we use, with the virial theorem, when we show that, for forces inversely proportional to power n of distance or for potential energies inversely proportional to power $(n-1)$, the time-averaged kinetic and potential energies are related by the formula

$$\bar{T} = \frac{1-n}{2}\bar{V}.$$

In the kinetic theory of material gases, of course, we usually assume that the pressure produced by the gas is purely a result of collisions between the molecules and the container walls, but this is merely an

---

[62] *Principia*, II, Proposition 23, Scholium.

operational convenience, not a necessary precondition; pressure terms of any kind, whatever their origin, are an expression of the action of force or potential energy. A gas in steady state is dynamically equivalent to a system with constant expansive force in all directions, and a system of this kind necessarily requires a virial relation of the form

$$\bar{V} = 2\bar{T}.$$

The pressure of the gas will therefore be really due to the potential energy of the molecules (V), even where it is derived (as $2T$) from a kinetic theory. Of course, even the modern kinetic theory, with gases made up of infinitely small, perfectly elastic molecules subject to no intermolecular forces, could itself be regarded as a mainly mathematical model, as these assumptions are a long way from being physically true.

In cases such as this, Newton is aware that certain fundamental truths are intrinsically abstract and are independent of the model used to develop them. The success of the mathematical model doesn't imply the truth of a physical one. Newton quite clearly disclaimed any fundamental belief in the static model of a gas, even though this led to the correct experimental result, perhaps because he realised that the derivation of Boyle's law did not depend on the finer details of the model.

> All these things are to be understood of particles whose centrifugal forces terminate in those particles that are next them, or are diffused not much farther. We have an example of this in magnetical bodies. Their attractive virtue [force] is terminated nearly in bodies of their own kind that are next them. The virtue of the magnet is reduced by the interposition of an iron plate, and is almost terminated at it: for bodies farther off are not attracted by the magnet so much as by the iron plate. If in this manner particles repel others of their own kind that lie next them, but do not exert their force on the more remote, particles of this kind will compose such fluids as are treated of in this Proposition. If the virtue of any particle diffuse itself every way *in infinitum*, there will be required a greater force to produce an equal condensation of a greater quantity of the fluid. But whether elastic fluids do really consist of particles so repelling each other, is a physical question.[63]

It is, of course, true that Newton never really suggested a truly alternative or more sophisticated model of gas behaviour, though in *De aere et aethere* he considered that heat expansion of a gas involved more than vibratory motion, but his purpose was not to derive complicated mechanical

---

[63] *ibid.*, Newton's description of the components of a magnetic body as attracting only their nearest neighbours is interestingly close to the forces between nucleons in the nucleus.

hypotheses of such things as gases but to show how their properties could be explained in abstract terms as a result of the actions of forces of attraction and repulsion between particles. In fact, the modern kinetic theory of gases cannot be justified on the grounds that it predicts the gas laws. It can only be justified on the basis of Brownian motion, as first observed in 1828, in which smoke particles or pollen grains move with a kind of jittery motion as though being buffeted by smaller gas or liquid molecules. Fatio de Duillier's aether seems to require something similar for aether particles.

Newton considered the possibility that gases did not necessarily represent the most rarefied state of which matter was capable. He thought at one time that 'aether' may be what we would call a fourth state of matter or a material substance even more rarefied than that of a gas.

> And just as bodies of this Earth by breaking into small particles are converted into air, so these particles can be broken into lesser ones by some violent action and converted into yet more subtle air which, if it is subtle enough to penetrate the pores of glass, crystal and other terrestrial bodies, we may call the spirit of air, or the aether. ... I believe that everyone who sees iron filings arranged into curved lines like meridians by effluvia circulating from pole to pole of the [load-]stone will acknowledge that the magnetic effluvia are of this kind. So also the attraction of glass, amber, jet, wax and resin and similar substances seem to be caused in the same way by a most tenuous matter of this kind...[64]

(At a later date, when he had rejected the Cartesian material aether, he told David Gregory that: 'Furrows among filings placed around an iron magnet are not made by effluvia but because the bits of the filings are themselves magnetically excited and arrange themselves longitudinally and through the poles'.[65])

His discussion of Fatio's aether theory involved a system of the same kind as modern gas theory describes for an ideal gas:

> The unique hypothesis by which gravity can be explained is however of this kind, and was first devised by the most ingenious geometer Mr. N. Fatio. And a vacuum is required for its operation since the more tenuous particles must be borne in all directions by motions which are rectilinear and very rapid and uniformly continued and these particles must experience no resistance unless they impinge upon denser particles.[66]

---

[64] *De aere et aethere*, *Unp.*, 227–228.
[65] Gregory, Memorandum, Cambridge, 5, 6, 7 May 1694, *Corr.* III, 338 (Latin 336).
[66] Draft addition for *Principia*, III, Prop. 6, Coroll. 4 and 5, 1690s; *Unp.*, 315.

This model may be compared with statements which suppose aether 'to be of the same nature with Aer but exceedingly more subtile'.[67]

Fatio also records, in writing to Newton on 24 February 1690, 'My Theory of Gravity is now I think clear of objections, and I can doubt little but that it is the true one. You may better judge when you see it'.[68] And two days later, according to the Royal Society Journal Book, 'Mr. Fatio read a Copy of a Letter of his to Mr. Hugenius, giving an account of the reason of gravity'.[69] He stated:

> That in the whole Universe there is dispersed a very subtle Matter, infinitely little and exceedingly agitated in right line; the Reflections of each particle whereof against gross Bodies and against any great mass of matter floating in them, lessening somewhat the swiftness of the particles. This he conceived would drive down all manner of Bodies towards one another or towards any greater Mass,

'in accordance with Newton's inverse-square law'.[70] It is interesting, in connection with Fatio's confident belief that Newton was impressed by his aethereal mechanism, that David Gregory subsequently added a note to a memorandum he had originally made on 28 December 1691, to the effect that 'Mr Newton and Mr Hally laugh at Mr Fatios manner of explain gravity'.[71] It had almost certainly failed the test that had previously made the Cartesian aether untenable: of preserving Kepler's laws of planetary motion in a resisting medium.

5.4. Resistance in Fluids

Most of Book II of the *Principia* is concerned with the special subject of resistance in fluids, and partly resulted from an investigation into possible theories of the aether. Newton realised early on that the stability of systems acting under universal gravitation would be seriously compromised by resistive or dissipative forces. He knew that resistance was significant with respect to the long-term stability of the Solar System. In addition, the mechanism of gravity still needed to be established. If an aether was involved, he would have to work out to what extent this allowed his action-at-a-distance mechanism to function without significant disruption. Book II is certainly not merely aimed at eliminating the Cartesian vortex theory,

---

[67] Draft Q 22 for *Opticks*, quoted Bechler (1974, 207).
[68] *Corr.* III, 390–391.
[69] *Royal Society Journal Book*, 26 February 1690, quoted *Corr.* III, 69.
[70] French text in Gagnebin (1949, 115); English text, *Unp.*, 206.
[71] *Corr.* III, 191.

which could be relatively easily disproved by comets going in orbits in every direction and in many planes. It is concerned mainly with establishing how far he could take the theory that replaced it.

Newton's studies of fluids seem to have been intended mainly to provide evidence which would confirm or deny the existence of a universal aether, but they nevertheless led to many results applicable to fluids in general: for example, the first analysis of wave motion, the formula for the velocity of sound, the law of dynamic similarity in the motions of bodies of the same shape and the law of viscosity for Newtonian fluids. Among other things, as already discussed in Section 5.2, Newton reduced the laws of hydrostatics to a branch of his more general dynamics. A 'magisterial use of hydrostatics' was also evident in his deduction of the figure of the Earth and its oblateness in Book III, Proposition 19.[72] Some of the results in *Principia*, Book II were not examined in detail theoretically, or put to the test experimentally, until relatively recent times.[73]

A number of general statements about resistance are made in both the *Principia* and in other works: 'The resistance of spherical bodies in fluids arises partly from the tenacity, partly from the attrition, and partly from the density of the medium'.[74] 'Viscosity is either a deficiency of fluidity (which is located in the smallness, and thus the separability of parts, understood as parts of last composition) or a deficiency of slipperiness or smoothness preventing the lowest parts from sliding over others'.[75] The effect of heat on resistance is discussed in Query 28 of the *Opticks*:

> Heat promotes Fluidity very much by diminishing the Tenacity of Bodies. It makes many Bodies fluid which are not fluid in cold, and increases the Fluidity of tenacious Liquids, as of Oil, Balsam, and Honey, and thereby decreases their Resistance. But it decreases not the Resistance of Water considerably, as it would do if any considerable part of the Resistance of Water arose from the Attrition or Tenacity of its Parts. And therefore the Resistance of Water arises principally and almost entirely from the *Vis inertiae* of its Matter.

---

[72] Drazin (1987, 351).
[73] This didn't stop judgements being made, and some of the ones that emerged before proper evaluations were made are still being quoted as authoritative.
[74] *Principia*, II, Proposition 14, Scholium. Néményi (1963) suggests cohesion, viscosity and inertia as approximate equivalents of tenacity, attrition and density.
[75] Pitcairne with Newton at Cambridge, 2 March 1692; *Corr.* III, 211.

Newton also told David Gregory in 1694 that '... the resistance of a medium whose internal particles are in motion is greater than that of one whose particles are at rest...'.[76]

Newton went to great lengths in trying to quantify resistance using theoretical models as well as experiments. Corollary 2 to *Principia*, Book II, Proposition 38 begins with the rather obvious statement that: 'The greatest velocity, with which a globe can descend by its comparative weight through a resisting fluid, is the same as that which it may acquire by falling with the same weight, and without any resistance', but continues with the less obvious 'and in its fall describing a space that is to four third parts of its diameter as the density of the globe is to the density of the fluid'. Resistance proportional to velocity was an interest as early as *Quaestiones quaedam philosophicae*: 'In the descention of a body There is to be considered the force which it receives every moment from its gravity (which must be least in a swiftest body) & the opposition it receives from the aire (which increaseth in proportion to its swiftnesse)'.[77] A classic example is Stokes' law, in which a falling sphere of radius a experiences a resistive force equal to $6\pi\eta a v$, where v is the velocity and η is the coefficient of viscosity. But Newton regarded the linear relation as 'more a mathematical hypothesis than a physical one'.[78] Terminal velocity is alluded to in *An Hypothesis of Light*: '... bodies let fall in water are accelerated till the resistance of the water equalls the force of gravity', as well as in the first Corollary to *Principia*, Book II, Proposition 3, and in Proposition 40 in the same Book, where it is a crucial component of the experiments on resistance to free fall.[79]

Book II has been the least regarded part of the *Principia*, and, though Newton was an important pioneer in fluid mechanics, he has received very little credit for this. In fluid dynamics, despite statements to the contrary, his principles and insights are generally true, and more important than his attempts at calculation. He is aware that mathematically correct results may not necessarily be a close match to reality, and that working models should not be regarded as physical hypotheses. In view of the long stream of criticism that the book has generated, it is something of a surprise to find

---

[76] D. Gregory, *Mathematical Notes from Newton*, 5, 6 and 7 May 1694; *Corr.* III, 333.
[77] QQP, 'Of Attomes', 12^v, NP.
[78] *Principia*, Book II, Proposition 4, Scholium.
[79] *Hyp.*, *Corr.* I, 370.

that a close analysis of the 53 numbered Propositions shows that most are actually *correct*; only two or three are open to question.[80]

Book II is an extraordinary collection of techniques and models created to solve particular problems. Here, Newton was working at the very frontier of what was possible in a highly intractable subject. The relatively high rate of success is a tribute to his remarkable ingenuity, and can be referred to his special talents as a problem solver.[81] In the construction of the book, sometimes the first iteration was correct, sometimes it required a second effort. The experimental data is of very high quality. Among the most successful sections are those which deal with hydrostatics, waves and simple harmonic motion (VIII, VI and V). In addition, the book contains some results of special mathematical interest, including the use of the calculus of variations in finding the solid of least resistance and the first published description of the method of moments.

A typically important insight is his use of similarity principles, in Section VII. Here, implying a law of 'dynamic similarity' for motions of bodies of the same shape, he assumes that different spheres in different media will have similar motions as long 'the value of a certain key parameter remains the same for both' (Proposition 32 and Corollaries 1–2, and Corollary 1 to Proposition 33). Also significant in this context are the corollaries to Propositions 5 and 7 in Section II. This type of approach, one of the most important in all studies of resisted motion, led ultimately to the discovery of the Reynolds number, and allowed the use of scale models in the experimental testing of such things as lift and drag coefficients of aircraft.[82]

Sections I and II of Book II discuss motion resisted by forces respectively proportional to the velocity and the square of the velocity, separating out the horizontal and vertical components. Newton produced treatments of these conditions which are now standard treatments of resisted motion in one and two dimensions. Once again, in contrast to all previous attempts at analysing such aspects of resisted motion, Newton offers a completely comprehensive consideration of all the important cases. Ehrlichson believed

---

[80] As we will see below, this does not include any from Section VII. Of the others, Whiteside has criticised Proposition 27 as a gross approximation, though Newton may well have known that it was. Proposition 51 is innovative and important qualitatively, but derives the wrong relation between time period and radial distances for vortices, whereas Proposition 52 finds the correct relation but from a seemingly incorrect starting-point and methodology.
[81] For Newton as a uniquely gifted problem solver, see Iliffe and Smith (2016, 17, 30).
[82] Smith (1999a, 190); Smith believes the idea should be more 'celebrated' than it is.

that Newton used calculus versions of the equations of motion in these particular propositions in working out the geometrical models that he needed for his exposition in the printed text, and several seem to indicate that he had some kind of prior knowledge of which would be the most effective.[83] Propositions 1–4, with their Corollaries, deal with resistance proportional to velocity in a uniform gravitational field. Proposition 1 considers the simplest case, of resistance to purely inertial motion. asserting that 'If a body is resisted in the ratio of its velocity, the motion lost by resistance is as the space gone over in its motion'. Propositions 2–4 consider horizontal, vertical, and both horizontal and vertical motions under such conditions. In Proposition 2, for example, the speed ratio increases in a geometric progression, that is exponentially. In Proposition 3, there is also an exponential relationship but the speed only increases until the body reaches terminal velocity. Proposition 4 considers resisted projectile motion under vertical gravity with resistance proportional to velocity.

Section II, Propositions 5–9, and their Corollaries, consider respective horizontal and vertical motions where resistance is proportional to velocity squared and the differential equations are no longer linear. Leibniz, who converted Proposition 9 into his own version of calculus, without significant difficulty, incorrectly claimed a generalisation of these results in his *Tentamen* of 1689.[84] Proposition 10, with the added complication of a variable density, has a major result in the identification of the approximate ballistic curve — a distorted hyperbola — with a demonstration in the process that it must have an asymptote.[85] Section III combines components of resistance proportional to velocity and velocity squared, again separated into horizontal motion without external force (Propositions 11 and 12 and Corollaries) and vertical motion under gravity (Propositions 13, with Corollary and Scholium, and Proposition 14, with Corollary).

Section IV is concerned with motion under the influence of a centripetal force with resistance from a medium with nonuniform density, varying inversely as the distance r from an 'immovable centre'. The orbits Newton deals with are equiangular or logarithmic spirals. Proposition 15 (with 9 corollaries) considers the effect of air-drag on a body descending under a centripetal force varying as the square of the density and then in an arbitrary manner. For an orbit that is an equiangular spiral, the density must

---

[83] Ehrlichson (1991, 1996).
[84] Leibniz (1689), 82–86. Whiteside, *MP* VI, 71.
[85] Whiteside reproduces the derivation of the 'logarithmic trajectory' (*MP* VI, 19) in *MP* VI, 85–87. See also *De motu corporum in medii resistentibus*; *Unp.*, 262–267, translation 287–292.

vary inversely as r. 'The analysis leading to this conclusion', which relies on geometrical properties of the equiangular spiral established in Lemma 3, is, for Chandrasekhar, 'a veritable *tour de force*',[86] while D. C. King-Hele and D. M. C. Walker state that: 'Connoisseurs of independent discovery may like to know that Newton's result was independently re-derived by Morduchow and Volpe in 1973'.[87] Colin Pask points out its relevance and that of the succeeding propositions to the problem of launching artificial satellites.[88]

Proposition 16 (with 3 corollaries) generalises to a centripetal attraction varying as the inverse nth power of r. In the special case of an inverse-cube attraction, the resistance will disappear, while for $n > 4$ it will become negative. The Scholium adds that the varying density condition only applies to the motion of bodies 'that are so small that the greater density of the medium on one side of the body above that of the other is not to be considered'. Newton supposes the resistance of the medium, '*caeteris paribus*', or 'other things being equal', 'to be proportional to its density'. The idea of an inverse square law with resistance leading to an orbit spiralling into the centre, may be compared with the behaviour of orbiting accelerating charges in later physics, for example, in precursors to the Bohr atom, where the assumed loss of energy through bremsstrahlung is similar in principle to the energy loss due to resistance. The purpose of Proposition 17 is 'To find the centripetal force and the resisting force of the medium, by which a body, the law of velocity being given, shall revolve in a given spiral', while Proposition 18 aims, 'The law of centripetal force being given, to find the density of the medium in each of the places thereof, by which a body may describe a given spiral'.

Propositions 1–18, which make up the first four sections of Book II (and were originally intended for Book I), together form the first comprehensive account of motion under resistance proportional to both velocity and velocity squared.[89] Newton's ultimate aim here was to relate these to the respective viscous and inertial components of resistance which he believed could be identified for real fluids. In this work, he described a genuine phenomenon for resisting media which was otherwise only appreciated at a much later period. In such media, at low velocity the Young's modulus or some other

---

[86] Chandrasekhar (1995, 545).
[87] King-Hele and Walker (1987), quoted in Chandrasekhar (1995, 539).
[88] Pask (2013, 356), referring to Blitzer (1971).
[89] Westfall says that, 'Almost without precedent, Newton created the scientific treatment of motion under conditions of resistance, that is, of motion as it is found in the world' (1971, 493).

such property determines what the medium does, but only density or inertia is important at high velocity.[90]

Newton was actually correct in stating that, over the range of the experiments he subsequently carried out, 'the resistance is predominately from the inertia of the fluid'; however, there is no 'purely inertial mechanism' by which this happens, and no general law of resistance of the kind Newton required has been found, even for the simplest solid bodies.[91] It does not arise from a 'combination of independent physical mechanisms', of internal friction or viscosity and inertia of the fluid, responding to different powers of velocity, in the form $a_1 v + a_2 v^2$, as Newton postulated.[92] The mechanisms only exist as an intricate combination which cannot be separated. Nevertheless, engineers still find it convenient to use combinations such as $a_1 v + a_2 v^2$ as an empirical approximation to the resistive force in the case of laminar flow, and a 'reasonably good fit' to the data.[93] (A constant third term, representing surface friction, a_0, which Newton added only in the third edition of the *Principia*, is excluded from modern accounts as a result of the 'no slip' condition in which the fluid particles in immediate contact with a body's surface move with the body.[94]) In addition, the results of Newton's resistance experiments, in combination with his tentative theorising, led to a number of significant developments, while the experiments suggested that, in practice, only a single power of velocity (v^2) was significant over most of the cases he investigated.

A key result was the proportionality of resistance to density (ρ), velocity squared (v^2) and the main cross-section (A) of the moving object, now known as Newton's law of fluid friction: '... the greater parts of the systems are resisted in a ratio compounded of the duplicate [squared] ratio of their velocities, and the duplicate ratio of their diameters, and the simple ratio of the density of the parts of the systems'. This is derived by an argument from first principles, which, like the derivation of Boyle's law, is almost a kind of dimensional analysis.[95]

---

[90] Bernal (1972, 235).
[91] Smith (2005, 134, 150).
[92] Smith (2005, 150).
[93] Smith (2005, 152). The alternative (and better) method is to take the drag coefficient as a function of the Reynolds number, which gives the fundamental ratio of viscous to inertial effects (see below).
[94] Smith (1999a, 189).
[95] *Principia*, II, Proposition 33, where it is derived for highly rarefied fluids and for elastic fluids like air. He also found it true for fluids like water where the molecules were more closely packed. In the first edition of the *Principia*, a Proposition 34 was added saying

Newton explains:

> For the resisting Power of fluid Mediums arises partly from the Attrition of the Parts of the Medium, and partly from the *Vis inertiæ* of the Matter. That part of the Resistance of a spherical Body which arises from the Attrition of the Parts of the Medium is very nearly as the Diameter, or, at the most, as the *Factum* of the Diameter, and the Velocity of the spherical Body together. And that part of the Resistance which arises from the *Vis inertiæ* of the Matter, is as the Square of that *Factum*. And by this difference the two sorts of Resistance may be distinguish'd from one another in any Medium; and these being distinguish'd, it will be found that almost all the Resistance of Bodies of a competent Magnitude moving in Air, Water, Quick-silver, and such like Fluids with a competent Velocity, arises from the *Vis inertiæ* of the Parts of the Fluid. And for these Reasons the Density of fluid Mediums is very nearly proportional to their Resistance.[96]

Newton's correct identification of the important parameters is equivalent to the formula for the drag equation $R = 1/4 \, \rho A v^2$ or $R = 1/2 \, C_D \, \rho A v^2$. The drag coefficient C_D is then defined as $2 \times$ resistance force/(fluid density $\times$ frontal area of the body $\times$ velocity^2). Newton's experiments with falling bodies in water and air, with this resistance formula, led to his finding an average resistance coefficient $z = 0.5$, compared with the modern value $z = 0.48$.[97]

In general, the drag coefficient is not a constant but depends on the dimensionless parameter, defined in the nineteenth century as the Reynolds number, $R_e = \rho v l/\mu$, where l is a characteristic length for the object moving through a medium with coefficient of viscosity μ. The drag coefficient decreases progressively as the Reynolds number increases, but not linearly. At low speeds and low Reynolds number, say of order 1, C_D decreases linearly with velocity and the resistance force is proportional to v. This is the region in which Stokes' law applies. However, over a considerable range of Reynolds number from about 1,000 to 100,000, the drag coefficient is approximately constant and the resistance is largely proportional to v^2. Newton's experimental work took him from very low values of the Reynolds

---

that the results of Propositions 32 and 33 would also apply if the particles had 'extremely great lubricity'. This was eliminated in the second edition, shifting the numbering of the later Propositions in Book II.

[96] Q 28. In Stokes' law the viscous force has the direct proportionality to diameter and velocity, but it cannot be attributed to a separate 'Attrition of the Parts of the Medium'.

[97] Néményi (1963). An earlier set of pendulum experiments had indicated the ρ and A dependence for the v^2 component but were too unreliable to give good values for the drag coefficient.

number up to about 80,000, and so from $R \propto v$ to $R \propto v^2$, with a more complicated region in between which he modelled, in his pendulum experiments, assuming $R \propto v^{3/2}$. It was the large region of relative stability in the drag coefficient/Reynolds number relation that allowed Newton to complete a relatively successful pioneering investigation in an area that researchers still find challenging today.

Newton had two models of resistance, which depended on the kind of structure of the fluids involved, though both obeyed the proportionality to $A\rho v^2$. The first (presented in Propositions 32–35) was an impact theory, which applied to highly 'rarefied' media, whose particles were extremely spaced out and exerted no forces on each other, and to elastic fluids, presumably like air, in which there were repulsive forces between the particles and also on the bodies on which they impacted. Julián Simón Calero, who emphasises Newton's role as the originator of this 'impact' theory, has stressed how it was only with the *Principia* that 'the fluids find themselves submitted to a theory', and that the results attracted 'the interest of all the authors' on the subject 'over at least a century'.[98]

The impact theory subsequently found significance in its application to the problem of aircraft flight. Here, in addition to establishing the connection between aerodynamic force and the product of fluid density, reference area, and squared velocity, using the impact theory, Newton indirectly made the first attempt at analysing the effect on the force of the angle of incidence, or, as it is now called, the angle of attack. Proposition 34 states that: 'If in a rare medium, consisting of equal particles freely disposed at equal distances from one another, a globe and a cylinder described on equal diameters move with equal velocities in the direction of the axis of the cylinder, the resistance of the globe will be but half as great as that of the cylinder'. The fluid here was assumed to be a 'rarefied and frictionless medium', composed of particles impacting on the body's surface, 'giving up their components of momentum normal to the surface, and then travelling downstream tangentially along the body surface'.[99] Newton knew that a real fluid didn't behave like this, but the result is a correct mathematical derivation for the conditions assumed.

Here, Newton 'frames' an implied "sine-square" law of resistance on a surface element' to show that the resistance of a sphere moving rapidly in a straight line in the resisting medium defined in the previous Proposition is 'exactly half that on the sphere's transverse (circular) section'.[100] Whiteside

---

[98]Calero (2008, 113).
[99]Anderson (1997, 39).
[100]Whiteside, *MP* VI, xxxii.

says that 'Newton's elegant equivalent appeal to the standard result (first demonstrated by Archimedes in his *On Conoids and Spheroids*, 20–22) that the volume of the frustum of a paraboloid of revolution is half that of the circumscribing cylinder somewhat obscures the basic simplicity of his deduction'.[101] In Proposition 35, Newton considers the consequences which follow for the motion of a cylinder and a sphere in a resisting medium given different sets of conditions, in particular when the collisions of the particles of the medium with the object are completely elastic, completely inelastic, and partially elastic. The resistance in the second case for either sphere or cylinder will be only half as great as in the first, and somewhere in between these values in the third. Corollaries 2–5 once again provide a resistance proportional to $\rho A v^2$, while Corollary 7, and Corollary 3 to Proposition 38 (which is based on a continuous rather than rarefied medium), describe how to find the velocity acquired, distance travelled and resistance generated by a sphere of known density and given initial velocity in a resisting medium also of known density.[102]

The proof of Proposition 34 implied that 'the impact force exerted by a fluid on a segment of a curved surface is proportional to $\sin^2\theta$, where θ is the angle between a tangent to the surface and the free-stream direction'. Essentially, supposing each element of the body's surface to be at an angle θ to the direction of fluid flow, the pressure on each element will be proportional to $\sin \theta$, and if the resistance on the element is due only to the normal component of the pressure from the fluid, $p \sin \theta$, it will be proportional to $\sin^2\theta$. Subsequently, when this factor was included in the expression for the aerodynamic force, and applied to a flat plate, with θ as the angle of attack, as was done subsequent to Newton's exposition, heavier-than-air-flight seemed to become practically impossible. Newton was consequently criticised for his 'erroneous' demonstration, even though he never made this application, nor did he assume that air behaved like the 'rarefied' medium that the calculation assumed, when his experiments on falling spheres (see below) suggested that it did not.[103] According to Calero, however: 'in the atmosphere at altitude of 100 km (generally considered to

---

[101] *ibid.*, 470.
[102] Westfall (1971, 493), considered the latter the 'climax' of Newton's work on resisted motion.
[103] Giacomelli (1963, 311 ff). Smith (2005, 128) asserts that both Proposition and Scholium 'make rigorously correct claims about such rarified fluids'. Newton was 'mathematically exploring possibilities with the goal of enabling the empirical world to choose among them' (129). See also Anderson (1997, 39).

be the edge of space) the air density is so low that the molecules have to run a long distance before colliding with another molecule. Therefore, faced with a moving body the gas behaves just as Newton forecast, with individual impacts between molecules'.[104]

Newtonian positions have shown an astonishing capacity for revival, possibly because they are often based on extremely abstract reasoning rather than being linked to specific applications. While Whiteside, in editing the *Mathematical Papers*, gave the impression that Proposition 34 was by that time of historical interest only from the physical point of view, this is now no longer the case.[105] The demonstration is not erroneous for the medium Newton described, and, ironically, according to John D. Anderson,

> Newton's sine-squared law has had a rebirth in modern aerodynamics, namely, for the prediction of pressure distributions on the surfaces of hypersonic vehicles. The physical nature of hypersonic flow, where the bow shock wave lies very close to the vehicle surface, closely approximates the model used by Newton — a stream of particles in rectilinear motion colliding with the surface and then moving tangentially over the surface. Hence the sine-squared law leads to reasonable predictions for the pressure distributions over blunt-nosed hypersonic vehicles, an application that Newton could not have foreseen.[106]

Newton did, however, foresee that his resistance model would be most successful for objects moving at very high speed, as Corollaries 2–5 to Proposition 33 clearly indicate.

Section VII uses a relative motion principle or principle of reciprocity, by which the same forces result on a stationary body under the impact of moving fluid particles as would occur for a body moving in a stationary fluid. Stated explicitly in Proposition 34 (which actually combines two of

---

[104] Calero (2008, 91).

[105] Whiteside, *MP* VI, 463, n. 23 refers to the 'artificiality' of the assumptions leading to the sine squared law. Drazin (1987) makes the specific claim that the sine-squared law only has 'historical significance' (352).

[106] Anderson (1997, 40). The effect is similar to that produced by the solar wind. Comte and Diaz (2004, 2–3), in their study of the Least Resistance problem, say that, 'even though the Newton resistance model is only a crude approximation, it appears to provide good results in many contexts, dealing, for instance, with a rarefied gas in hypersonic aerodynamics', and they say that such 'distinguished specialists in this area' as 'von Karnan, Ferrari, Lighthill and Sears have used this model'. Hui (2006, 2) says that it provides 'a rather crude approximation to real physics; however, it appears to provide good results in the following situations: for a body in a rarefied gas with low speed, for bodies which move in an ideal gas with high Mach number, and for slender bodies'. See also Buzzatto and Frediani (2009, 34).

Newton's principles: reciprocity and dynamic similarity), this has become a fundamental principle used in fluid mechanics, for example, in using wind tunnels to test stationary scale models of aircraft.

The famous Scholium, which follows, on the solid of least resistance, is also completely valid mathematically for the rarefied medium which it assumes, though it has no practical value in ship design, as Newton proposed. In the first edition of the *Principia*, however, Newton followed his theoretical calculation of the solid of least resistance with a suggestion that his experimental studies of resistance might be extended to find 'the aptest shape for ships' at 'little cost', by using scale models of ships' hulls in tanks with a steady flow of water, which is a perfectly valid suggestion, later followed up.[107] John Craig had apparently suggested the problem of ship design to him during a visit to Cambridge in 1685. The experimental method was subsequently adopted and necessarily led to results which could not be obtained by simplified mathematical models.

Not only was Newton's method in the calculation original, his result was also surprising, and has been described as 300 years ahead of its time. According to V. I. Arnol'd:

> The extremum has a break and Newton knew about it. A smooth ellipsoid offers more resistance to motion through a fluid than a body of more angular section. The break appears in Newton's picture but is absent from the picture in some later editions of the *Principia* because, until the late twentieth century, when the problem became topical in astronautics, the mathematicians did not understand Newton.[108]

While the treatment of resistance in highly rarefied media in the first part of Book II, Section VII (Propositions 32–35) is now recognised as intrinsically correct, the treatment of the resistance in 'continuous media' in the second part (Newton's second theory of resistance, Propositions 36–39) is considerably more problematic, for the incompressible, inviscid fluid that he assumed produces exactly zero resistance, as d'Alembert later found — though not if there is any amount of viscosity.[109] Newton's theoretical model actually followed his experimental results, and he changed it in important ways in the second edition of the *Principia*. Significantly, Newton did not propose that his model was necessarily a realistic description of how a continuous material medium actually resisted the movement of a material

---

[107] Cohen (1999, 183).
[108] Zdravhovska (1987). See also *Newton and the Great World System* for the mathematical significance.
[109] d'Alembert (1752).

object. He regarded it simply as a means of calculating the component of resistance provided by the medium's inertia, and proportional to v^2, and such a model, called a 'perfect fluid', is still used by some fluid dynamicists to solve particular problems.

For the incompressible, inviscid fluid, with particles close packed and in direct contact with their neighbours, Newton proposed to investigate the inertial loss of momentum of objects of various shapes by successive impact with the fluid particles. As a preliminary, he set up an experiment on the speed of a jet of water issuing from a hole at the bottom of a cylindrical tank, which was reported in Proposition 36 (originally 37). The aim was to find the inertial force which would result from removing a column of liquid and which, he assumed, would act as the inertial component of resistance on a material cylinder of the same dimensions. Here, he reasoned that if the total height of the water of density ρ was h and cross-section A, and the velocity of outflow V, then the total loss of momentum was $\rho h A V$. However, if the water travelled $2s$ in time t under final velocity V, then, under its instantaneous velocity v, it would travel distance $2sv/V$ in the same time. The mass of water moved would be $2\rho s A v/V$ and the momentum transferred $2\rho s A v^2/V$. Equating $\rho h A V$ and $2\rho s A v^2/V$ gives $v^2/V^2 = h/2s$, meaning that the velocity acquired would be equivalent to that produced by falling half the height of the tank, $v^2 = gh$. When Newton tested this by measuring the volume rate of flow out of the hole, his experiment totally confirmed his prediction. Proposition 36/37, in which this was discussed, was also remarkable for leading to the first statement of reaction propulsion, while the diagram of the apparatus he used has been described as displaying 'a forerunner of a modern wind tunnel'.[110]

Within a few years, however, in 1690, Newton's confidence was shaken, when Fatio de Duillier drew his attention to an experiment, first performed by Torricelli, in which a pipe attached to the outlet sent a column of water almost up to the height from which it had started, leading to the conclusion that the velocity acquired was equivalent to that produced by falling the whole height of the tank, $v^2 = 2gh$. This was the value that would be predicted from 'Galileo's theorem', or in modern terms the conservation of energy, rather than the conservation of momentum argument that Newton had previously used. For the second edition of the *Principia*, Newton decided it was necessary to revise the whole section. The problem was that volume

---

[110] Smith (2001, 269).

rate of flow experiments still indicated the correctness of $v^2 = gh$.[111] He didn't resolve the problem until 1712.

The revised Proposition 36 has been the most heavily criticised item in the *Principia* almost from the time of publication until the present, largely because attention has been focused on the 'model' that Newton seemingly used to illustrate his idea. Yet, it is in fact an extraordinary achievement, a classic example of Newton's abstract style of reasoning. Approaching seventy, in the middle of conflicts with Leibniz and Flamsteed and troubled by problems at the Mint, with the second edition of the *Principia* impending, Newton reconciled two conflicting theories and two conflicting experiments using a theoretical concept equivalent to streamline flow, and the experimental discovery of the *vena contracta* principle, a 'very important hydraulic phenomenon', in which the jet has a point of minimum cross-section near but just outside the container wall. The contraction of the jet then leads to a flow rate reduced by a factor of $\sqrt{2}$ from the calculated value.[112]

In Newton's explanation, the water flows in a cataract, defined by boundaries looking exactly like modern streamlines, beginning with the surface of the fluid where it touches the tank walls and curving inwards until they reach the hole. Like modern streamlines, they meet only at a distance of about a hole's length outside the hole. The curvature of the flow makes it possible to reconcile vector momentum and scalar energy calculations, and the factor of 2 in the velocity squared is resolved physically when the streamlines from opposite directions meet and produce a sudden acceleration in the jet stream. It was an abstract resolution, not dependent on any model, and the theoretical resolution was matched by the quality of the experimental discovery of the *vena contracta*. Comparison with a modern value of the 'efflux coefficient' of 0.62, suggest that Newton's measurement of the width of the stream at a distance of almost half an inch, for a whole of diameter 25/40 inches, was accurate to less than 1/40 of an inch.[113]

Yet, for all its success in achieving this resolution of a surprisingly difficult problem, the Proposition has been variously regarded as a bluff,

---

[111] It is important to note that Newton's conclusions of 1687, both theoretical and experimental, were not *erroneous*, as is often claimed, merely incomplete, as he would subsequently find.
[112] Cascetta (1995, 233).
[113] Smith (2005, 139); Calero (2008, 102). It is remarkable that the Proposition in its two versions was responsible for as many as *four* separate innovations: reaction propulsion, a precursor of the wind tunnel, streamline flow and the *vena contracta*.

a fudge and a blunder by numerous commentators. Many earlier writers seriously thought that the water would fall straight down from the surface, and couldn't believe in Newton's construction of curved flow lines. Brougham and Routh, who produced the first systematic account of the *Principia* in the nineteenth century, thought it 'open to very serious objections'.[114] Truesdell, the originator of the non-existent 'fudge factor' in Newton's work, characteristically regarded this and the whole of Section VII, as perfect examples of 'bluff' and 'fudge'.[115] Westfall, normally so astute in his analysis of Newton's dynamics, was exceptionally harsh in his analysis of the Proposition, on the basis that the final velocity of the efflux could be achieved in one line using conservation of energy principles,[116] but Newton's aim was also to describe how the motion occurred, and how it could be reconciled simultaneously to the conservation of momentum, which seemingly gave different results. The seriousness with which he treated the Proposition can be seen from its exceptional length and the inclusion within it of 5 Cases and 10 Corollaries. If Newton had intended it as a 'bluff', he would have despatched it far more quickly and with much less attempt at detailed explanation.

The reason for the criticisms, however, is obvious. Newton chose to illustrate his novel concept of what is essentially streamline flow by a seemingly bizarre analogy between the relatively inactive parts of the water and a block of melting ice, the ice melting as the surface of the cataract descended through it. The 'ice' was simply a metaphor, not a literal description of the fluid's microstructure, yet commentators unused to Newton's abstract style of reasoning fixed their attention on his metaphors and working models as though he intended them to be 'hypotheses' in the Cartesian style, which could be 'refuted' when better physical models were proposed, and did this for the whole of Section VII.[117]

---

[114] Brougham and Routh (1855, 264).

[115] Truesdell (1968a, 201).

[116] Westfall (1971, 501–505) and (1980, 707–712). Westfall was heavily influenced here, as he was in the whole 'fudge factor' story, by Truesdell. As with Truesdell, his condemnation of Proposition 36 extended to the whole of Propositions 37–40, even to the experiments on resistance, whose brilliance has now been firmly established by Smith and others. But Proposition 36, which is totally coherent when read with close attention, has yet to be rehabilitated or studied in detail.

[117] Truesdell, who characteristically homes in on the 'cataract of melting ice' (it was actually a cataract passing *through* melting ice), typically gives no specific analysis of the technical details to support his criticism, and, as on some other occasions, relies purely on bluster. He also completely disregards Newton's experimental discovery of the *vena contracta*. Amazingly, his account is still quoted as though authoritative.

As Newton saw it, Propositions 37–39 were aimed at calculating the inertial part of the resistance force in continuous media. In principle, experiments would then be able to extract the other resistance components from the data. The calculations were not intended to justify the continuum model or to assess its validity. Since they only concerned the inertial component, they could be set up in an almost hypothesis-free abstract manner. Proposition 37 developed the analysis of the efflux experiment to conclude that a solid cylinder moving 'forward in a compressed, infinite and non-elastic fluid, in the direction of its length' would generate a resistive force on its 'transverse section' equivalent to the force which would destroy or generate its whole motion in the same time as it took to move four times its length, 'as the density of the medium to the density of the cylinder, nearly'. In principle, the force required to remove the whole momentum of a body of mass m, cross-sectional area A and length l travelling at speed v is $mv/\text{time} = mv/(4l/v) = mv^2/4l = 1/4$ density $\times Av^2$. If ρ is the density of the medium, the equivalent resistive force exerted by the medium is $1/4\ \rho A v^2$.

Comparing this to the drag equation $R = 1/2\ C_D\ \rho A v^2$ would seem to suggest that Newton's 'model' had 'predicted' a drag coefficient for a cylinder of $C_D = 1/2$, whereas measured values are in the region $C_D = 1$. Newton never did any drag experiments on cylinders, but he was aware of the effect of shape on resistance, as well as surface conditions,[118] and the Scholium which follows the Proposition points out that it only concerns the resistance arising 'from the magnitude of the transverse section of the cylinder, neglecting that part of the same which may arise from the obliquity of the motions'. He then invokes 'the obliquity of the motions with which the parts of the water, pressed by the antecedent extremity of the cylinder, yield to the pressure, and diverge on all sides, retards their passage through the pieces that lie round that antecedent extremity, toward the hinder parts of the cylinder, and causes the fluid to be moved to a greater distance'. This, he says, will increase the resistance almost in the same ratio by which it diminishes the efflux of water in Proposition 36. If we take this as the factor 2 seen in that Proposition, the effective C_D would become closer to the observed value. Even so, the Proposition was not intended as a 'prediction' to be tested; it was simply a means of assessing the inertial component

---

[118] This is additionally apparent in the Scholia following Lemma 7 and Proposition 39.

of resistance on a cylinder considering only the effect on the transverse section.[119]

Cylinders were clearly difficult to observe in a consistent manner. Spheres provided a much simpler type of object for experimental investigation. For spherical objects in rarefied media, the 'resistance is to the force by which the whole motion of the globe may be destroyed or produced in the time in which the globe can describe two thirds of its diameter; with a velocity uniformly continued, as the density of the medium to the density of the globe', assuming globe and particles of the medium to be perfectly elastic, and 'this force, where the globe and particles are infinitely hard and void of any reflecting force, is diminished one half'.[120]

For continuous media, in which Newton supposed the sphere pressed only against adjacent particles, which then pressed on the particles next to them, and so on, the resistance was diminished by another half, and the measure required eight thirds, rather than two thirds, of the diameter of the globe (Proposition 38). Comparing this result to that of the cylinder, we see that the respective volumes are $\pi r^2 l$ and $4\pi r^3/3$, or Al and $2AD/3$, where A is the cross-sectional area of each object and D the diameter of the sphere. Using the same reasoning we find that the globe requires $8D/3$ stopping distance rather than $4l$, leading again to $C_D = 1/2$ if the drag equation applies. This would be the basis of Newton's experimental investigations.

However, to devise a theory that would explain the facts of these experimental investigations, Newton had been obliged to suppose that, in continuous media, the shape did not determine the resistance coefficient. So, Lemmas 4–7 claimed that, in water, cylinders, spheres and spheroids of 'equal breadth' would all experience the same resistance (to the transverse section) as a disc of the same diameter, whether the objects moved or the water, and, according to Proposition 38, Corollary 1, this would be 'in a ratio compounded of the duplicate ratio of the velocity and the duplicate ratio of the diameter, and the ratio of the density of the mediums', as in the case of rarefied media.

Whatever the status of Newton's working hypothesis, he used it as the basis of a highly successful series of experiments on spheres falling in different media, which are recorded in *Principia*, Book II, Proposition 40 and its

---

[119] A fascinating story related by George Gamow describes how Proposition 37, in which the resistive force depends only on the length of the cylinder and not on the initial velocity, was used in World War II to explain how the depth of a bomb's penetration into the ground had no dependence on the height from which it was dropped (Simony, 2012, 269).
[120] *Principia*, Book II, Proposition 35, Scholium.

Scholium, and the theory was (as always) necessary to provide a framework in which the experiments could be conceived. Proposition 38 provided the basic structure for the interpretation of the results; though, in two of the experiments, he also used a version (Proposition 39), with a correction factor, taken from Corollary 2 to Proposition 37, to allow for the fact that the real length of the fluid was finite. To translate the measurements of distance x and time t into the equivalent of a resistance force was in itself nontrivial for it effectively meant solving the differential equation $m\ddot{x} = mg - b(\dot{x})^2$, for $x = 0$ and $v = 0$ at $t = 0$. According to modern textbooks, the solution for velocity is $v(t) = V \tanh(gt/V)$, where the terminal velocity $V = \sqrt{mg/b}$, while the solution for distance can be given as either $x(t) = (V^2/g) \ln(\cosh(gt/V))$ or $x = (V^2/2g) \ln(V^2/(V^2-v^2))$. Newton nowhere gives these solutions but the tables for theoretically expected values of velocity and distance in Proposition 40, which are accurate to eight significant figures, show that he must have known some version of them.[121]

The fourteen experiments, which included twelve by himself of spheres measured falling in a tank of water, and two of spheres falling in air from the cupola of St. Paul's Cathedral, in experiments carried out by Hauksbee in 1710 and Desaguliers in 1719 under Newton's direction, gave results on the resistance on spheres which were 'as accurate as any experimentally determined values before the twentieth century'.[122] The fact that Newton respected the integrity of data is shown in these high-quality resistance experiments, which defy his attempts to impose a neat theory on the phenomenon.[123] He also realised that he had no real prospect of finding a viscous 'component' of resistance in his data and downgraded the significance of his earlier pendulum experiments in relation to those involving falling spheres. While most of the experiments on spheres gave results close to his theoretical prediction of a drag coefficient, $C_D = 0.5$, some fell nontrivially on either side of this value. The cases where the results were larger seemed to involve higher velocities, and Newton thought that a significantly large velocity might cause a loss of pressure in the fluid at the rear of a falling sphere. Other cases where the resistance was too low coincide with those

---

[121] Pask (2013, 370–371), who cites Lamb (1945) as the source for the modern solutions.
[122] Smith (2005, 140). Those of 1710 used glass spheres, filled with air, water or mercury, and those of 1719 used spherical bladders, filled with various liquids, and balls of lead. Smith (2001, 276) gives the respective values for C_D in air as 0.504 to 0.538 for 1710 (Reynolds number 78,000) and 0.499 to 0.518 for 1719 (Reynolds number 40,000), and says they show 'remarkably good agreement with modern values'. Results for water were 'slightly high': 0.462–0.519.
[123] Smith (2005, 144).

in which the same thing happens with modern data. Though Newton made no separate comment on the lower values, George E. Smith regards their alignment with modern results as a strong indicator of the 'high quality of Newton's data'.[124]

The experiments showed that the resistance coefficient for a sphere was the same in two media apparently showing different responses to inertial mechanisms — a fact of considerable interest in itself. In such cases, we might believe that a more fundamental principle (such as conservation of energy) is the real reason for the 'coincidence', and that it manifests itself through different processes under different conditions, for example, in terms of viscosity where this is required. However, the mathematical model which explains the phenomenon can still be successful even if it doesn't reflect the 'physical' conditions that emerge at that level. We have already seen something of this kind with Newton's mathematical explanation of Boyle's law. It has links with the idea of the renormalisation group in twentieth century physics.

Earlier negative assessments of Newton's work in this area were based on a lack of appreciation of how science opens up a new area of research, in this case one which both Descartes and Galileo had previously rejected as impossible.[125] Later assessments have been much more favourable, despite the fact that Newton, in Propositions 36–39, was unable to match theory and experiment in the same fundamental way as in Books I and III of the *Principia*. Colin F. Gauld, for example, regards the combination of theory and experiment as an example of 'frontier' science, which can be followed by students as an object lesson in how scientists carry out real investigations in a previously unknown area.[126] For George E. Smith, the Newtonian theory was one that had value in becoming 'an instrument in further research',[127] whereas 'the century and a half of hydrodynamics between d'Alembert and Prandtl' was a much better candidate for 'bad science' than Newton's theory of inviscid fluids in Propositions 36–39.[128]

In fact, even though Newton's working hypothesis of separate viscous and inertial components for the resistance force, proportional to v and v^2,

---

[124] *ibid.*
[125] Descartes (1629); Galileo (1991, 252); both quoted by Smith (2005, 156). Westfall (1971, 493), describes Newton's work as 'Almost without precedent'.
[126] Gauld (2010). This was also true of the less successful pendulum experiments (Gauld, 2009).
[127] Smith (2005, 151).
[128] Smith (2005, 153).

was invalid as a *physical* explanation, both his calculations and experiments led directly to the approximate identification of the predominantly inertial effect over the range he studied, and also the predominance of the dependence on v^2. The fact that an idealised 'inviscid' fluid cannot actually deliver inertial resistance doesn't mean that it is an incorrect working model for calculating the inertial effect that is produced partly through the medium of viscosity. If a model produces 'correct' results, it generally has a degree of validity, at least in a heuristic sense. True 'coincidences' seldom happen in physics.[129]

5.5. Vortices

To a considerable extent, Book II of the *Principia* was aimed at removing Descartes' system of vortices from physics. As this was still the dominant explanation of the motion of the heavenly bodies in the solar system, and would remain so for at least another half century, even after Newton's demonstration of the validity of an inverse square law of gravity, it is not surprising that he gave it so much emphasis. It had, after all, taken a very considerable struggle for Newton to eradicate it completely from his own mode of thinking. He could hardly expect his contemporaries to fall into line without a systematic explanation of why the model that they had always accepted as the unchallenged starting-point of their investigations failed on almost every account. Strangely enough, this ubiquitous model was not based on any genuine systematic study of vortex motion, and Newton, as in so many other areas, had to break new ground in tackling this aspect of material dynamics in mathematical terms almost for the first time.

Vortices actually appeared early in Newton's work with his detailed observations of the vortices in water produced by a moving body, which include the small 'motes' which appear at the 'stagnation points' directly behind the body. These observations, with an implied extension to air, appear in the early 'Aristotelian' essay 'Of violent Motion' in his *Quaestiones*

---

[129] Parallel cases might be the use of a one-way light signal to determine simultaneity in special relativity, the derivation of time dilation from a system in uniform motion and the exclusion of backward motion in the use of Huygens' principle. It is a common error to suppose that heuristic models are 'incorrect' if superseded by more technically accurate descriptions. As Newton no doubt knew, almost all descriptions of real systems are heuristic to some degree, and the relative 'correctness' of a theory can be largely measured by the degree of heurism it incorporates. The key element in the resistance calculations is that they involve a 'book-keeping' exercise concerning the inertial process which has to be valid regardless of the actual physical mechanism by which it is maintained.

quaedam philosophicae.[130] Vortices are also involved in Newton's water bucket experiment and in his explanation of the Magnus effect. The water bucket experiment was mainly concerned with Newton's attempt to distinguish absolute and relative motions, but the patterns he observed there are widespread in fluid motion and have been observed in cloud structures on Saturn, in hurricanes, and in other natural vortex systems.[131] The Magnus effect, on the other hand, is an extraordinary example of how some of Newton's best physical ideas occurred only in contexts where they were introduced by analogy. Here, a very accurate analysis of the transverse force produced on a spherical object as a result of an air circulation came up in his correspondence on light only because he wished to refute a particular mechanical model of refraction.

> Then I began to suspect, whether the Rays, after their trajection through the Prisme, did not move in curve lines, and according to their more or less curvity tend to divers parts of the wall. And it increased my suspicion, when I remembered that I had often seen a Tennis-ball, struck with an oblique Racket, describe such a curve line. For, a circular as well as a progressive motion being communicated to it by that stroak, its parts on that side, where the motions conspire, must press and beat the contiguous Air more violently than on the other, and there excite reluctancy and reaction of the Air proportionately greater.[132]

The Magnus effect, as described in Newton's letter, is the phenomenon whereby a spinning object flying in a fluid creates a whirlpool of fluid around itself, and experiences a force perpendicular to the line of motion. The effect is essentially the same as that around an aerofoil, the basic structure required for flight in all modern aircraft. The Magnus effect is a particular manifestation of Bernoulli's theorem: fluid pressure decreases at points where the speed of the fluid increases. As we now explain it, the spinning sphere creates a thin boundary layer around itself due to viscous friction, and this produces a circular motion of the air in which the ball is spinning. The lift force turns out to have the same structure as Newton's force of fluid resistance $R = 1/4\ \rho A v^2$, with the constant of proportionality for a smooth ball in the region 0.1–0.3, compared with approximately 0.25 in fluid resistance. As in the example cited by Newton, it occurs widely in sporting

---

[130] QQP, 'Of violent Motion', f. 98^r, NP; Herivel (1965, 122–123). The actual observations recorded date from Newton's pre-Cambridge days.
[131] See, for example, Jansson *et al.* (2006). The curvature of a rotating liquid surface has also been used to create mirrors for reflecting telescopes.
[132] Newton to Oldenburg, 6 February 1672; *Corr.* I, 94.

contexts, but it also has many other applications. Remarkably, there is also an *optical* Magnus effect, connected with spin-orbit interaction in photons from a laser beam in the inhomogeneous medium of an optical fibre, when the polarisation of the photons switches from right- to left-handed circular; it is detected by an angular shift in the speckled pattern representing the modal noise.[133]

In striking out at the Cartesian system, Section IX of Book II was principally concerned with vortex motion, how it could be generated, and how it could be transferred via the dense Cartesian aether. This required knowledge of how internal friction or viscosity operates. Newton's experiments with falling spheres had not given him any insight into the viscous aspect of fluid resistance, so he had to rely on theory. As usual, his method of abstraction and his ability at choosing the most appropriate conditions for idealisation led to successful results. A clearly marked 'Hypothesis' at the beginning of the section defines Newtonian fluids, where linearity applies. Nearly all the successful work in this area has been on such fluids. The Hypothesis states Newton's law of viscosity: that, if Newtonian fluids flow in parallel layers, the shearing stress in them (or force per unit area between adjacent molecular planes) is proportional to the velocity gradient (or rate of change of velocity across them) — or equals the product of the velocity gradient and viscosity. Newton says: 'The resistance arising from the want of lubricity in the parts of a fluid is, other things being equal, proportional to the velocity with which the parts of the fluid are separated from one another'.[134] Here the 'want of lubricity' is equivalent to the shear stress τ, the frictional action on the liquid. The 'velocity with which the parts of the fluid are separated from one another' is the rate of strain experienced by a fluid element in the flow; mathematically, this becomes a velocity gradient dV/dn. If the constant of proportionality is the viscosity μ, then the Newtonian shear-stress law can be expressed by the relation $\tau = \mu(dV/n)$.

Nearly all gases are Newtonian fluids, and so obey this law, which thereby represented 'a major contribution to the state of the art of aerodynamics at the end of the seventeenth century'. 'Integration of the effects of the shear stresses over the entire surface of the body gives rise to the skin-friction

---

[133] Barkla and Auchterlonie (1970) have outlined the immediate post-Newtonian history of the effect, which included a demonstration for spherical bodies by Robins (1742), using quantitative experiments, followed by a subsequent denial of its validity by Euler (1777). Rayleigh (1877) was the first to associate it with Magnus, whose purely qualitative work (of 1853) was based on rotating cylinders rather than spheres.
[134] *Principia*, Book II, Section IX, Hypothesis.

drag exerted on the body. For slender, streamlined bodies, skin friction is by far the largest contributor to the total drag on the body; hence Newton's relationship for shear stress was of vital importance for aerodynamics'.[135]

The first two Propositions after the Hypothesis describe vortices made from spinning a cylinder and a sphere in a fluid. It is clear from Newton's comments that these Propositions and their many Corollaries are founded on experimental observation as much as mathematical theory.[136] Proposition 51 defines for the first time the problem of Couette flow, the laminar flow of a viscous fluid between relatively moving parallel plates,[137] in the process correctly identifying the frictional force as proportional to both the relative velocity of the layers and the extent of the area of contact. Here, it is between rotating parallel cylinders, a relatively simple system, much studied subsequently, which is illustrated by Newton with streamlines coaxial to the cylinders. In this system, the flow is created by the viscous drag on the fluid and a pressure gradient which is exerted parallel to the cylinders.[138] The Couette flow between two cylinders is one in which the Navier–Stokes equations produce an exact solution. It is such a classic case that it has been called the 'Hydrogen atom of fluid dynamics' and compared to the *Drosophila* of genetics.[139]

Newton also discusses the flow round a single cylinder, correctly saying that such a system will generate vortices in a fluid.[140] His observations showed that the cylinder did indeed transfer momentum to the layer of fluid next to it, and that each layer of fluid transferred momentum to the next layer, as he assumed by fluid friction, with successive layers acquiring a momentum determined only by their distances from the centre. According to Gavin Hamilton, Newton thus, 'in one simple experiment', 'revealed two concepts of overwhelming significance to the understanding of fluid dynamics — viscosity and laminar flow'.[141] His imperfect understanding of torque and angular momentum, and his incorrect use of a balance of forces

---

[135] Anderson (1997). In being the first 'to develop a theory of air resistance', Newton has been described as 'arguably' 'the world's first aerodynamicist' ('Aerodynamics', *New World Encyclopedia*).
[136] Proposition 51, for example, refers to experiments with cylinders made in 'deep standing water' (386).
[137] Smith (1999a, 193).
[138] See Corollaries 2–6.
[139] Tagg (1994); van Gils *et al.* (2012); Fardin *et al.* (2014, 3523).
[140] For parallel cylinders, G. G. Stokes showed in 1848 that vortices in the form of eddies would be produced by fast movement of the inner cylinder (Stokes, 1880).
[141] Hamilton (1980, 2). Laminar or streamline flow was, as we have seen, also significant in Proposition 36.

rather than torques, however, led him to imagine that the vortex would maintain an angular velocity ω inversely proportional to distance from the centre r, rather than distance squared, with time period T proportional to distance, and velocity v a constant at all distances.[142] This would imply that a fluid made up of identical particles also maintained the same kinetic energy in each.

Unlike cylinders, spinning spheres do not create vortices, as Newton assumed for Proposition 52. This Proposition correctly took into account the boundary condition for a viscous fluid, namely, adherence to the solid surface, but no complete solution has yet been found for its problem of the rotation of a sphere in a viscous fluid. However, it is important to recognise that Newton is not giving us *his own theory* but rather the one that he imagines Descartes would have to put forward to justify his explanation of celestial motions. In this case, though he may have once again used a balance of forces rather than torques in his derivation, he was led, perhaps by the fortuitous reduction from spherical to circular symmetry in the application to planetary motion, to the correct condition for irrotational vortex motion.[143]

Vortex motion lies between two limiting cases. The free or irrotational vortex (as in water draining through a circular hole) has the tangential velocity inversely proportional to the distance from the centre of the vortex ($v \propto 1/r$), or angular velocity inversely proportional to the square of this distance ($\omega \propto 1/r^2$). This means that, for all identical mass elements m, angular momentum (mvr) is conserved. This is the case considered in Proposition 52 and all its Corollaries. Newton says that 'the angular motions of the parts of the fluid about the axis of the globe are reciprocally as the squares of the distances from the centre of the globe' (Corollary 1). A consequence is that the periodic time ($T = 2\pi/\omega$) is directly proportional to r^2. In Corollary 7 he describes a globe contained in a spherical vessel of fluid revolving about a common axis in the same direction 'with the periodical times ... as the squares of the semi-diameters'. He says that 'the parts of the fluid will not go on in their motions without acceleration or

---

[142] Dobson (1999), referring to earlier work by Stokes. Newton's analysis is otherwise correct.

[143] It may be true that a 'correct' use of torque rather than force would lead to $\omega \propto 1/r^3$ rather than Newton's $\omega \propto 1/r^2$ (Dobson, 1999), but the whole process by which this happens is itself non-physical, and has no privileged status in defining a 'correct' method. The fact that free vortices require $\omega \propto 1/r^2$ is one of the reasons why spheres cannot produce vortices while cylinders can. Newton's results are stated for planar circular vortices, as required for a description of the planetary system, and are correct for this case.

retardation, till their periodical times are as the squares of their distances from the centre of the vortex. No constitution of a vortex can be permanent but this'.[144] Corollary 6 also shows that, for several globes producing vortices, the system will not remain stable, as 'the vortices will not be confined by any certain limits, but by degrees run mutually into each other; and by the mutual actions of the vortices on each other, the globes will be perpetually moved from their places...'

The other limiting case is the forced or rotational vortex in which all the fluid is forced to move with the same angular speed (ω) and $v = \omega r$. Here T is a constant. Angular momentum conservation occurs at the level of the entire system integrated over distances from 0 to R (the radius of the vortex) to give $1/2\ MR^2\omega$, where $1/2\ MR^2$ represents the moment of inertia. This would happen for fluid in a shallow cylindrical container rotated at constant ω. In Corollary 11, Newton refers to the driving motions of his globe in a spherical container ceasing 'and the whole system' being 'left to act according to the mechanical laws'; under these circumstances, 'the vessel and globe, by means of the intervening liquid, will act upon each other, and will continue to propagate their motions through the fluid to each other, till their periodic times become equal among themselves, and the whole system revolves together like one solid body'. He also describes the final rotational state of the water in the bucket experiment as being one of forced movement with the bucket. It seems that Newton recognised the existence of the limiting cases.

It would seem also that Newton, who, despite contrary statements, had a knowledge (if an imperfect one) of angular momentum and its conservation, and a concept equivalent to the moment of inertia from as early as the 1660s, had an idea of the special significance of the $T \propto r^2$ relationship for permanent vortex motion, and the eleven Corollaries to Proposition 52 suggest that he paid special attention to it, perhaps partly through experimental observation;[145] and Corollary 11 suggests that he understood also its relation to the other limiting case. The real point of these propositions, of course, is not the idealised mathematical analysis of such vortex generation, but the negative point that the vortices, however

---

[144] This is a key statement which implies that only this condition (however derived) is important for permanent or long-lasting free vorticity. Newton does not try to apply the idea of permanence to the cylindrical case, where his examples are mainly concerned with double cylinders.

[145] His derivations may have been contrived, as on other occasions, to generate results already achieved by experiment.

generated, cannot be reconciled with Kepler's laws of planetary motion. In none of these cases ($T \propto r^2$, T constant, and even $T \propto r$), whether real or imagined, is Kepler's condition for planetary motion $T^2 \propto r^3$ fulfilled, and Newton's point is that any attempt to construct a permanent vortex system in which this law obtains would have to be hopelessly contrived and unphysical.[146] The corollaries, in particular, show that this argument does not in any way depend on the correctness or otherwise of Newton's mathematical analyses relating to cylinders and spheres in fluids. The arguments are constructed on principles that Newton had found to apply for vortex motion in general.

They are all the stronger, in fact, for his realisation that vortices in any real fluid cannot persist indefinitely, 'some active principle' being needed to maintain them.[147] Since he was making a negative, rather than a positive, point about Descartes' proposed mechanism for planetary motions, he was under no obligation to create a system which was self-perpetuating, as Descartes required. In fact, in the particular mechanism he envisaged, whereby the fluid acts as though composed of a series of laminar shells surrounding the central body, each moving with a uniform angular speed proportional to the radius, motion would be continually transferred out towards the circumference and to the 'boundless extent' beyond.[148] In addition, the increased centrifugal force at the equator with respect to that at the poles would lead to transfer of material between them.[149] Though these suggest that the shell structure would be unstable, the spherical symmetry would ensure that the free vortex condition ($\omega \propto 1/r^2$) would hold, and the model, with the shells incorporating the independent units of mass m, can be taken as a first approximation to a derivation of this condition.

In Proposition 53, Newton also shows simply that bodies such as planets moved about in a fluid vortex must be of the same density as the vortex, and move 'according to the same law with the parts of the vortex, as to velocity and direction of motion'. A body that was more dense (and so made of more massive elements) would drift out of the vortex with a spiral motion because constant mv^2/r for orbital motion would make r bigger for the same v, while

---

[146] II, Proposition 52, Scholium, 392–394; General Scholium, 543. Interestingly, the vortex conditions T = constant and $T \propto r$ hold for galactic rotation (interior and exterior) as seen from a long distance, where the galaxy effectively becomes a single body approximating to uniform density, but this cannot, of course, be applied to the interior of the solar system.
[147] II, Proposition 51, Corollary 4.
[148] ibid., Corollary 3.
[149] ibid., Case 3.

a body that was less dense would drift in, because r would then be smaller for the same v.

In these terms Newton quite convincingly shows that vortical motion is not a plausible hypothesis for a planetary theory: 'the hypothesis of vortices is utterly irreconcileable with astronomical phenomena, and rather serves to perplex than to explain the heavenly motions',[150] though he had, of course, an even stronger argument in that 'comets are carried with very eccentric motions through all parts of the heavens indifferently, with a freedom that is incompatible with the notion of a vortex'.[151] An additional Scholium aims at a more exact argument using Kepler's second or area law for elliptical orbits. We imagine three concentric elliptical orbits with the Sun at their common focus. According to the laws of astronomy, the planets move more slowly at aphelion than at perihelion. According to fluid dynamics, however, the matter of the vortex supposedly carrying the planets will move more *swiftly* at aphelion than at perihelion. The continuity equation or conservation of fluid flow, used here, may be 'only really applicable to a two-dimensional plane flow', 'in which in all planes perpendicular to a certain direction an identical velocity distribution is found', as Néményi argued, but the area law, as another manifestation of the conservation of angular momentum, must necessarily yield the same contradiction with vortex motion for the planets as Newton had found for the relationship between T and r.[152]

Vortices may have been the key component in the Cartesian world system but they were not explained there on any dynamical principles. As in so many other cases, Newton was the first to provide a quantitative treatment of the subject. His analysis, though created purely to make a negative case, in fact included a number of fundamental principles related to vortex notion, while a whole battery of arguments showed that, in one way or another, Descartes' 'hand-waving' description of the world-system as a series of interlocking vortices could not be made compatible with a mathematical description of the universe, privileging Kepler's laws. Nowadays, largely because of the methodological revolution which Newton's work ultimately forced physics

---

[150] II, Proposition 53, Scholium.
[151] General Scholium, which also explained that the moons of Jupiter and Saturn would require the almost impossible situation of vortices within vortices. The first use of comets as an argument against vortices may be relatively early, possibly as early as 1682, when a comet appeared with retrograde motion. (The comet he had observed in 1664 was also retrograde, as he seems to have known by 1680–1681. Ruffner, 2012, 246, 250) However, Newton's doubts about vortices may date from much earlier as part of his general dissatisfaction with Descartes's ideas.
[152] II, Proposition 53, Scholium; Néményi (1963).

to adopt, his carefully worked-out and immensely rigorous derivation of the
fundamental law of gravity as immediate explanation of Kepler's laws and
his elimination of model-dependent assumptions would have made his case
unanswerable. Not so then, when the hypothesis-creating mechanistic mode
of theorising was still very much the dominant paradigm.

Newton's pioneering application of mathematical argument to vortex
theory made little impression on other prominent mathematicians who were
already persuaded by the Cartesian idea, even though that had been based
on only the vaguest of qualitative descriptions of vortex motion.[153] In
almost exact repetition of the events of 1672, two of the most prominent
Cartesians, Huygens and Leibniz, quickly contrived to present vortex-based
mathematical theories of the planetary system, the former acknowledging
Newton's work, the latter providing a deliberate, though unacknowledged,
inversion of it.[154] Newton could do nothing except reiterate his attack on
vortices in the General Scholium added to the second edition of the *Principia*
in 1713, with the clear statement that the area law or conservation of
momentum involved in the motion of each single planet, essentially fulfilling
the free vortex condition $T \propto r^2$ when the planet is taken as an independent
mass unit, is mathematically and physically incompatible with a vortex
maintaining the collective relation $T^2 \propto r^3$ *between* them. The Cartesian
intervention, however, prolonged the idea of the vortex model for another
fifty years, long after its usefulness as an explanatory model had expired,
and the relative success of this delaying tactic suggests that we should not
underestimate the overwhelming, and almost atavistic, contemporary belief
in the Cartesian system, and the immensity of Newton's achievement in
ultimately installing something so radically different in its place.

Newton, of course, saw that there must be minimal resistance in the
heavens on purely observational grounds. He referred to the well-known
'guinea and feather' experiment in stating that the resistance of a rarefied
medium is negligible:

> Liquors which differ not much in Density, as Water, Spirit of Wine, Spirit
> of Turpentine, hot Oil, differ not much in Resistance. Water is thirteen
> or fourteen times lighter than Quick-silver and by consequence thirteen or
> fourteen times rarer, and its Resistance is less than that of Quick-silver in
> the same Proportion, or thereabouts, as I have found by Experiments made
> with Pendulums. The open Air in which we breathe is eight or nine hundred
> times lighter than Water, and by consequence eight or nine hundred times

---

[153] Aiton (1972).
[154] Huygens (1690); Leibniz (1689).

rarer, and accordingly its Resistance is less than that of Water in the same Proportion, or thereabouts; as I have also found by Experiments made with Pendulums. And in thinner Air the Resistance is still less, and at length, by rarifying the Air, becomes insensible. For small Feathers falling in the open Air meet with great Resistance, but in a tall Glass well emptied of air, they fall as fast as Lead or Gold, as I have seen tried several times.[155]

This is particularly significant because, as he outlines in Proposition 40 of *Principia*, Book II, it relates to the motions of astronomical bodies in the heavens, while a draft addition to Book III, Proposition 6 argues that,

> ...if a certain subtle matter divided into least particles were uniformly scattered through the empty spaces of the heavens and filled as much as their thousandth part; and if the bodies of planets or comets or other globes were solid and destitute of all pores: these would lose a thousandth part of their motion before they would describe the length of three semidiameters. And it is reasonable that in a heaven more filled with matter they would lose a larger part of their motion, in proportion to the density, and even more if they were not solid bodies. Whence Comets, which pass through the planetary heavens... in all directions would soon lose all their motions, and balls of lead shot from guns would very quickly stop unless the spaces of the heaven and of air were nearly vacuous.[156]

It was an illustration of how far Newton had come from the Cartesian views of his contemporaries.

---

[155] Q 28.
[156] Draft addition to *Principia*, III, Prop. 6, Coroll. 4 and 5, 1690s; *Unp.*, 315–316. Newton also noted that a dense aether 'in the pores of bodies' would 'stop the vibrating motions of their parts wherein their heat & activity consists' (draft Q 28, CUL, Add. 3970.3, f. 246^r, NP).

Chapter 6

Microforces: Cohesion and Chemistry

6.1. Cohesion

By the time of the *Principia*, and probably long before that, Newton knew exactly how nature worked, from the smallest scale to the largest. The only problem was that he knew that he couldn't convince his contemporaries. Put in the simplest terms, it was a competition between forces of attraction and repulsion between particles of matter. Of course, the specific details were far from worked out, and still elude us, but physics has never deviated from this explanation. In Query 31 chemical reactions from Newton's alchemical investigations were used to support his argument that material substances were brought together and pulled apart by attractive and repulsive forces between their component particles. These forces involved 'active principles' which meant that matter could not be innately passive or inert as the Cartesians had supposed. The forces were not just chemical, but also determined the physical states of matter, and they required new laws of operation.

In Newtonian physics the second and third laws of motion (or equivalent formulations such as the conservation of momentum, energy or angular momentum) were necessary conditions, but specific application also required the fundamental laws which governed individual forces. The law of gravity had been identified for large-scale systems, where systematic observations had been available for centuries. Newton wrote in a draft conclusion to the *Principia*: 'Hitherto I have explained the system of this visible world, as far as concerns the greater motions which can easily be detected'.[1] What he needed to do now was to extend his methodology from the macro- to the microscale. Further development in physics would require the discovery of all the other laws of force operating on smaller scales.

---

[1] Draft *Conclusio, Unp.*, 333.

What might be called Newton's 'programme' is followed by Laplace and many others in the eighteenth and nineteenth centuries. This is to seek forces of other kinds, which are *analogous* to gravity and equally inexplicable. The idea that this must be the mechanism underlying the natural world is an aspect of Newton's Ockhamist procedure. He also referred to the principle of self-similarity at different scales, which he called the analogy of nature and which is the philosophy behind the renormalization group in more recent physics. If the analogy of nature applied, there could be many such forces.

Controversy about the idea of attraction as a supposedly 'occult' quality forced him into making a number of defensive statements, as in the *Principia*:

> I here use the word *attraction* in general for any endeavour, of what kind soever, made by bodies to approach each other, whether that endeavour arise from the action of the bodies themselves, as tending mutually to or agitating each other by spirits emitted; or whether it arises from the action of the aether or of the air, or of any medium whatsoever, whether corporeal or incorporeal, any how impelling bodies placed therein towards each other.[2]

His treatment would be mathematical, not physical in the hypothetical sense.

> I likewise call attractions and impulses, in the same sense, accelerative and motive; and use the words attraction, impulse, or propensity of any sort towards a centre, promiscuously, and indifferently, one for another; considering those forces not physically, but mathematically: wherefore the reader is not to imagine that by those words I anywhere take upon me to define the kind, or the manner of any action, the causes or the physical reasons thereof, or that I attribute forces, in a true and physical sense, to certain centres (which are only mathematical points); when at any time I happen to speak of centres as attracting, or as endued with attractive powers.[3]

It was more important, in the first instance, to find the laws of attraction than to inquire about the cause.

> How these Attractions may be perform'd, I do not here consider. What I call Attraction may be perform'd by impulse, or by some other means unknown to me. I use that Word here to signify only in general any Force by which Bodies tend towards one another, whatsoever be the Cause. For we must learn from the Phænomena of Nature what Bodies attract one

---

[2] *Principia*, Book I, Proposition 69, Scholium.
[3] *Principia*, Definition 8.

another, and what are the Laws and Properties of the Attraction, before we enquire the Cause by which the Attraction is perform'd.[4]

The active principles causing the attractions and repulsions were not 'occult qualities', as some alleged, but 'general laws of nature'.

It seems to me, farther, that these Particles have not only a *Vis inertiae*, accompanied with such passive Laws of Motion as naturally result from that Force, but also that they are moved by certain active Principles such as is that of Gravity, and that which Causes Fermentation, and the Cohesion of Bodies. These Principles I consider, not as occult Qualities, supposed to result from the specifick Forms of Things, but as general Laws of Nature, by which the Things themselves are form'd; their truth appearing to us by Phænomena, though their Causes be not yet discover'd. For these are manifest Qualities, and their Causes only are occult.[5]

Newton was concerned with one of these forces, cohesion, from his earliest entries to *Quaestiones quaedam philosophicae*.[6] It was a crucial case, for cohesion was almost impossible to explain using Cartesian or mechanistic impact physics. Newton referred to it in practically every major work he wrote, linking it with capillary action and surface tension. The capillary rise of water in a narrow glass tube was a favourite example. The experiment by Hauksbee on the capillary action on a drop of oil of oranges[7] was clearly of special importance, for he described it several times. He found that, at a separation of 10^{-7} inches, the force would be strong enough to hold up a column of the oil 52,500 inches high. Newton calculated the force of capillarity at 10^{-7} inches because this was the order of magnitude which he had derived from his optical studies for the sizes of material corpuscles. Though introduced by way of illustration, this value was specifically intended to represent the typical distance over which cohesive forces could be expected to act, and to represent the value of force at the *molecular* level. 10^{-7} inches introduces the nanoscale into physics, as it considers material bodies of a size equivalent to 2.5 nanometres. Newton, among many other things, is the first nanoscientist.

The strength but short range of the force involved probably led Newton to imagine that there were perhaps even greater attractive forces at an even shorter range within these atoms and that, to be short range, as he had already explained in the *Principia* and elsewhere, the forces must

---

[4] Q 31.
[5] *ibid.*
[6] QQP, 'Conjunction of bodys', f. 90^v, NP.
[7] Chapter 4, Section 4.4.

be proportional to inverse powers of the distance greater than 2. In the *Principia*, Newton puts cohesion on a mathematical-theoretical basis:

> If a body be attracted by another, and its attraction be vastly stronger when it is contiguous to the attracting body than when they are separated from each other by a very small interval; the forces of the particles of the attracting body decrease, in the recess of the body attracted, in more than the squared ratio of the distance of the particles.[8]

> If the forces of the particles of which an attractive body is composed decrease, in the recess of the attractive body, in a triplicate or more than a triplicate ratio of the distance from the particles, the attraction will be vastly stronger in the point of contact than when the attracting and attracted bodies are separated from each other, though by never so small an interval.[9]

6.2. Attraction and Repulsion in the Different States of Matter

To a large extent, Newtons discussion of examples of attraction and repulsion was descriptive. He was trying to make a case for the generality of the phenomena, rather than attempting to quantify it or subject it to mathematical analysis. However, the striking exactness of observation in a multiplicity of specific cases reveals a characteristically Newtonian drive towards a comprehensive and unified account of the seemingly disparate phenomena in as abstract a manner as possible. Newton discusses repulsion in liquids and solids in several places: 'And just as magnetic bodies repel as well as attract each other, so also the particles of bodies can recede from each other by certain forces. Water does not mix with oil, or lead with iron or copper, on account of the repulsion between the particles'.[10] 'For there is a certain secret principle in nature by wch liquors are sociable to some things & unsociable to others'.[11]

Penetration of one material into another was not a question of porosity but of attractions or repulsions between specific types of particles: '... liquors and spirits are disposed to pervade or not pervade things on other accounts then their subtlety... Some fluids (as oil and water) though their pores are in freedom enough to mix with one another, yet by some secret principle

---
[8] I, Proposition 85.
[9] I, Proposition 86.
[10] Draft *Conclusio*, *Unp.*, 336.
[11] Letter to Boyle, 28 February, 1679; *Corr.* III, 292.

of unsociableness they keep asunder...'[12] 'Thus water will not mix with oil but readily with spirit of wine or with salts. It sinks also into wood which quicksilver will not, but quicksilver sinks into metals, which ... water will not. So aqua fortis dissolves ☿ [mercury] not ☉ [gold]; aqua regia ☉ [gold] and not ☿ [mercury], etc'.[13]

> Just as the magnet is endowed with a double force, the one of gravity and the other magnetic, so there are various forces of the same particles arising from various causes. There must be one force whereby the particles of oil attract each other, and another whereby they repel the particles of water. That the forces by which particles attract one another, however, decrease rapidly with distance from the particles may be gathered from the small quantity of a dissolving menstruum which the particles of a dissolved body can attract and retain.[14]

Solution of salts in water suggested a greater attraction of the solute particles for the water molecules than for each other:

> If a very small quantity of any Salt or Vitriol be dissolved in a great quantity of Water, the Particles of the Salt or Vitriol will not sink to the bottom, though they be heavier in Specie than the Water, but will evenly diffuse themselves into all the Water, so as to make it as saline at the top as at the bottom. And does not this imply that the Parts of the Salt or Vitriol recede from one another, and endeavour to expand themselves, and get as far asunder as the quantity of Water in which they float, will allow? And does not this Endeavour imply that they have a repulsive Force by which they fly from one another, or at least, that they attract the Water more strongly than they do one another? For as all things ascend in Water which are less attracted than Water, by the gravitating Power of the Earth; so all the Particles of Salt which float in Water, and are less attracted than Water by any one Particle of Salt, must recede from that Particle, and give way to the more attracted Water.[15]

The penetration of mercury into dense solid metals must be accounted for by attractions between the mercury and metal particles, for mercury will not penetrate lighter materials for which its particles have no attraction:

> Mercury penetrates gold, silver, tin and lead, but it does not penetrate wood and bladders and salts and rock; water on the contrary penetrates wood, bladders, salts and rock but does not penetrate metals: the cause of this is not the greater or lesser subtlety of the parts of water or mercury,

---

[12] *Hyp.*, *Corr.* I, 368.
[13] Letter to Boyle, 28 February 1679; *Corr.* II, 292.
[14] Draft *Conclusio*, *Unp.*, 336–337.
[15] Q 31.

but is, I suspect, to be attributed to the mutual attraction of particles of a like kind, less to the repulsion of likes.[16]

Examples could be found involving many different types of material:

And is it not for want of an attractive virtue between the Parts of Water and Oil, of Quick-silver and Antimony, of Lead and Iron, that these Substances do not mix; and by a weak Attraction, that Quick-silver and Copper mix difficultly; and from a strong one, that Quick-silver and Tin, Antimony and Iron, Water and Salts, mix readily? And in general, is it not from the same principle that Heat congregates homogeneal Bodies, and separates heterogeneal ones.[17]

'In every solution by menstruum the particles to be dissolved are attracted by the parts of the menstruum more than by each other reciprocally'.[18] 'A tincture of cochineal with spirits of wine liberated in a large quantity of water even in a small dose nevertheless dyes the whole of the water, obviously because the particles of cochineal are then attracted by the water more than by each other reciprocally'.[19]

The contrasting properties of aqua fortis and aqua regia as solvents for silver and gold provided a specially striking case of such 'selective attraction':

When *Aqua fortis* dissolves Silver and not Gold, and *Aqua regia* dissolves Gold and not Silver, may it not be said that *Aqua fortis* is subtil enough to penetrate Gold as well as Silver, but wants the attractive Force to give it Entrance; and that *Aqua regia* is subtil enough to penetrate Silver as well as Gold, but wants the attractive Force to give it Entrance? For *Aqua regia* is nothing else than *Aqua fortis* mix'd with some Spirit of Salt, or with Sal-armoniac; and even common Salt dissolved in *Aqua fortis*, enables the *Menstruum* to dissolve Gold, though the Salt be a gross Body. When therefore Spirit of Salt precipitates Silver out of *Aqua fortis*, is it not done by attracting and mixing with the *Aqua fortis*, and not attracting, or perhaps repelling Silver? And when Water precipitates Antimony out of the Sublimate of Antimony and Sal-armoniac, or out of Butter of Antimony, is it not done by its dissolving, mixing with, and weakening the Sal-armoniac or Spirit of Salt, and its not attracting, or perhaps repelling, the Antimony?[20]

The presence of a 'mediating' substance could frequently make particle types attractive to each other which were normally 'unsociable': 'When any metal is put into common water the water cannot enter into its pores to

---

[16] Draft *Conclusio, Unp.*, 336.
[17] Q 31.
[18] Pitcairne with Newton at Cambridge, 2 March 1692; *Corr.* III, 211.
[19] *ibid.*, 210.
[20] Q 31.

act on it and dissolve it. Not that water consists of too gross parts for this purpose, but because it is unsociable to metal'.[21]

> But a liquor which is of it self unsociable to a body may by the mixture of a convenient mediator be made sociable. So molten lead which alone will not mix with copper or with regulus of Mars, by the addition of tin is made to mix with either. And water by the mediation of saline spirits will mix with metal. Now when any metal is put into water impregnated with such spirits, as into aqua fortis, aqua regia, spirit of vitriol or the like, the particles of the spirits as they floating in the water, strike on the metal, will by their sociableness enter into its pores and gather round its outside particles, and by advantage of the continual tremor the particles of the metal are in, hitch themselves in by degrees between those particles and the body and loosen them from it, and the water entering into the pores together with the saline spirits, the particles of the metal will be thereby still more loosed, so as by that motion the solution puts them into, to be easily shaken of and made to float in the water: the saline particles still encompassing the metallic ones as a coat or shell does a kernel...[22]

Mechanisms could be suggested for the expulsion of gases and vapours when reactions involved the production of heat:

> In solutions, when particles rush together violently they grow hot by concussion, and by the vibratory motions of heat particles are driven off; if these particles are coarser they go off as air, vapours and exhalations; if least of all as light. The particles when joined so cohere that they can be distilled, or sublimed together or remain fixed, yet often in new associations they abandon one another and the deserted body is precipitated.[23]

The particles of solids and liquids, which attract each other in those states of matter, repel each other in the gaseous state:

> So also the particles of liquid and solid bodies attract each other when contiguous, but when separated by heat or fermentation they flee from each other and constitute vapours, exhalations, and air. And on the other hand coming together by the action of cold or fermentation they cohere and revert into liquids and compact bodies. For that remarkable philosopher, the Honourable Mr Boyle, teaches us that true air arises from the fermentation of dense bodies and returns into them.[24]

There is a degree of repulsion also between the separated parts of hard bodies: 'And in the same manner we see why the separated parts of hard

---

[21] Letter to Boyle, 28 February 1679; *Corr.* II, 291–292.
[22] *ibid.*, *Corr.* II, 292.
[23] Draft *Praefatio*, *Unp.*, 306.
[24] Draft *Conclusio*, *Unp.*, 338.

bodies can only with great difficulty be brought together so that they touch each other completely. Whence they do not cohere as they did before'.[25]

Mutual repulsion of the particles makes it difficult to bring lenses together: 'For if you place on a somewhat convex lens, very highly polished, such as the objective of a very long telescope, a second lens, which is smaller, plane, and equally well polished, you will find that some effort is required to bring them into contact, and that when the pressure is removed they will spring apart spontaneously, the more so as the upper lens is less heavy...'[26]

The repulsion can be seen from optical effects:

> I placed on a bi-convex objective lens from a fifty-foot telescope a plane glass of less weight, and in the middle of the glasses about the point where they almost touched there appeared certain coloured circles. Then lightly pressing upon the upper glass, so that it came closer to the lower one, there emerged new coloured circles in the middle of the former circles; and then increasing the pressure there emerged in the middle of the white circle a sort of spot darker than the other colours. This spot was the indication of the contact of the pieces of glass. For the surfaces of the glasses which elsewhere reflected light from themselves, reflected almost no light at all at that spot, but offered very little hindrance to the unimpaired passage of the light just as if the glasses were continuous there, and being joined together by fusion were deprived of intermediate surfaces. Therefore the upper glass could not be brought by its weight into contact with the lower one. That was effected by the greater force of the hands pressing down. Indeed, with all the force with which I compressed those glasses I could scarcely make them touch at all. For on removing my hands they did not cohere in the least.[27]

The minimal separation of the glasses could be determined from the fringe pattern:

> I have sometimes observed the Diameter of that Spot to be between half and two fifth parts of the Diameter of the exterior Circumference of the red in the first Circuit or Revolution of Colours when view'd almost perpendicularly; ...whence it may be collected that the Glasses did then scarcely, or not at all touch one another, and that their Interval at the perimeter of that Spot when view'd perpendicularly was about a fifth or sixth part of their Interval at the circumference of the said red.[28]

---

[25] Draft *Conclusio*, *Unp.*, 339.
[26] *De aere et aethere*, *Unp.*, 222.
[27] Draft *Conclusio*; *Unp.*, 339.
[28] *Opticks*, Book II, Part 1, Observation 8.

The same applies to plane glasses:

> When plane glasses are used, even though they are not more than half an inch wide, you can so compress them that the colours appear, but the whole strength of one man is scarcely or not quite sufficient to bring them into complete contact, so that the black spot appears. (However, the intervening air because of the minuteness of the space becomes so rare that where the colours have appeared the glasses cannot be separated without the use of a considerable force. The external air exerts a greater pressure than the internal air, with its elastic force weakened by rarefaction, is able to overcome.)[29]

And polished marbles: 'From the same repelling Power it seems to be that ... two polish'd Marbles, which by immediate Contact stick together, are difficultly brought so close as to stick'.[30]

Repulsion is also observed in the parts of broken bodies, molten metals in an iron vessel, and powders:

> In the same way the parts of glass, steel or any broken body cannot even when pressed with the greatest force be reduced to their former contact so that they cohere as before, for were complete contact restored without doubt their cohesion would be restored (also). Further, lead and tin when melted and poured into an iron vessel do not attain contact in such a way that the casting adheres to the iron.[31]

Sometimes the repulsion could be overcome using water as an intermediary:

> And thence it happens that ... particles of dry powder cannot be brought together by a compressive force so that they cohere as the parts of a solid body do. But if water is poured on so that (the particles) compose a continuous body, and then the water is evaporated, they are brought into contact and through that contact they form a body by cohesion; when it is dried by fire so that the fluid softening it is driven off, it is converted into a firm and hard body.[32]

The particles of gases and vapours mutually repel each other once they are extracted from liquids and solids:

> Also the particles of vapour and of air, so long as they are in water and solid bodies, and not yet assuming the form of vapour or air, cohere with contiguous particles. But as soon as they are separated by the motion of heat or fermentation and recede beyond a certain distance from the bodies and from each other, they fly apart (as we said) and in receding from each

---

[29] *De aere et aethere*, Unp., 222.
[30] Q 31.
[31] *De aere et aethere*, Unp., 222.
[32] Draft *Conclusio*, Unp., 340.

other compose a medium which has a strong tendency toward expansion. For in the cylindrical cavities of narrow glass tubes they flee from the glass, so that air is more rare there than in free spaces (as a certain person has ingeniously remarked, following an hypothesis of his own)...[33]

Because of the relatively large spheres of activity of its individual particles, air appears to have a relatively low rate of diffusion through container walls compared with liquids such as water and oil: '...the Air can pervade the bores of Small Glasse pipes, but not yet so easily as if they were wider, & therefore stands at a greater degree of rarity then in the free æreall Spaces, & at so much a greater degree of rarity as the pipe is Smaller...'[34]

> Furthermore, that air although fluid and subtle seems to creep through the pores of bodies with greater difficulty than water or oil, so that it may be contained in a bladder which a liquid easily pervades, results from the fact that its particles, keeping their distance from all (others), are reluctant to approach the sides of the pores. They are indeed very small; but as to the faculty of permeating narrow places, (their) subtlety ought to be estimated from the whole sphere which each claims for itself.[35]

Surface tension shows that the particles on the surfaces of liquids exert forces: '...fluids near their Superficies are less pliant & yeelding then in their more inward parts, & if formed into thin plates or Shells, they become much more Stiff & tenacious then otherwise. Thus things wch readily fall in water, if let fall upon a bubble of water, they do not easily break through it, but are apt to Slide downe by the sides of it, if they be not too big and heavy'.[36] 'Also powders floating on liquids avoid contact with the liquid and are submerged with difficulty even when they are fairly heavy, like filings of metal. Similarly, flies and other small creatures are wont to walk on the yielding water without wetting their feet'.[37]

That separate and distinct bodies avoid contact is shown by bubbles:

> The matter may be better understood from this experiment. I covered a fairly large bubble ... with a glass vessel so that the external air should not by its motion disturb the colours appearing in the bubble. The top of the bubble was ringed round with concentric circles of different colours; these circles widened out; they crept down the sides towards the lower part of the bubble while new circles arose towards the top. At length a whiteness appeared towards the top and soon in the midst of it a spot of dark blue

---

[33] Draft *Conclusio*, *Unp.*, 337.
[34] *Hyp., Corr.* I, 366–367.
[35] *De aere et aethere*, *Unp.*, 225.
[36] *Hyp., Corr.* I, 373.
[37] *De aere et aethere*, *Unp.*, 222–223.

colour in whose centre a dark spot then appeared and afterwards in its centre a blacker spot which hardly reflected the light at all, and then at last the bubble usually burst. These colours are of the same kind and have the same origin as the colours between glasses. For the watery skin of the bubble between the internal and external air serves in this case to exhibit colours just as the layer of air between the two glasses does. Just as when the layer of air was made thinner by compression of the glasses new colours arose until the black spot appeared, so when the skin of the soap-bubble becomes thinner at the top by the continual flowing down of the upper parts to the lower part, new colours appear successively and ultimately the black spot is observed. And as the skin of the soap-bubble is thinnest at the top where the black spot appears but notwithstanding this is not wholly lacking in thickness, so from the black spot in the middle of the glasses it cannot be concluded that the layer of intermediate air near that spot is destitute of all thickness. Therefore separate and distinct bodies vehemently avoid mutual contact.[38]

6.3. Double Forces

Newton identified various effects of repulsion in matter but tended to ascribe them to a single kind of repulsive force. Two, in particular, could be ascribed to identifiable repulsive forces: one occurred at relatively small distances, as when two object-glasses were pressed together, and another when particles which were widely separated, as in gases, tended to flee from one another by thermal motions.[39] Opposing these was the force of attraction producing cohesion when particles could be brought close enough together to form solids or liquids.

It is notable that Newton's discussion of repulsion considered the possibility that observed effects of repulsion were in some cases due to insufficient attraction; particles of salt and vitriol in solution repel each other 'or at least attract the water more strongly than one another'. Particles of matter would always be subjected to some sort of overall force; generally, he believed, when attractive forces ceased, repulsive forces would begin. 'Since Metals dissolved in Acids attract but a small quantity of that Acid, their attractive Force can reach but to a small distance from them. And as in Algebra, where affirmative Quantities vanish and cease, there negative ones begin; so in Mechanicks, where Attraction ceases, there a repulsive Virtue

---

[38]Draft *Conclusio*; *Unp.*, 339–340.
[39]The repulsion in the case of gases can be considered as an 'entropic' force, that is one that results from the system's attempt to achieve maximum entropy. It is an emergent property that does not directly result from the microforces relating to the molecular structure of the gas, but is nevertheless a very real effect.

ought to succeed'.⁴⁰ 'But the expulsive force of particles seems to be exercised at greater distances, and even to be less in their immediate neighbourhood; but at greater distances, where the attractive force has decreased makes its effects felt'.⁴¹

Ultimately, double forces were at work, dominating at different distance scales: 'When the particles of air, exhalations, vapours, salts in water, and particles of dyes are at a distance they repel each other; but when they are brought to touch each other the particles at last cohere and by cohering are turned into solid bodies again. The contiguous particles of all bodies cohere, and their distant ones frequently repel one another'.⁴²

> Such operations may be supposed to be effected by double forces: one of these impels adjacent particles towards one another; this is the stronger but decreases more quickly with distance from the particle. The other force is weaker but decreases more slowly and so at greater distances exceeds the former force; this drives the particles away from each other. It is certain that particles of iron are sometimes impelled towards the magnet, sometimes again repelled from it, and that the more strongly the more closely they approach the magnet; and that the particles of all bodies are impelled towards the Earth by the force of gravity just as lighter bodies are impelled towards glass and amber by the electric force.⁴³

Such forces were fundamental to all phenomena:

> Thus almost all the phenomena of nature will depend on the forces of particles, if only it be possible to prove that forces of this kind do exist. And although the names of attractive and repulsive forces will displease many... ...nevertheless they are simple and easy to conceive, and of the same kind as the natural philosophy of the cosmic system which depends on the attractive forces of greater bodies.⁴⁴

Newton generally avoided speculation on the nature of the attractions and repulsions, but one of his suggestions concerning repulsion has an interesting contemporary connection.

> Many opinions may be offered concerning the cause of this repulsion. The intervening medium may give way with difficulty or not suffer itself to be compressed. Or God may have created a certain incorporeal nature which seeks to repel bodies and make them less packed together. Or it may be in the nature of bodies not only to have a hard and impenetrable nucleus but

---

[40] Q 31.
[41] Draft *Conclusio*; Unp., 337.
[42] Draft *Praefatio*; Unp., 306.
[43] *ibid.*, 307.
[44] Draft *Conclusio*; Unp., 345.

also a certain surrounding sphere of most fluid and tenuous matter which admits other bodies into it with difficulty.[45]

In the modern theory of matter, of course, the electron cloud surrounding the nucleus in the atom acts as a source of repulsion to prevent solid matter collapsing in on itself under the attractive power of the van der Waals force of cohesion.

The forces even affected light:

And that there is such a Virtue, seems to follow from the Reflexions and Inflexions of the Rays of Light. For the Rays are repelled by Bodies in both these Cases, without the immediate Contact of the reflecting or inflecting Body. It seems also to follow from the Emission of Light; the Ray so soon as it is shaken off from a shining Body by the vibrating Motion of the Parts of the Body, and gets beyond the reach of Attraction, being driven away with exceeding great Velocity. For that Force which is sufficient to turn it back in Reflexion, may be sufficient to emit it'.[46]

The analogy of nature implied that they were similar at all scales. A 'simple and fully consonant' nature must use the same principles in 'regulating the motions' of the smaller bodies as of the larger,[47] '... performing all the great Motions of the heavenly Bodies by the Attraction of Gravity which intercedes those Bodies, and almost all the small ones of their Particles by some other attractive and repelling Powers which intercede the Particles'.[48]

Frequently, he denied that his theories implied certain possibilities while giving the impression that he may be prepared to consider them. One such case concerned the existence of actual point-centres for attractive forces; another concerned the possibility of a magnetic origin for polarization. Here, he left his options open to await future developments, while stating that his current theories did not depend on the assumption of such facts as hypotheses. The spheres of attraction and repulsion within particles which Newton considered in the 1690s resemble the series of alternate spheres of attraction and repulsion imagined by Bošković and others after him,[49] but, unlike the eighteenth century natural philosophers who followed him, he avoided producing explicit models, and included nothing about

---

[45] *De aere et aethere*, *Unp.*, 223. This is also interesting in connection with the 'small atmosphere' due to the electric spirit lying close to the surface in an electrifiable body like glass.
[46] Q 31.
[47] Draft Conclusion for *Opticks*, Book IV, c 1690; Westfall (1971, 379; 1980, 521).
[48] Q 31.
[49] Bošković (1746, 1748, 1749).

such speculations in his published works. He tended to concentrate largely on the opposition between cohesion and thermal expansion. The matter was treated more extensively by his disciples. After Stephen Hales had introduced a new aspect of repulsion in the last book to receive Newton's imprimatur as President of the Royal Society,[50] Desaguliers clarified the issue in 1739 with the explicit statement that the attractive force of cohesion was opposed within matter by an even shorter-range repulsive force (the one that corresponded with Newton's early speculation on the 'sphere of most fluid and tenuous matter').[51] There were therefore two separate causes of repulsion: the short-range force and the tendency to thermal expansion.

6.4. Alchemy and Chemistry

The great Cambridge economist, John Maynard Keynes, was concerned that part of Newton's legacy would be lost to Cambridge University, and possibly to scholarship in general, when the remaining parts of the Portsmouth collection, which had been refused by the university in the nineteenth century, were advertised for sale to the highest bidder in 1936.[52] He attended the auction, bought as many of the manuscripts as he could, and presented them subsequently to King's College. He was startled, however, when he took the trouble to read them, for his conclusion was that:

> Newton was not the first of the age of reason. He was the last of the magicians, the last of the Babylonians and Sumerians, the last great mind which looked out on the visible and intellectual world with the same eyes as those who began to build our intellectual inheritance rather less than 10,000 years ago. Isaac Newton, a posthumous child born with no father on Christmas Day, 1642, was the last wonderchild to whom the Magi could do sincere and appropriate homage.[53]

Perhaps he should not have been so surprised. David Brewster, Newton's nineteenth century biographer, had looked into some of these papers, and was shocked to find that Newton had not only been the great rationalist of the *Principia* and the great experimenter of the *Opticks*, but had also been 'the copyist of the most contemptible alchemical poetry' and had seriously studied and annotated works that were 'the obvious product of a fool and a knave'.[54] This kind of thing may have shocked the Victorians

---

[50] Hales (1727).
[51] Desaguliers (1739); Heilbron (1982, 57–58).
[52] Taylor (1936).
[53] Keynes (1947).
[54] Brewster (1855, II, 301).

but it became perfect copy for some of the biographers of the twentieth and twenty-first centuries. For one modern biographer, for example, Newton could be described as 'The Last Sorcerer'.[55]

More recent studies have changed this perspective entirely. Newton's alchemy was neither sorcery nor magic, but a perfectly rational study which cannot be separated from his work in chemistry, and many insights that Newton derived from this work were to be influential in the special contributions he made in the recognisably rational part of his science.[56] Whatever its origins, alchemy for Newton was a form of chemistry, like that of Robert Boyle, and he only started investigating alchemical texts after he had already developed considerable expertise in 'rational' chemistry, as we can see from the early *Chemical Dictionary* which precedes any of the alchemical texts.[57] In fact, he may well have had an opportunity to work at the practical side of the subject when he lodged with Apothecary Clark in the years before going to Cambridge.

'Chymistry', as it was called, was not differentiated from alchemy in the works of such seventeenth century scientists as Boyle and Newton. Newton certainly wrote letters to men who maintained large correspondence networks, like Collins and Oldenburg, which referred to (al)chemical topics without giving any impression that this was anything other than the province of a natural philosopher of the age. His alchemical notes even contain translation from alchemical symbols and names to chemical ones, though Newton also believed that something more than ordinary chemistry was involved.[58] In search of the hidden wisdom that he thought they might contain, he always showed greatest interest in the most recondite parts of the texts he examined, and, when, rightly or wrongly, he subsequently interpreted them as processes he could reproduce in the laboratory, he invariably used the most impeccable scientific procedures for the experiments. There is evidence that this scientific 'code breaking' was being practised by Newton right from the beginning of his alchemical studies. The exotic language in the manuscripts was derived from his sources and we have no proof that it had the same symbolic meaning to him as it had to the authors on which he drew.

Newton certainly wrote a number of what are clearly alchemical documents, though nothing that can be considered a worked-out 'treatise' like

---

[55] White (1997).
[56] Newman (2002).
[57] *c* 1667–1668, Bodleian Library, Oxford, MS Don. B. 15.
[58] Hall (1992, 186–187).

those of his sources. The nearest equivalent to a systematic exposition of alchemical principles in his work comes in the relatively short document, *Praxis*, written near the end of his alchemical career in 1693, although this is not a finished work and contains a great deal of crossing out and substitutions of text, as well as some direct translations of alchemical into chemical language.[59] Although he was interested in materials from the whole range of alchemical authorities, some of those which had the strongest influence on him came from near contemporaries in the seventeenth century. This was a period when alchemy was not only very active but also, through practitioners like Robert Boyle, approaching more closely to recognisable 'science'. The modern-sounding descriptions and explanations of chemistry that Newton was eventually able to generate and incorporate into his thoroughly 'scientific' treatise *Opticks* were partly a result of a long effort at systematic abstraction in the usual Newtonian manner, and partly a result of the fact that the alchemical tradition was itself committed to rationalising its findings using more general principles, which could themselves give hints to finding even more rationalised principles, whether or not the specific details were finally retained.

Some of Newton's ideas, such as the transformation of water, come from Jan Baptist van Helmont, a Flemish contemporary of Descartes, whose *Ortus medicinae* (1667) was in Newton's library. Van Helmont's ideas on the structure of matter, based on mercury, sulphur and salt, the principles derived from the mediaeval Arabic alchemists via the influential sixteenth century iatrochemist Paracelsus, included vitalistic corpuscles with powers allowing them to ferment and vegetate.[60] They were developed in the mid-seventeenth century by George Starkey (Eirenaeus Philalethes, in his alchemical disguise), who also certainly influenced Newton,[61] as did the earlier seventeenth century alchemist, Michael Sendivogius, who had been another source for Starkey.[62] Like Boyle, with whom he exchanged ideas, Newton was certainly interested in transmutation, especially in relation to the corpuscular theory.[63] It was an inevitable consequence of the view of matter which his metaphysics required, though whether he believed that anyone could claim success in this area is much more problematic.

---

[59] *Praxis*, 1693, Dobbs (1991, 293–305).
[60] Newman (2002, 358–369, 359–363).
[61] KCC, Keynes MSS 25, 34, 36, 51, 52, 221.
[62] *Sendivogius Explained*, KCC, Keynes MS 55, 1680s.
[63] Boyle (1676). Newton to Oldenburg, 26 April 1676, *Corr*. II, 1–3.

Alchemical language shades into more rational language as he attempts to explain symbolic chemical writing of earlier alchemists in terms of what he has observed in the laboratory and in models of matter as composed of structured particles with internal forces at work. The vitalistic corpuscular theory, with internal structure, of the alchemists was transformed by Newton into a structured matter theory involving forces and recognisably material particles. The alchemists' active semen of every structure, a 'speck of light' which metals needed for transmutation, morphed in his work into force-based active principles. As William R. Newman has written: 'The precise observations that he made of chemical affinity and his speculations about the invisible structure of matter are as 'scientific' as any other parts of his work'.[64] Though, for some, alchemy may have been a kind of religion rather than a science, Newton's alchemy was not more significantly religious than any other work of his.[65]

In fact, in Newton's hands, *chemistry* became an exact tool of quantitative analysis, with small quantities measured to great precision. Such work was a foretaste of the reaction stoichiometry that a century later would lead Antoine Lavoisier to the chemical revolution of the later eighteenth century. He had, for instance, a number of analytic tests, and saw that the amount of a substance which combined with another was limited; that substances tended to combine with like ones, but a combination with a contrary substances could be facilitated by the mediation of a third. For Newton, the alchemists' 'mediation' became catalytic. In his *De natura acidorum*, he claimed that the alchemists' 'Sulphur' meant acid and that 'Mercury' meant earth, or, in more modern terminology, base.[66] So the alchemists' three 'principles', mercury, sulphur and salt, would, in this reading, be equivalent to acid, base and salt, though it may be that the meanings of these terms should not be so narrowly defined. The contemporary manuscript containing the alchemical treatise *Praxis* similarly identifies the four earlier 'elements' of Empedocles and Aristotle — fire, air, water and earth — from which the alchemists' 'principles' descended, with 'Sulphur acidum', 'Sp^{tus} mercurij',

---

[64] Newman (2002, 358–369, 359–367).

[65] Newman (2002, 367). There is no need either to assume that it had anything to do with the fulfilment of psychological needs that Carl Gustav Jung saw as characteristic of earlier alchemy (Dobbs, 1975, 26–34).

[66] Westfall (1975). Examples of analytic tests, Add MS 3975, p. 125; precision experiments, Add MS 3973, ff 13–13^r; small quantities, Add MS 3975, p. 143; limited degree of combination between substances, Add MS 3975, p. 107. cf. p. 120; like interacts with like, Add MS 3973, f 12. Add MS 3975, p. 107.

'Aqua pont', and 'Sal fix', or, in more modern terms, acid, base, water and salt.[67]

Significant statements which Newton made show his transformation of ideas based on alchemy into his own more scientific thinking. In the key text, the document beginning *Of Natures Obvious Laws*, from the early 1670s, he writes: 'Nothing can be changed from what it is without putrefaction... no putrefaction can bee without alienating the thing putrefyed from what it was'.[68] And: 'all the operations in vulgar chemistry (many of which to sense are as strange transmutations as those of nature) are but mechanicall coalitions or seperations of particles as may appear in that they returne into their former natures if reconjoned or (when unequally volatile) dissevered, & that without any vegetation...'[69] The transmutations remained even after he had abandoned the aethereal mechanisms of *Of Natures Obvious Laws* and the *Hypothesis of Light*, during the writing of the *Principia*, for actions at a distance between material particles: 'Any body can be transformed into another, of whatever kind, and all the intermediate degrees of quality can be induced in it'.[70] '...all things arise from imposing diverse forms and textures on a certain common matter and... all things are resolved again into the same matter by the privation of forms and textures'.[71]

William R. Newman has shown that *Of Natures Obvious Laws* immediately postdated a work in Latin in the same manuscript, *Humores minerales* (*Mineral Humours*), written under the influences of Michael Sendivogius, Johann Grasseus and Bernhard Varenius (whose *Geographia generalis* Newton had just edited), which saw metals in acid solution sinking down into the Earth, but rising as fumes when vaporised by terrestrial heat. The fumes then mixed with the descending liquids and vaporised them, thus maintaining a perpetual circulation. However, the mixing of liquids and vapours caused a separation of the component alchemical 'principles' of the metal, which then led to a new generation of metals and metallic ores within the Earth.[72]

---

[67] Babson MS 420, p. 2; Dobbs (1991, 164). The three principles had a somewhat indirect, and not necessarily fixed, relationship to the four elements. One way of describing this would make sulphur a combination of the properties of fire and air; mercury a combination of the properties of air and water; and salt, a combination of those of water and earth.
[68] *Of Natures Obvious Laws*, NPA.
[69] *ibid.*
[70] *Principia*, first edition, 1687, III, Hypothesis 3; Koyré (1965, 263).
[71] Manuscript version of Hypothesis 3 for *Principia*, III, 1687; Westfall (1971, 388).
[72] Newman (2009).

In *Of Natures Obvious Laws*, Newton distinguished between nitre and sea salt by their textures rather than chemical structures, nitre having components more easily separated, and developed 'a quite original' treatment of how they are formed by the interaction of water and the metallic fumes rising from below the Earth's surface, nitre meeting a vapour and sea salt water in more liquid form. In treating salt formation in this way, Newton was not implying that it was not *chemical*, but that it was a more mechanical process than the formation of metals in the Earth. Nitre, he knew, could be reconstituted or 'redintegrated' after a chemical reaction. Boyle, who subscribed to a belief that the corpuscles of the Paracelsian principles in chemical compounds remained unchanged during their formation, had described, in his *Certain Physiological Essays* (1661), an example of 'redintegration', in which molten saltpetre had been broken up into its constituents, using burning charcoal. The expulsion of gases during the reaction left a residue of 'fixed nitre' (potassium carbonate), to which the addition of 'spirit of nitre' (nitric acid) led to a new creation of saltpetre.[73]

Despite the appearance of irrevocable change in chemical reactions, the original substances could be reconstituted, showing that the seeming change in physical properties was not fundamental but due to the way the parts were arranged:

> Yet those grosser substances are very apt to put on various external appeanes according to the present state of the invisible inhabitant as to appear like bones flesh wood fruit etc Namely they consisting of differing particles watry earthy saline airy oyly spirituous etc, those parts may bee variously moved one among another according to the acting of the latent vegetable substances & be variously associated & concatenated together by their influence.[74]

However, that was not the entire story:

> There is ... besides the sensible changes wrough in the textures of the grosser matter a more subtile secret & noble way of working... and the immeadiate seate of these operations is not the whole bulk of matter, but rather an exceeding subtile & inimaginably small portion of matter diffused through the masse which if it seperated there would remain but a dead & inactive earth.[75]

---

[73] *Of Natures Obvious Laws*, NPA; Boyle (1661). Newman (2009, 2010).
[74] *ibid.*
[75] *ibid.*, Dobbs (1991, 256–270), Westfall (1971, 307).

Newton was making a distinction here between 'vulgar chemistry' which he saw as mechanical concatenation of the parts of 'grosser matter', and the 'subtile secret & noble way of working' which concerned things which were *individual species* and not concatenated ones — elemental in effect, and particularly metals. These were not put together by vulgar chemistry, but could be generated only by a kind of replication from their own kind (a multiplication in alchemical language), like real living things, with the existing material of this kind acting as template or protoplast. Here, 'vegetation is the only naturall work of metalls', not the 'gros mechanicall transposition of parts'. It was certainly a process analogous to the one involving real vegetation, and something like the growth of crystals. Newman has described it as 'a goal-directed process' directed by seeds deep within matter.[76]

Newton writes in a draft version of Query 25:

> For particles of one & the same nature draw one another more strongly then particles of different natures do. And therefore in the bowels of the earth particles of the same nature are apt to assemble in the same masses & those of different natures in different masses. And when many particles of the same kind are drawn together out of the nourishment they will be apt to coalesce in such textures as the particles which drew them did before because they are of the same nature as we see in the particles of salts which if they be of the same kind always crystallize in the same figures.[77]

Newton seems to be saying that metals are formed in this way in the Earth. He also seems to be attributing it to a very subtle level in nature, possibly aethereal. He is emphatic in separating out the two processes. 'Natures actions are either vegetable or purely mechanicall', examples of which include 'grav. flux. meteors. vulgar Chymistry'. He says of the latter process:

> All these changes thus wrought in the generation of things so far as to sense may appear to bee nothing but mechanisme or severall dissevering & associating the parts of the matter acted upon & that becaus severall changes to sense may bee wrought by such ways without any interceding act of vegetation. ... Nay all the operations in vulgar chemistry (many of which to sense are as strange transmutations as those of nature) are but mechanicall coalitions or seperations of particles as may appear in that they

---

[76] Newman (2009, 48).
[77] Draft Q 25, CUL, Add. 3970.3, f. 235ᵛ NP. Interestingly, it is possible that there are metallic crystals many km long at the centre of the Earth, which are responsible for its magnetic field. The metallic bond is noticeably between atoms of the same kind.

returne into their former natures if reconjoned or (when unequally volatile) dissevered, & that without any vegetation.⁷⁸

The differences are fundamental:

So far therefore as the same changes may bee wrought by the slight mutation of the textures of bodys in common chymistry & such like experiments may judg that such changes made by nature are done the same way that is by the sleighty transpositions of the grosser corpuscles, for upon their disposition only sensible qualitys depend. But so far as by vegetation such changes are wrought as cannot bee done without it wee must have recourse to som further cause And this difference is vast & fundamental because nothing could ever yet bee made without vegetation which nature useth to produce by it.⁷⁹

'Vegetation' is like creating a copy from a template which has to exist already:

1 All vegetables have a disposition to act upon other adventitious substances & after them to their one temper & nature. And this is to grow in bulk as the alteration of the nourishment may bee called groth in vertue & maturity or specificateness. 2 When the nourishment has attained the same state with the species transmuting the action ceaseth 3 And then is that body thus maturated able in like manner to act upon any new matter & transform it to its owne state & temper. Hence the more to the lesse mature is as agent to patient.⁸⁰

It is driven from something deep within matter: 'The portion fully mature in all things is but very small, & never to bee seene alone, but only as tis inclothed with watry humidity. The whole substance is never maturated but only that part of it which is most disposed The maine bulk being but a watry insipid substance in which rather then upon which the action is performed'.⁸¹

It is interesting that Newton, with whatever background may be assumed to be behind this composition, seems to be making a distinction of some kind between a process that would create compounds from elementary materials and a process that would create the more elementary materials, including metals, themselves, the first being effectively chemistry as we now know it, and the second a process not yet achieved in the laboratory, but probably taking place somewhere within the Earth. Though the connection is very remote, something

---

⁷⁸ *Of Natures Obvious Laws*, NPA.
⁷⁹ *ibid.*
⁸⁰ *ibid.*
⁸¹ *ibid.* This has similarities with the description from *De aere* of material particles as consisting of a hard nucleus surrounded by a repulsive sphere of more tenuous matter (see 6.3).

a little like what Newton might have been imagining takes place to create elementary substances in the process of nuclear fusion in the centres of stars and supernovae.[82] It is certain that the distinction between the two processes remained in Newton's work, even after he abandoned the idea of a Cartesian aethereal circulation in favour of material corpuscles governed by forces acting at a distance, but we have not yet uncovered what form the generation of metals might have taken. What is important from a modern perspective is that Newton, whatever his starting point, is unable to prevent himself synthesizing and abstracting, and, the more abstracted the ideas become, the closer they become to the way we think today. Clearly, his ideas are not always the ones that we use, but the abstract categories frequently are, and he is also successful in detecting their presence in earlier authors. It doesn't seem as if he consciously reconstructs them — it is just part of an automatic process — but the re-imagining of ideas expressed in totally different forms and for completely different purposes by his predecessors became a significant and fascinating aspect of Newton's creative process.

The significant aspect of the 'more subtle secret and noble way of working' is that, with a considerable amount of help from the alchemical tradition, it is a logical consequence of Newton's metaphysical programme. Matter cannot logically be composed of parts that can never be interchanged with each other. However, all the laboratory processes to which he has ever had any access, and everything that he has been able to extract from the hosts of earlier authorities he has studied, seem to work out only as versions of 'vulgar chemistry', mere redistributions of the parts of 'grosser matter'. Some things, perhaps metals, which have a long history of being obtained from mines deep within the Earth, seem to be preserved through all these redistributions. The restructurings only occur at a particular level; the products are not random; they are restructurings of existing structures, which 'vulgar chemistry' is unable to create, just as it is unable to create life. Since chemistry is unable to create these structures, something more subtle must. Newton sought to find this process through more than a quarter of a century of intense experimentation and on a few occasions may have felt he was close to a breakthrough, but he never found anything beyond chemistry. His last major effort in this direction seems to have been in 1693, the year in which he had his breakdown. It has often been conjectured that the two were linked. William Stukeley claimed that there was a fire at this time in

---

[82]The Earth, of course, has radioactive materials in its core and crust, particularly metals transforming into other metals.

which he lost a manuscript treatise on chemistry.[83] Soon after this his work in this area ceased abruptly, and he is not known to have pursued alchemy, at least in experimental form, after he left Cambridge for the Mint.

Many of Newton's 'alchemical' experiments are a form of metallurgy, involving such things as eutectic alloys and the star regulus of antimony, and they had practical results in the invention of at least two alloys, one a speculum for telescope mirrors, compounded of copper and arsenic, another a fusible alloy of lead, tin and bismuth (Newton's metal), with a melting point of 97°C, for use in thermometric work.[84] Later, at the Mint, he wrote papers on the metallurgy of tin and on the refinement of gold, but these may not have been related to his own experiments. They seem likely to have been a distillation of accumulated knowledge from the practical experience of workers in the trade.[85]

In the kind of alchemy that stemmed from mediaeval Arabic sources, 'mercury' was a sort of metallic principle common to all the known metals, including gold, silver, iron, copper, zinc, tin and lead, whereas 'sulphur' was the component which made the metals distinct from each other. Many experiments in alchemy were attempts at extracting this 'true' mercury from metals, metallic ores or metallic compounds. In his earlier career, between about 1669 and 1675, Newton made a search for the 'mercury' of metals in a series of experiments, recorded in his chemical notebook (Cambridge University Library, Add MS 3975) which has been documented and analysed by the historian Betty Jo Teeter Dobbs. In his first experiments ('Mercury' extracted the wet way), Newton began with an observation recorded by Boyle in the second edition of his *Physiological Essays* (1669). Effectively mercury metal was dissolved in nitric acid to produce mercuric nitrate solution with the emission of nitrogen dioxide. Then, adding lead precipitated mercurious oxide as a white powder, which clearly was not the 'true' mercury. Newton, then, extended the experiment by trying other metals, tin, copper, iron and silver, finding out which precipitated which from the solution. Though this

---

[83] Stukeley to Conduitt, 15 July 1727, Keynes MS 136, p. 10. A fire is also referred to in the *Diary of Adam de la Pryme* (1870). This would be in repetition of the equally putative fire-breakdown sequence of 1677. Westfall (1980, 277).

[84] For mirrors: *Phil. Trans.*, 7, no. 81, 40004, 1672; Brewster (1855, II, 361, 535). For fusible alloy: 'Scala graduum caloris', *Philosophical Transactions*, 22, no. 270, 824–829, March–April 1701 (Latin); 'A Scale of the Degrees of Heat', translated by Henry Jones in 1749 in the third edition of *Philosophical Transactions Abridg'd*, 1749; 1809 edition, 4, 572–575; both reproduced in Cohen, *Papers and Letters* (1958, 259–268).

[85] Hall (1992, 332, 334); *Report on a new design of furnace* (1708) (on tin), NP; *Corr.* V, 82–88 (on gold).

had been begun in the spirit of alchemy, the result was, in fact, chemistry, and was eventually included in the *Opticks* in this form.[86]

Newton was again inspired by Boyle in his second set of experiments ('Mercury' extracted the dry way), this time by *The Origine of Formes and Qualities* (1668). Mercury sublimate or mercuric chloride was baked together with various metals in a crucible and sal ammoniac or ammonium chloride was added. The result was the production of ordinary mercury or mercury amalgams with the other metals. Newton seems to have thought he had obtained the 'mercuries' of the various metals, perhaps some component of their inner structures, but not the 'true' or 'philosophick' mercury. Again, a version of this appeared in the *Opticks*.[87]

At this point, Newton switched to a new investigation on the 'star regulus of antimony'. Metallic antimony was extracted from the ore, stibnite, antimony sulphide, but then known as 'antimony', by reducing it with another metal, especially iron, when it was known as 'regulus of iron' because it was believed that iron remained in the final product. Depending on the conditions, a crystalline form of the metal could be produced with a star-like appearance, with branches radiating from a central point. Other patterns could imitate life-like forms, such as ferns. If extracted with a non-metal reducing agent, it was known as 'regulus of antimony'. Newton's *Index Chemicus*, an enormous work compiled over a long period, describes how to make it with saltpetre and tartar. It also describes how antimony can be used to refine gold. In the laboratory, Newton experimented on producing star reguli with five different metals. However, he also took notes on the star regulus from the pseudonymous alchemist Basil Valentine, who likened its structuring to a 'magnetic' property (not to be confused with real magnetism), and identified another 'magnet' or 'Chalybs', referred to by Michael Sendivogius, with antimony. Sendivogius thought that metals grew from 'seeds' in the Earth, no doubt influencing Newton's views in *Of Natures Obvious Laws*, and thought that the 'Chalybs' (which he also called 'Steel') drew the seeds out of gold. Newton thought that antimony in this form might be able to draw out the 'philosophick' mercury from metals.[88]

---

[86] Dobbs (1975, 139–141). Chemical notebook, CUL, Add MS 3975. Q 31. Boyle (1669, 202–203).
[87] Dobbs (1975, 141–146). Chemical notebook, CUL, Add MS 3975. Q 31. Boyle (1668).
[88] Dobbs (1975, 146–161). *Index Chemicus*. Keynes MS 30, NPA. Chemical notebook, CUL, Add MS 3975, f. 42^r, *Essay on the Preparation of 'Star Reguluses*, Dobbs (1975, 249–250). Keynes MS 19. Keynes MS 55.

A fourth series of early experiments by Newton, concerned with 'The Net', was inspired by the *Arcanum philosophiae opus* of the not strictly alchemical early seventeenth century writer Jean d'Espagnet, which had used figurative language about a 'finely woven net' for catching 'fat and scaly flickering fishes'. Newton interpreted the fishes as being of two kinds, representing the 'mercury' and 'sulphur' components of metals, with the net as another way of separating them. Through an unknown number of experimental trials, he managed to create a metallic 'net-like' structure by alloying copper with the star regulus of antimony, prepared from stibnite and iron, but there is no indication of how he came upon this particular combination. Later on he created new interpretations for both the star regulus and the net.[89]

A final series involved a number of amalgamations of mercury with the various star reguli of antimony he had already created, formed by heating the materials at relatively high temperatures, once again in an attempt to find the 'philosophick' mercury. This time, he may have used real mercury as a more likely source for the 'philosophick' mercury than the other metals that he had tried in the second set of experiments. In these experiments, he also generally used a third ingredient, tin, lead or more often a metal with a symbol, later associated with 'quintessentia' (quintessence), which may have been bismuth. Here we see the alchemical idea of 'mediation' being used to facilitate the process, possibly to help separate the components. Mediation, as we have seen, led to the chemical concept of catalysis.[90]

Certainly, many of these experiments have an origin in alchemical beliefs and practices, but it is notable how Newton always transforms them in some way to fit into his own picture. The later experiments are at present less well understood but it is likely that some of the chemical consequences ended up as components of the Queries in the *Opticks*, especially the long Query 31. The many chemical examples discussed in that Query show clear evidence of having been closely observed in the laboratory and probably on many occasions. Whatever their original inspiration, they have been thoroughly assimilated into Newton's model of a theory of matter governed by attractive and repulsive forces. How this 'transition', if that was what it was, actually happened has yet to be established, but clearly Newton investigated many of the most arcane alchemical writings at the same time as

---

[89] Dobbs (1975, 161–163). Chemical notebook, CUL, Add MS 3975. Keynes MS 19. Keynes MS 55.
[90] *ibid.*, Chemical notebook, CUL, Add MS 3975.

having a totally rational and experimentally-based knowledge of laboratory chemistry.

Incidental references show how Newton used his chemical background also in physical investigations. In his attempt to overcome the problem of chromatic aberration, for example, he studied the refracting properties of a large number of substances. Oils, he observed, had very high refractive indices, like diamond. He proposed, therefore, that diamond must be 'an unctuous [oily] Substance coagulated', which, in modern terms, means consisting mostly of carbon.[91] By classifying it with such high refractive index organic materials as camphor, olive oil, linseed oil, spirit of turpentine and amber, he implied that it was, like them, inflammable or 'sulphureous', because it represented a more condensed version of the same component which caused them to be inflammable. In this context, of course, 'sulphureous' means a substance which is combustible *like* material sulphur, rather than one which *contains* sulphur as an ingredient, in the same way as 'phosphorescent' implies a substance which glows *like* phosphorus, rather than one which contains it, a distinction which Brewster, among others, would fail to recognise.[92] Further to his speculations about diamond, Newton asked in Query 7: 'Is not the strength and vigor of the action between Light and sulphureous Bodies observed above, one reason why sulphureous Bodies take fire more readily, and burn more vehemently than other Bodies do?' The connections made here between the chemical and physical properties of these materials may be considered 'one of the first beginnings of the study of the relations between optical properties and composition', a subject which became of considerable subsequent interest to chemists.[93]

In another example, though Newton's discovery of the composition of white light was made independently of his work in chemistry, his two early essays 'Of Colours' in which it is first described occur in chemical contexts in his manuscripts, and William R. Newman has shown that the chemical background may well have influenced the way he developed his interpretation. In particular, Newman believes that Newton's analysis and synthesis of white light could be considered as an example of optical 'redintegration'. Newton's first essay 'Of Colours' shows that he observed the differing refrangibilities of various colours by looking at objects through

---

[91] *Opticks*, Book II, Part 3, Proposition 10. Hazen (2003, 9) paraphrases it as 'a solidified oil'.
[92] Brewster (1855, 211). 'Sulphur', to Newton, often means the 'principle' rather than the material.
[93] Berry (1954, 86).

a prism and then attributed dispersion to the corpuscles moving at different velocities; the second essay introduced the projection of the elongated spectrum and the subsequent recovery of white light by synthesis, implying also the new discovery that colours were immutable.[94]

6.5. Chemistry and Active Principles

Newton's unique and special contribution was to integrate the many strands of the alchemical tradition, together with material derived directly from more conventional chemical observation, into a theory of forces between material corpuscles, arranged in hierarchical structures, and acting across large amounts of empty space within the matter itself. As was usual with him, ideas derived from various sources were subjected to a degree of unconscious merging and abstracting which made them compatible with more advanced and more abstracted notions worked out in other areas, such as mechanics and gravity. The basic ideas were emerging at the time of the *Principia*, and were communicated to Fatio at an early date before surfacing in the Queries added to successive editions of the *Opticks*. At the same time, disciples such as John Keill, James Keill and John Freind started producing their own applications of the Newtonian theory to the structure of matter and chemical reactions.[95] Though attacked in *Acta eruditorum* in 1710, Freind's work, in particular, was significant in finding support in France, where it contributed to the characteristic way in which French chemistry would develop during the eighteenth century. One French natural philosopher, the Comte de Buffon, even went so far as to proclaim that chemical affinity was produced by the same 'general law by which celestial bodies act upon one another'. Others, such as Guyton de Morveau, attempted to establish specific affinities in quantitative terms.[96]

Chemical phenomena are linked to active principles or the forces between particles and the motions of particles in a number of drafts written by Newton around 1705 for Query 23 of the Latin *Opticks* of 1706. The chemical examples, like the physical ones, are not merely anecdotal but directed towards an explanation which minimises mechanistic hypotheses.

> By the actions of the same spirit some particles of bodies can the more strongly attract one another, others less strongly, and thence can arise

---

[94] Newman (2010), QQP, *Of Colours*, CUL Add. MS 3996), NP. *Of Colours*, c 1665–1666 (CUL, Add. MS 3975). See also Newman (2016).
[95] John Keill (1702); James Keill (1708); Freind (1709).
[96] Buffon (1749–1767); Thackray (1970, 159, 214).

varying congregations and separations of particles in fermentations and digestions, especially if the particles are agitated by slow heat. Heat in fact, brings together homogeneous particles on account of greater attraction and separates heterogeneous particles on account of lesser attraction.[97]

In *De natura acidorum*, 1691/92, he writes:

Mercury can pass, and so can aqua regia, through the pores that lie between particles of the last order, but not others. If a menstruum could pass through those others or if the particles of gold of the first, or perhaps of the second composition of gold could be separated; that metal might be made to become a fluid, or at least more soft. And if gold could be brought once to ferment and putrefy, it might be turned into any other body whatsoever. And so of tin, or any other bodies.[98]

Newton, here, of course, refers to a kind of transformation beyond the chemical, but he is now practising a form of physics without an aether, and, in this form of physics, a transformation will require penetrating into the hierarchical and porous structure of matter at a deeper level than ever before.[99] (Interestingly, this defined the way in which it would be ultimately done, though, of course, using techniques very different from those attempted by Newton.[100])

Consideration of the states of matter led Newton to the processes of distillation and sublimation, and it is possible that it was in considering such processes in the light of long-term speculation from the alchemical tradition that he first developed the idea of chemical composition and the qualitative distinction between the cohesion due to chemical attraction and that due to physical association, and possibly first began to speculate on a hierarchy of attractions within matter. Distillation loosens the particles within a volatile liquid and separates them out into vapour, but it will not separate out the particles of strong acids once they have been chemically compounded, even though the acids themselves are volatile; consideration of sublimation

---

[97] Draft c 1705 for Q 23 of Latin *Opticks* of 1706, CUL, Add MS 6970, 604^r; McGuire (1968, 177).
[98] *De natura acidorum*, 1691/92; a conflation of the versions in *Corr.* III, 211 and Cohen, *Papers and Letters* (1958, 258).
[99] According to Newman (2016), 'Real transmutation, which Newton has in mind when he speaks of vegetation, had long been thought of in alchemy as something that occurs at a deeper microstructural level of matter' (475); 'Newton saw the possibility of transmutation only at the extreme nano-stage of corpuscular hierarchy' (476). It would, of course, have been well beyond the nano-level of modern terminology because this was already the order of his optical experiments.
[100] At the mechanical level, however, they were the same because nuclear transmutation uses the same processes of violent collision, absorption and emission of particles under attractive and repulsive forces as Newton imagined must be occurring in ordinary chemical reactions.

extends the process to solids. Observations of the processes of distillation and sublimation thus allowed Newton to make a direct comparison of the two presumed force of attraction and to show that the force of chemical attraction is incomparably greater than the force of physical cohesion in either liquids or solids.

Newton's examples on distillation included the actions of sal ammoniac. 'And because all the particles of a compound are greater than the component particles, and larger particles are agitated with greater difficulty, so the particles of sal ammoniac are less volatile than the smaller particles of the spirits of which they are composed'.[101] Newton's explanation here is valid even in modern terms. 'Sal ammoniac', which in modern chemistry, is ammonium chloride, NH_4Cl, is created from the combination of ammonia, NH_3, and hydrogen chloride, HCl, and the particles of ammonium chloride are certainly 'greater than the component particles' and 'agitated with greater difficulty'. Newton had a special interest in sal ammoniac, using deliquescence as an an analytic test for its presence.[102] He also used it with *aqua fortis* to dissolve antimony, which was then precipitated with fresh water. After the solution had been filtered and then distilled, he obtained a dry white salt and a saltish faeces. By extraordinarily accurate measurements he found that the proportions of water, *aqua fortis*, sal ammoniac, antimony, dry volatile salt, and faeces were in the ratios of 64, 32, 16, 25, 5, and 1 parts by mass. Westfall remarks that 'This was real quantitative analysis'.[103]

Another distillation occurred with sal alkali.

> If spirit of vitriol (which consists of common water and an acid spirit) be mixed with sal alkali or with some suitable metallic powder, at once commotion and violent ebullition occur. And a great heat is often generated in such operations. That motion and the heat thence produced argue that there is a vehement rushing together of the acid particles and the other particles, whether metallic or of sal alkali; and the rushing together of the particles with violence could not happen unless the particles begin to approach one another before they touch one another. The force of whatever kind by which distant particles rush towards one another is usually, in popular speech, called an attraction. For I speak loosely when I call every force by which distant particles are impelled mutually towards one another, or come together by any means and cohere, an attraction. Moreover, if the solution of a metal in spirit of vitriol be distilled, that spirit which by itself will ascend at the heat of boiling water does not ascend before the material

---

[101] Draft *Conclusio*; *Unp.*, 335.
[102] CUL, Add MS 3973, ff 35–37, 36^v. Westfall (1975).
[103] *ibid.*, 38^v. Westfall (1975).

is heated almost to incandescence, being held down by the attraction of the metal.[104]

Further examples included arsenic and oil of vitriol (sulphuric acid).

When Arsenick with Soap gives a Regulus, and with Mercury sublimate a volatile fusible Salt, like Butter of Antimony, doth not this shew that Arsenick, which is a Substance totally volatile, is compounded of fix'd and volatile Parts, strongly cohering by a mutual Attraction, so that the volatile will not ascend without carrying up the fixed? And so, when an equal weight of Spirit of Wine and Oil of Vitriol are digested together, and in Distillation yield two fragrant and volatile Spirits which will not mix with one another, and a fix'd black Earth remains behind; does not this shew that Oil of Vitriol is composed of volatile and fix'd Parts strongly united by Attraction, so as to ascend together in form of a volatile, acid, fluid Salt, until the Spirit of wine attracts and separates the volatile Parts from the fixed?[105]

When Oil of Vitriol is mixed with a little Water, or is run *per deliquium*, and in Distillation the Water ascends difficultly, and brings over with it some part of the Oil of Vitriol in the form of Spirit of Vitriol, and this Spirit being poured upon Iron, Copper, or Salt of Tartar, unites with the Body and lets go the Water; doth not this shew that the acid Spirit is attracted by the Water, and more attracted by the fix'd Body than by the Water, and therefore lets go the Water to close with the fix'd Body? And is it not for the same reason that the Water and acid Spirits which are mix'd together in Vinegar, *Aqua fortis*, and *Spirit of Salt*, cohere and rise together in Distillation; but if the *Menstruum* be poured on Salt of Tartar, or on Lead, or Iron, or any fixed Body which it can dissolve, the acid by a stronger Attraction adheres to the Body, and lets go the Water?[106]

Much the same occurs with all strong acids.

So also spirit of nitre (which is composed of water and an acid Spirit) violently unites with salt of tartar; then, although the spirit by itself can be distilled in a gently heated bath, nevertheless it cannot be separated from the salt of tartar except by a vehement fire. And similar reasoning applies to all sufficiently strong acids and the bodies dissolved in them. In the same way if oil of vitriol be mixed with common water, a vehement heat arises from the impetuous rushing together of the particles, and then the united particles so cohere that water, although by itself very volatile, cannot be wholly distilled off before a large part of the oil is driven to ascend by the heat of the water at boiling point, or rather more. And by a similar cohesion

---

[104] Draft *Conclusio*; *Unp.*, 333–334.
[105] Q 31.
[106] *ibid.*

the phlegmatic water and the acid spirit, of which aqua fortis is composed, hold on to each other and distill over together.[107]

Other reactions of interest include those of spirits of soot and sea-salt, which together create sal ammoniac, and of spirits of salt with antimony, which produces butter of antimony.

> And is it not also from a mutual Attraction that the Spirits of Soot and Sea-Salt unite and compose the Particles of Sal-armoniac, which are less volatile than before, because grosser and freer from Water; and that the Particles of Sal-armoniac in Sublimation carry up the Particles of Antimony, which will not sublime alone; and that the Particles of Mercury, uniting with the acid Particles of Spirit of Salt compose Mercury sublimate, and with the Particles of Sulphur, compose Cinnaber; and that the Particles of Spirit of wine and Spirit of Urine well rectified unite and, letting go the Water which dissolved them, compose a consistent Body; and that in subliming Cinnaber from Salt of Tartar, or from quick Lime, the Sulphur by a stronger Attraction of the Salt or Lime lets go the Mercury, and stays with the fix'd Body; and that when Mercury sublimate is sublimed from Antimony, or from Regulus of Antimony, the Spirit of Salt lets go the Mercury, and unites with the antimonial Metal which attracts it more strongly, and stays with it till the Heat be great enough to make them both ascend together, and then carries up the Metal with it in the form of a very fusible Salt called Butter of Antimony, although the Spirit of Salt be almost as volatile as Water, and the Antimony alone almost as fix'd as Lead?[108]

Hall sees this as close to 'an essentially modern view of a replacement reaction $HgCl + Sb \rightarrow Hg + SbCl$', requiring only forces of chemical affinity or relative attractions, and with no 'philosophic mercury' involved.[109] The passage in general provides a coherent account of the chemical combinations of various atoms and radicals, and the one between mercury and sulphur producing cinnabar actually describes a combination between the atoms of two substances now known to be elementary producing a compound material: $Hg + S \rightarrow HgS$.

The descriptions of these various interactions in Query 31 is paralleled by an earlier discussion in a draft *Conclusio* for *Principia*:

> And the acid spirit in butter of antimony, because it is copious and by itself extremely volatile, cohering with the metallic particles takes them with it in distillation, so that both ascend together by a very violent heat in the form of a fusible salt. In the same way the spirits of urine and salt by cohering

---

[107] Draft *Conclusio*, *Unp.*, 334.
[108] Q 31.
[109] Hall (1992, 190).

compose sal ammoniac; and that salt by cohering with other bodies carries up the particles of these in sublimation. And sal alkali liquefies by attracting vapour from the air, and retains the attracted liquor so firmly that the two can scarcely be separated by distillation.[110]

Through his analysis of distillation and sublimation, Newton had certainly observed the difference in kind between chemical attractions and the weaker physical forces of cohesion, but chemical and physical effects were not made separate subjects for investigation by him as they are for us. His discussion of chemical matters is entirely before any real classification of chemical substances and lacks the precision of terminology which we now consider essential; but we can recognise so many of the modern distinctions, at least in embryo form, in Newton's work that we are forced to conclude that he was an innovator here too.

Chemical reaction is another form of physical restructuring brought about by forces acting on the component particles. 'Moreover these very great forces by which the particles of bodies attract each other mutually and cohere will effect remarkable results in fermentation, putrefaction and chemical reaction'.[111] 'Through forces of this kind solvents, salts, spirits and other bodies either act upon one another or not; come together either more swiftly or more slowly; either do not mix easily or are separated from each other with difficulty or do not cohere'.[112]

The Queries contain many examples of exothermic and endothermic reactions, in which heat is emitted or absorbed.

> For when Salt of Tartar runs *per Deliquium*, is not this done by an Attraction between the Particles of the Salt of Tartar, and the Particles of Water which float in the Air in the form of Vapours? And why does not common Salt, or Salt-petre, or Vitriol, run *per Deliquium*, but for want of such an Attraction? Or why does not Salt of Tartar draw more Water out of the Air than in a certain Proportion to its quantity, but for want of an attractive Force after it is satiated with Water? And whence is it but from this attractive Power that Water which alone distils with a gentle luke-warm Heat, will not distil from Salt of Tartar without a great Heat? And is it not from the like attractive Power between the Particles of Oil of Vitriol and the Particles of Water, that Oil of Vitriol draws to it a good quantity of Water out of the Air, and after it is satiated draws no more, and in distillation lets go the Water very difficultly? And when Water and Oil of Vitriol poured successively into the same Vessel grow very hot in the mixing, does not this

---

[110] Draft *Conclusio*; *Unp.*, 334–335.
[111] *De vi electrica*; *Corr.* V, 368.
[112] Draft *Praefatio*; *Unp.*, 305–306.

Heat argue a great Motion in the Parts of the Liquors? And does not this Motion argue, that the Parts of the two Liquors in mixing coalesce with Violence, and by consequence rush towards one another with an accelerated Motion? And when *Aqua fortis*, or Spirit of Vitriol poured upon Filings of Iron dissolves the Filings with a great Heat and Ebullition, is not this Heat and Ebullition effected by a violent Motion of the Parts, and does not that Motion argue that the acid Parts of the Liquor rush towards the Parts of the Metal with violence, and run forcibly into its Pores till they get between its outmost Particles, and the main Mass of the Metal, and surrounding those Particles loosen them from the main Mass, and set them at liberty to float off into the Water? And when the acid Particles, which alone would distil with an easy Heat, will not separate from the Particles of the Metal without a very violent Heat, does not this confirm the Attraction between them?[113]

In some cases, flame is produced.

When Spirit of Vitriol poured upon common Salt or Salt-petre makes an Ebullition with the Salt, and unites with it, and in Distillation the Spirit of the common Salt or Salt-petre comes over much easier than it would do before, and the acid part of the Spirit of Vitriol stays behind; does this not argue that the fix'd Alcaly of the Salt attracts the acid Spirit of the Vitriol more strongly than its own Spirit, and not being able to hold them both, lets go its own? And when Oil of Vitriol is drawn off from its weight of Nitre, and from both the Ingredients a compound Spirit of Nitre is distilled, and two Parts of this Spirit are poured on one part of Oil of Cloves or Carraway Seeds, or of any ponderous Oil of vegetable or animal Substances, or Oil of Turpentine thicken'd with a little Balsam of Sulphur, and the Liquors grow so very hot in mixing, as presently to send up a burning Flame; does not this very great and sudden Heat argue that the two Liquors mix with violence, and that their Parts in mixing run towards one another with an accelerated Motion, and clash with the greatest Force? And is it not for the same reason that well rectified Spirit of Wine poured on the same compound Spirit flashes; and that the Pulvis *fulminans*, composed of Sulphur, Nitre, and Salt of Tartar, goes off with a more sudden and violent Explosion than Gun-powder, the acid Spirits of the Sulphur and Nitre rushing towards one another, and towards the Salt of Tartar, with so great a violence, as by the shock to turn the whole at once into Vapour and Flame? Where the Dissolution is slow, it makes a slow Ebullition and a gentle Heat; and where it is quicker, it makes a greater Ebullition with more Heat; and where it is done at once, the Ebullition is contracted into a sudden Blast or violent Explosion, with a Heat equal to that of Fire and Flame. So when a Drachm of the above-mention'd compound Spirit of Nitre was poured upon half a Drachm of Oil of Carraway Seeds *in vacuo*, the Mixture immediately made

---

[113] Q 31.

a flash like gun-powder, and burst the exhausted Receiver, which was a Glass six Inches wide, and eight Inches deep.[114]

The explosive reaction between spirit of nitre and oil of caraway seeds was first reported by Frederick Slare in 1694, and became quite celebrated.[115]

Colour changes occur during chemical reactions, with different chemicals producing different colours of flame, for example, the blue flame of sulphur, the green of copper, the yellow of tallow, and the white of camphor. Flames are also produced in different ways, for example, by heat throwing off the volatile parts of ignited bodies, leaving the 'fixed parts' as ashes; 'an auxiliary flame can ignite volatiles away from their source and the resulting flame flashes back to the origin of the volatiles'.[116] Flame can be distinguished from vapour, with flame, unlike vapour, being red hot. Bodies do not 'only shine by vehement heat' because hot and cold flames can be distinguished.[117]

The chemical reactions are produced by strong 'active principles' that act as new sources of motion within the particles of matter.

> Now the above mentioned motions are so great and violent as to show that (in fermentations) there is new motion in the world generated from other principles than the usual laws of motion ... bodies (which almost rest) are put into (new) motions by a (very potent active principle which acts upon them only when they approach one another and which is much more potent than are the passive laws of motion arising from the vis inertiae of the matter).[118]

'And if there be another principle of motion there must be other laws of motion depending on that principle'.[119]

The process that we call 'oxidation' is due to the actions of a 'sulphureous spirit', a component of air which is responsible for combustion. 'In coals and ignited vapours, however, heat seems to be excited and conserved by the action of a sulphureous spirit... And inasmuch as air abounds in suphureous spirits, it also makes ignited matter grow hot and because of the subtlety of the spirit is required for the maintenance of fire'.[120] The 'sulphureous spirit' was not a substance which was in itself 'sulphureous', but one which

---

[114] Q 31.
[115] Slare (1694, 212–213).
[116] Simms (2004, 67–68).
[117] Simms (2004, 60–61, 68–69, 72).
[118] Draft Q 23/31; McGuire (1968, 171).
[119] Draft Q for *Opticks*; McGuire (1968, 171).
[120] Draft *Conclusio*; *Unp.*, 343.

caused a 'sulphureous' substance to combust, a kind of complementary principle:

> That sulphureous spirit meeting with the fixed parts of vegetable salts looses their spirits, however, and united with those fixed parts forms that fixed salt which is usually found in ashes and is called sal alkali, and by action of that sulphureous spirit runs per deliquium. For such are the properties of oil and spirit of vitriol. They attract a humour from the air and meeting with the fixed parts of common salt and salt of nitre (which are both vegetable salts), they loose their spirits and expel them. But nitre deflagrated in the open air with sulphur or with any sulphureous body is converted into sal alkali, whereas if nitre alone is mixed with clay and distilled in a closed glass, no sal alkali is extracted from the caput mortuum.[121]

The 'sulphureous' or 'acid' part of the air responsible for 'oxidation' has connections with ideas of Newton's contemporaries Hooke, Boyle, and especially John Mayow, about a particulate component of the air ('nitro-aerial particles') responsible for combustion and respiration. 'For the Air abounds with acid Vapours fit to promote Fermentations, as appears by the rusting of Iron and Copper in it, the kindling of Fire by blowing, and the beating of the Heart by means of Respiration'.[122] Interestingly, the connection between 'oxidation' and 'acidity' was maintained in Lavoisier's choice of the name 'oxygen' or 'acid former' for the component of air responsible for these processes, even though oxygen plays no part in determining the nature of acids.

The most obvious manifestation of the significance of oxygen in combustion observed at the time (though not, of course, fully recognised) was the explosion of gunpowder, where the heating of the component nitre (or potassium nitrate) liberates large quantities of the gas to generate an explosive reaction. Concerning this, Newton writes:

> What has been said before is confirmed by gunpowder which is composed of nitre, sulphur and charcoal. The powdered charcoal readily takes fire and ignites the powdered sulphur. The acid fume of the sulphur invades the nitre and by this encounter makes it hot, and looses its spirit, which spirit driven out by heat and converted into vapour expands itself violently. If sal alkali be mixed in due proportion with it, and the powder dried by fire to drive off all the humour that this salt commonly attracts to itself, the powder now deflagrates with a more violent explosion after the manner of fulminating

---

[121] Draft *Conclusio*; *Unp.*, 343–344.
[122] Q 31. One of Newton's early letters on his reflecting telescope (30 March 1672, in reply to Adrien Auzout) refers to an 'Acid spirit, wherewith the Atmosphere is impregnated' corroding metallic telescope mirrors.

gold. For the sulphureous spirit meets more willingly and violently with sal alkali than with salt of nitre, and by that meeting agitates all the material more and more and makes it grow more intensely hot. Indeed, it is to be believed that the culinary fire is excited and maintained at a slow and gradual rate by the material causes of fire which act more vigorously in these compositions, for those causes abound less in it.[123]

6.6. Basic Chemical Principles

Acids, bases and salts are outlined in some passages, and suggestions of chemical radicals; the differences between physical and chemical changes are described, in addition to those between mixtures and compounds. The reaction acid plus base produces salt lies behind a statement such as: 'Between acid particles and fixed particles deprived of acid by fire, there is very great attraction. Out of these are compounded particles less attractive'.[124] Newton also says that the greatest degree of attraction occurs between 'acid particles and fixed particles deprived of acid by fire' (i.e. particles of bases).[125] Detail is provided in *De natura acidorum*:

> The particles of acids are coarser than those of water and therefore less volatile; but they are much finer than those of earth, and therefore much less fixed than they. They are endowed with a great attractive force and in this force their activity consists by which they dissolve bodies and affect and stimulate the organs of the senses. They are of a middle nature between water and [terrestrial] bodies and they attract both. By their attractive force they surround the particles of bodies be they stony or metallic, and they adhere to them very closely on all sides, so that they can scarcely be separated from them by distillation or sublimation. When they are attracted and gathered together on all sides they raise, disjoin and shake the particles of bodies one from another, that is, they dissolve the bodies, and by their force of attraction by which they rush to the [particles of] bodies, they move the fluid and excite heat and shake asunder some particles to such a degree, as to turn them into air and generate bubbles; and this is the reason of dissolution and violent fermentation... But the acid, suppressed in sulphureous bodies, by attracting the particles of other bodies (for example, earthy ones) more strongly than its own, causes a gentle and natural fermentation and promotes it even to the stage of putrefaction in the compound. This putrefaction arises from this, that the acid particles which have for some time kept up the fermentation do at length insinuate themselves into the minutest interstices, even those which lie between the parts of the first composition, and so, uniting closely with those particles,

---

[123] Draft *Conclusio*; *Unp.*, 344.
[124] Latin MS (CUL, Add MS 3970, f. 604ʳ); McGuire (1968, 177–178).
[125] Draft Conclusion to the *Principia*, translated in Cohen and Whitman (1999, 292).

give rise to a new mixture which may not be done away with or changed back into its earlier form.[126]

Many examples of the basic process are provided.

The particles of *sal alkali* consist of acid and earthy combined in just this way, but these acid ones prevail with such a strong attractive force that they are not separable from the salt by fire; moreover they precipitate metals dissolved [in them] by attracting from them the acid particles by which they were previously dissolved.

If the acid particles are joined with the earthy ones in a lesser proportion they are so closely held by the latter that they are, as it were, suppressed and hidden by them. For they no longer excite the organs of sense, nor do they attract water, but they compose bodies which are sweet and do not readily mix with water; that is, they compose fatty bodies such as we find in *mercurius dulcis*, ordinary sulphur, luna cornea and copper which has been corroded by mercury sublimate.[127]

To Archibald Pitcairne, who visited Newton in 1692 at the time when he was involved in a serious attempt at making progress in his alchemical experiments, he said: 'By distillation nitre is converted largely into acid spirit... But nitre burnt with charcoal turns largely into salt of tartar since fire applied in this way drives the particles of acid and earth into itself and binds them together more strongly'.[128] And also: 'Particles of acid combined with fine earth make up salt, and salt dissolves in water for the reason that water attracts acid. Moreover acid causes alteration and thus makes nitre, vitriol etc. But some kind of acid so attracted by an earth that it cannot be separated by force of fire constitutes an alkali'.[129]

Query 31 supplies detail on some instances:

When Mercury sublimate is re-sublimed with fresh Mercury, and becomes *Mercurius Dulcis*, which is a white, tasteless Earth scarce dissolvable in Water, and *Mercurius Dulcis* re-sublimed with Spirit of Salt turns into Mercury sublimate; and when Metals corroded with a little acid turn into rust, which is an Earth tasteless and indissolvable in Water, and this Earth imbibed with more acid becomes a metallick Salt; and when some Stones, as Spar of Lead, dissolved in proper *Menstruums* become Salts; do not these things shew that Salts are dry Earth and watry Acid united by Attraction, and that the Earth will not become a Salt without so much acid as makes it dissolvable in Water? Do not the sharp and pungent Tastes of Acids

---

[126] *De natura acidorum*, 1691/92; *Corr.* III, 209–210.
[127] *De natura acidorum*, 1691/92; *Corr.* III, 209.
[128] Pitcairne with Newton at Cambridge, 2 March 1692; *Corr.* III, 210.
[129] *ibid.*, *Corr.* III, 212.

arise from the strong Attraction whereby the acid Particles rush upon and agitate the Particles of the Tongue? And when Metals are dissolved in acid *Menstruums*, and the Acids in conjunction with the Metal act after a different manner, so that the Compound has a different Taste much milder than before, and sometimes a sweet one; is it not because the Acids adhere to the metallick Particles, and thereby lose much of their Activity? And if the Acid be in too small a proportion to make the Compound dissolvable in Water, will it not by adhering strongly to the Metal become unactive and lose its Taste, and the Compound be a tasteless Earth? For such things as are not dissolvable by the Moisture of the Tongue, act not upon the Taste.

In one of his comments to Pitcairne, Newton says: 'Vitriol is the sediment of metal torn away by acid of sulphur and combined with it'.[130] Previously, in the early letter he wrote to Boyle in 1679, he had described the process in almost mechanical terms:

If into a solution of metal thus made, be poured a liquor, abounding with particles, to wch ye... saline particles are more sociable than to ye particles of ye metal, (suppose with particles of salt of Tartar:) then so soon as they strike on one another in ye liquor, ye saline particles will adhere to those more firmly then to ye metalline ones, & by degrees be wrought of from those to enclose these. Suppose A a metalline particle, enclosed with saline ones of spirit of nitre, and E a particle of salt of tartar, contiguous to two of the particles of spirit of Nitre b & c, & suppose ye particle E is impelled by any motion towards d so as to roll about ye particle c till it touch ye particle d: the particle b adhering more firmly to E then to A, will be forced of from A. And by ye same means ye particle E as it rolls about A will tear of ye rest of ye saline particles from A, one after another, till it has got them all or almost all about it self. And when ye metallic particles are thus divested of ye nitrous ones which as a mediator between them & ye water held them floting in it: the Alkalizate ones crouding for ye room ye metallic ones took up before, will press these towards one another & make them come more easily together: so yt by the motion they continually have in ye water they shall be made to strike on one another, & then ... they will cohere & grow into clusters, & fall down by their weight to ye bottom, wch is called precipitation.[131]

Later descriptions, though less detailed, tended to be in the same vein.

There are also quantitative measures of combination and affinity, and what later became known as the electrochemical series of metals, based partly on his earliest experiments at finding the 'Philosophick Mercury'.

---

[130] Pitcairne with Newton at Cambridge, 3 March 1692; *Corr.* III, 212.
[131] Letter to Boyle, 28 February 1679; *Corr.* II, 292–293.

Though incorporating material from some of his earliest experiments in chemistry, the idea of the series found its best-known expression in Query 31:

> When Salt of Tartar *per deliquium*, being poured into the Solution of any Metal, precipitates the Metal and makes it fall down to the bottom of the Liquor in the form of Mud: Does not this argue that the acid Particles are attracted more strongly by the Salt of Tartar than by the Metal, and by the stronger Attraction go from the Metal to the Salt of Tartar? And so when a Solution of Iron in *Aqua fortis* dissolves the *Lapis Calaminaris*, and lets go the Iron, or a Solution of Copper dissolves Iron immersed in it and lets go the Copper, or a Solution of Silver dissolves Copper and lets go the Silver, or a Solution of Mercury in *Aqua fortis* being poured upon Iron, Copper, Tin, or Lead, dissolves the Metal and lets go the Mercury; does not this argue that the acid Particles of the *Aqua fortis* are attracted more strongly by the *Lapis Calaminaris* than by Iron, and more strongly by Iron than by Copper, and more strongly by Copper than by Silver, and more strongly by Iron, Copper, Tin, and Lead, than by Mercury? And is it not for the same reason that Iron requires more *Aqua fortis* to dissolve it than Copper, and Copper more than the other Metals; and, that, of all Metals, Iron is dissolved most easily, and is most apt to rust; and, next after Iron, Copper?

Through E. F. Geoffroy this kind of thinking actually influenced the development of chemistry, just as the ideas of Newton and the eighteenth century Newtonians on forces and particles actually influenced the development of atomism and field theory in nineteenth century physics. There is a case for saying that the chemical revolution of the eighteenth century began with Geoffroy's affinity tables in 1718.[132] We have for too long viewed the history of chemistry in terms of classification and systematics, but in a longer term perspective the theoretical study of chemical forces inaugurated by Newton and his eighteenth century followers is at least as important.

The special nature of chemistry is, in principle, a pragmatic finding rather than a fundamental one. It relies on the fact that, in addition to the fact that matter is structured, there is also a special level of structure within matter that produces a special category of material interaction. That is why it needed a period of systematization wholly separate from the high energy physics approach pioneered by Newton before it could

---

[132] Geoffroy (1718). The historian of chemistry, Henry Leicester, says that: 'The influential 1923 textbook *Thermodynamics and the Free Energy of Chemical Reactions* by Gilbert N. Lewis and Merle Randall led to the replacement of the term "affinity" by the term "free energy" in much of the English-speaking world'. (1956, 206).

be incorporated into science at the same level. Newton, though he was influenced strongly by Boyle, and used many chemical examples, was not really writing about chemistry at all. He by-passed chemistry to make a direct attack on the microphysics of matter which was ultimately responsible for chemical operations, and, seen in a twenty-first century perspective, his work on matter was far more advanced than that of the instigators of the eighteenth century chemical revolution, and now appears astonishingly prescient.

6.7. Newton and the Chemical Revolution

The eighteenth century tradition stemming from Newton was more physical than chemical, but it introduced affinity and the electrochemical series of metals. Ultimately affinity made possible Dalton's atomic theory. Newton's careful quantitative analysis or stoichiometry also prefigured those of Joseph Black and Lavoisier, though it didn't influence them. Of course, Boyle was the significant figure in chemistry at the time and wrote much more on it than Newton. The more classifying aspect of his work bore fruit in the chemical revolution of the late eighteenth century. But Newton was probably the most significant figure of his time in beginning the process which successfully integrated chemistry at a more fundamental level into physics a century later.

Of course, Newton and his contemporaries did not fully recognise the categories of chemical thought which were the direct result of that revolution — the differences between elements, mixtures and compounds, physical and chemical changes, acids, bases and salts, atoms and molecules — though he had some understanding of most of them, yet his ideas, if seen in terms of the most contemporary physics, are amazingly correct. It is sometimes said that it is anachronistic to view Newton's work in the light of contemporary ideas which turn out to be similar, but this is a result of misunderstanding the nature of the 'recursive' or abstract analytical aspect of his thinking. It is surely a result of his facility at finding the right categories for organizing his thoughts at a fundamental level that Newton was so often correct in his speculations, even when he was guessing at many removes from his actual experience. And this may well be as much relevant to the development of scientific method as the axiomatically rigorous and axiomatic approach of the *Principia* or the systematic and inventive experimentation of the *Opticks*.

The ultimate development of the Newtonian approach to chemistry brings us back to the alchemy with which we began, for to establish the 'secret and noble way of working' beyond the 'vulgar chymistry', and the 'interactions of gross particles', we need to find a route to a transmutation that is beyond chemistry and involves new forces. Newton's last major effort on alchemy was apparently the experiment on 'multiplication' of gold and other metals recorded in his treatise *Praxis*, in 1693.[133] Though at first it seemed a success, the euphoria died away quickly and was followed by something resembling a nervous breakdown.

Remarkably, the 'philosopher's stone' was closer at hand than he may have thought. Newton had always been interested in mines and the extraction of metals. The first letter he ever addressed to an individual (though it may never have been sent) advises his Cambridge colleague and lifelong friend Francis Aston to make various observations about mining and metallurgy as practised in the various parts of central Europe he was about to visit, including the mines of gold, copper, iron, vitriol, antimony and other minerals at Schemnitium (Schemnits or Selmenzbanya) in Hungary. Newton also asked Aston to examine some of the processes described in the various alchemical tracts collected in the Hungarian Count Michael Maier's *Symbola Aureae Mensae Duodecim Nationum*, published at Frankfurt in 1617.[134]

Now, during his time at the Mint, when he was involved in maintaining a currency based on the durability of the precious metals, gold and silver, the government was, for political reasons, obliged to buy large quantities of tin from the Cornish mines, that were surplus to requirements, at a price well above the going rate. Newton was forced to stockpile the tin and to try to sell it on the open market.[135] He actually produced a draft, in 1708, on its metallurgy, though this may have been based on information from commercial producers rather than on his own experiments.[136] Very probably the Cornish miners had come across a dark-looking rock which had been

---

[133] *Praxis*, 1693; Dobbs (1991, 293–305).

[134] Newton to Francis Aston, 18 May 1669, *Corr.* I, 9–11. Newton writes: 'Observe the products of nature in severall places especially in mines with ye circumstances of mining & of extracting metalls or mineralls out of theire oare and refining them and if you meet wth any transmutations out of one species into another (as out of Iron into Copper, out of any metall into quicksilver, out of one salt into another or into any insipid body &c) those above all others will bee worth your noting being ye most luciferous & many times lucriferous experiments too in Philosophy'. Also: 'Whither at Schemnitium [Schemnits = Selmenzbanya] in Hungary (where there are Mines of Gold, copper, Iron, vitrioll, Antimony, &c) they change Iron into Copper by dissolving it in a Vitriolate water they find in cavitys of rock in the mines'.

[135] Westfall (1980, 606, 615–616); Hall (1992, 331–332); Gjertsen (1986, 362–363).

[136] Hall (1992, 332). *Report on a new design of furnace* (1708, NP).

showing up in the silver mines of the Erzgebirge Mountains, Germany since at least the sixteenth century, not too far from the area that Aston went to visit in 1669. In 1727 F. E. Brückmann described a sample from Jáchymov in Bohemia, naming it pitchblende. Almost a century after Newton, Martin Klaproth, lecturer in chemistry to the Prussian Royal Artillery in Berlin, examined pitchblende from the Johanngeorgenstadt deposit in Germany and identified it as the ore of a new metal, uranium, which he named after the newly-discovered planet Uranus. Pitchblende turned out to be an oxide of uranium (UO_2), and the French chemist Eugène Péligot isolated the first sample of the metal in 1841.[137]

In 1896, in Paris, a hundred years again from Klaproth's discovery, Henri Becquerel found that pitchblende was radioactive, fogging a photographic plate that had not been exposed to light. A young research student in Cambridge, Ernest Rutherford, showed that the invisible radiation emitted had two distinct components, which could be separated by their relative power of penetration of different thicknesses of metal foil; they were named alpha and beta. Thorium compounds seemed to indicate the presence of a third, which was later identified and named gamma.[138] Back in Paris, Marie Curie found that pitchblende was more radioactive than uranium, leading to the conclusion that it must contain an even more radioactive component. After a heroic struggle, with tons of pitchblende, she isolated two new elements, polonium and radium, with the help of her husband, Pierre Curie. Rutherford, now in McGill University, Montreal, showed, with Frederick Soddy, that the radioactive process involved a change in the chemical identity of the element uranium through a long series of decay products, including polonium and radium.[139]

Rutherford realised that radioactivity was nature practising its own brand of alchemy, and he even as much as published a book called *The Newer Alchemy* in 1937,[140] though, when he and Soddy first realised the implication of their discovery, he is reported to have exclaimed: 'For Mike's sake, Soddy, don't call it *transmutation*. They'll have our heads off as alchemists'.[141] Rutherford also received the Nobel Prize for chemistry rather than for physics, leading him to remark on the fact that the fastest transformation

---

[137] Brückmann (1727); Klaproth (1789); Péligot (1841).
[138] Becquerel (1896); Rutherford (1899).
[139] Curie *et al.* (1998); Rutherford and Soddy (1902).
[140] Rutherford (1937).
[141] Howorth (1958), 83–84; Badash (1978); Trenn (1977, 42, 58–60, 111–117).

he knew of was his own transformation from a physicist to a chemist.[142] It was eventually realised that the three types of radiation he had identified were manifestations of three physical forces as different from each other as anything in the whole of nature. While the gamma radiation was the familiar electromagnetic radiation from the electric force, the alpha and beta emissions were the results of short-range forces, the strong and weak nuclear interactions, deep within the atom, which both changed its chemical nature. The weak force, in addition, changed the *physical* composition of matter, as the only force of nature which could change the nature of a fundamental particle.

The two forces thus change chemistry at an elemental level. The strong force can lead to chemical change by the indirect process of rearranging the particles or fermions within the nucleus and removing part of it, so that the electron structure responsible for the chemistry has to change as well. The weak force leads to chemical change of the same kind by changing the characteristics of the fundamental nuclear fermions. The weak interaction is unable to tell one fermion from another, and this is the reason why it causes transitions between neutrons and protons. Both forces cause transmutation, the strong interaction transmuting atoms, while the weak interaction transmutes the ultimate particles.

For Rutherford, who moved to Manchester in 1907, and back to Cambridge in 1919, radium became the philosopher's stone, for the alpha particles which it emitted could be used to penetrate deeper into the centre of matter than any chemical reaction had managed, first to discover the tiny central nucleus (1911), containing most of the atom's mass, and the place of origin of the radioactive forces, and then to effect the first ever deliberate transformation of the atoms of one element into those of another (1919). Thirteen years later, technology had developed to the point where Rutherford's Cambridge colleagues Cockcroft and Walton could accelerate charged particles using just the electric force with sufficient energy to produce a wholly artificial transmutation.[143] Accelerators effectively harness the power that Newton saw was locked up in the atom, and emerges naturally with radioactivity. They are the alternative man-made method of transmutation.

---

[142] Rutherford (1908): 'I have dealt with many different transformations with various periods of time, but the quickest that I have met was my own transformation in one moment from a physicist to a chemist'. Jarlskog (2008).
[143] Rutherford (1911a, 1911b); Rutherford (1919); Cockcroft and Walton (1932).

The Cornish mines, which started extracting uranium commercially during the nineteenth century, apparently did play a part in this story, for it is believed that Marie Curie used radium extracted from South Terras mine for experiments on radiation. The 1913 share prospectus for the company Société Industrielle Du Radium Ltd., which extracted radium under Dr. Marcel Leon Pochon as superintendent, included reports on sampling at the Cornish mine; one from the Curie laboratory concerned the radioactivity of the water in the mine, the material dumped and samples taken from the deeper parts of the mine before they were flooded.[144]

Newton could not achieve an alchemical transmutation himself with the techniques and materials at his disposal, but his beliefs were all confirmed: that one chemical could be transmuted into another by a process that was beyond ordinary chemistry; that this process involved forces right at the centre of matter that had not yet been discovered, but which were stronger and more short-range than the processes so far observed involving the electric force; that matter was mostly empty space with a hierarchical structure leading up to the level at which such properties as the colours of objects became manifest; and that the future of physics lay with the investigation and exploitation of the electric force.

---

[144]Smale (1993); Gillmore *et al.* (2001).

Chapter 7

Nonscience

7.1. Alchemy and Ancient Wisdom

A good deal of Newton's known writings are concerned with subjects that we would now call 'nonscientific'. In many ways, however, this is an arbitrary distinction. Newton's intellectual pursuits can only be considered a unified endeavour in the search for the fundamental truth about the world. It is clear that everything that he did had a very strong basis in a theologically-inspired metaphysics which he developed at a relatively early period and never abandoned. In addition, Newton seems to have applied a strong element of rationality, not to say rationalising, to every area he investigated. We also know that he rejected, for whatever reason, the totally irrational beliefs held by many in his day on such things as demons, ghosts, witchcraft and astrology. Where modern thinkers might object to Newton's processes and conclusions is not so much in the methodologies he uses, but in his belief that his sources contained good evidence and could be treated scientifically. Often his nonscientific work could be regarded as an application of good scientific technique to dubious data. Even so, he had some notable successes, even in these areas.

We have already seen that his pursuit of alchemy was not as irrational as many earlier commentators believed, and, like the work of several illustrious contemporaries, such as Robert Boyle, cannot be divorced from the entirely rational pursuit of chemistry. In Newton's case, it led to purely chemical results and was pursued with a totally 'scientific' approach to experimentation. The 'alchemical' documents that he seems to have composed should be seen as an attempt at making sense of a tradition that he believed contained scientific truths hidden within figurative language, which, at least with some of the later authors in the alchemical tradition, seems to be a valid interpretation. Essentially, though Newton's writings on alchemy may look at first sight like nonscience, in fact they are not because his aims were to use scientific methodology to discover fundamental truths about nature, and the results he actually achieved, rather than sought, were channelled into the world as a form of chemistry.

For all the sensationalism accompanying 'revelations' about Newton's work as an adept, there is, in fact, no example of an extensive alchemical treatise composed by Newton, and there is no evidence whatsoever that he was interested in 'multiplication' or making gold for anything other than 'scientific' reasons. He was interested only in secrets of nature that might be hidden in alchemical texts, which he hoped would include access to the method of transmutation that clearly must occur at some level in the real world. As it stands, his own work in the laboratory can always be interpreted in terms of chemistry or metallurgy in its outcomes, and only ever emerged in these forms. He did not seek to create or preserve mysteries written in arcane language for the adepts, but to solve them. The compiling of an immense *Index Chemicus*, which went through three different states, suggests how seriously he sought to bring to order the immense number of disparate facts that previous writers had uncovered.

The work called *Clavis*, or *The Key* (Keynes MS 18), once thought to be a Newtonian original and an indication that he wrote alchemical treatises in addition to studying alchemical authors, is a Latin translation of a treatise composed by Eirenaeus Philalethes (*alias* George Starkey) in 1651.[1] When 'B.R', whom he knew must be Robert Boyle, published an account of a philosophical 'mercury' which could heat gold in the Royal Society's *Transactions* in 1676, and asked for advice on whether he should reveal the recipe, a possible route to the sought-after 'multiplication', Newton wrote a letter to Oldenburg expressing extreme scepticism as to the process, and proposing instead that it was a result of the 'sociability' of the metallic particles. He argued, however, that Boyle should not publish the process in any case because of the 'immense dammage to the world if there should be any verity in the Hermetick writers', their processes suggesting 'other things beside the transmutation of metals (if those great pretenders bragg not) which none but they understand'.[2] Phrases like 'great pretenders' do not suggest great confidence in the authors he had spent so long examining

---

[1]Starkey (1651) in Dobbs (1975, 251–255). Dobbs (1975, 178), and Westfall (1980, 370), were inclined to think it might be authentic, but Karen Figala (1977, 107), thought it might be by Starkey, and Newman (1987) finally proved it.
[2]Boyle (1676); Newton to Oldenburg, 26 April 1676, *Corr.* II, 1–2. Newton's warning of the danger in making this knowledge freely available is interesting in the light of later events.

or 'in the reality of the cosmic mystery Humphrey [Newton] imagined' him 'to be seeking', no more, in fact, than 'the possibility of it'.[3] Later he tried to warn off Locke from pursuing the quest, suggesting that, unlike his friend, he was not 'perswaded of it', and promising 'one argument against it which I could never find answer to'.[4] According to Hall: 'There was nothing magical in alchemy as Newton conceived it, no method of acting upon material substance by supernatural powers'.[5]

In addition, although some of the alchemical authors he quoted and was interested in used astrological language, there is not the slightest evidence that Newton ever had any belief whatsoever in astrology himself. He told his nephew John Conduitt on 31 August 1726 that, as a student at Cambridge, in 1663, he had acquired a book on astrology at Stourbridge Fair, which seems to have been Richard Saunders's recently published *Palmistry, the secrets thereof disclosed*. Trying to make sense of the diagrams and calculations in the book, he then acquired books on mathematics by Euclid, Descartes and van Schooten and was 'soon convinced of the vanity & emptiness of the pretended science of Judicial astrology'.[6]

In the case of demons, ghosts and witchcraft, his mature views, as developed from the 1680s, were against the orthodox thinking of his day, but totally in line with his own metaphysics and his understanding of the mind–matter problem. For him: 'The spirits of God of fals Prophets & of Antichrist are [in 1 John 4] plainly taken not for any substantial Spirits but for ye good or evil dispositions & true or fals perswasions of mens minds...'. The same was true of the spirit of Antichrist, the Dragon of the Apocalypse, demons of all kind, devils as fallen angels, and Satan himself, as portrayed, for example, in *Paradise Lost* and many other works. 'From this figure of putting serpents for spirits & spirits or Daemons for distempers of ye mind, came ye vulgar opinion of ye Jews & other eastern nations that mad men & lunaticks were possessed with evil spirits or Daemons'. Christ only used such language as a Prophet to comply with the traditional Jewish way of speaking. Demons were therefore not real entities, but 'figures for disordered psychotic states'. Temptation by the devil meant being seduced by your own inner urges (your own 'inner demons', as we would say today) not by some external supernatural agent. For Newton, the orthodox view of demons and

---

[3] Hall (1992, 198).
[4] Newton to Locke, 6 August 1692, *Corr.* III, 218–219.
[5] Hall (1992, 200). This is true but Hall seems to have been over-sceptical about the alchemical (i.e. transmutational) nature of the quest (Newman, 2016).
[6] On astrology, to John Conduitt, 31 August 1726, *MP* I, 15–19.

devils was a heretical view imported by the Jews and early Christians from pagan cultures, and had as much credibility as believing that pagan gods were real beings. In a century when, in Britain and America, people were being put to death for being witches, Newton asserted that 'to beleive that men or weomen can really divine, charm, inchant, bewitch or converse with spirits is a superstition of the same nature wth beleiving that the idols of the gentils were not vanities but had spirits really seated in them'.[7] A famous story, recorded by the diarist Abraham de la Pryme, has him rebuking a number of scholars of St. John's who had gathered outside a house opposite the College which was reputed to be haunted: 'Oh! yee fools, will you never have any witt, know yee nott that all such things are meer cheats and impostures? Fy, fy! go home, for shame'.[8]

Less 'rational' to modern ways of thinking was his belief in a 'prisca sapientia' (sacred wisdom), an ancient source of wisdom which had been lost.[9] It was undoubtedly the driving force behind his intense study of alchemical authors, and his special interest in the more abstruse aspects of their works. The belief in an ancient wisdom was a common one among Renaissance scholars; it is found, for example, in Copernicus and Natale Conti, and still surfaces regularly today. There may be a reason for it in the self-similarity of theories, certainly at the fundamental level. Simple ideas are at the heart of nature and repeat themselves so often that one can get the impression that earlier people had the same ideas as later ones. Sometimes they have the same *basis*, though with subtle differences in meaning, accounting for the success of people like Newton in absorbing nonscientific ideas like alchemy and theology, and even ancient history, into science. It is hard, for example, to avoid a concept similar to atoms, which certainly existed in ancient times, but which could easily have been lost if a few ancient texts had not been preserved. In the case of atoms, for

---

[7]Snobelen (2004); *Treatise on Revelation*, Yahuda MS 9, mid-late 1680s, NP. Snobelen warns the reader against believing that Newton's stance represents an 'incipient rationalism' (177); it was rather an *exegetical* option (181). It is nonetheless remarkable for its age, and certainly heretical. Newton never set out to be a rationalist or to put forward arguments that (with all their compromising) would appeal to 'reasonable men' — he was an extremist. However, his very extremism led him to create systems that excluded contradictions and anomalies, and demanded consistency. His arguments, hence, often end up by appearing more 'rational' than those derived from (compromised) 'rational principles'.
[8]de la Pryme (1870, 42), cited in Westfall (1980, 502). The incident occurred in spring 1694.
[9]Royal Society, Gregory MS 247; some material published in Gregory (1702); McGuire and Rattansi (1966).

example, it is clear that there are only two options for matter. It must be either continuous or discrete, and with such a stark dichotomy it is almost inconceivable that either of the two options would remain unexplored. Again, we know for certain that *some* ancient wisdom has been lost, for example, Euclid's books on porisms, which Newton tried to restore, and we can only imagine what must have been destroyed in the great library of Alexandria and by the attrition of the centuries.

To restore ancient wisdom in this sense is a perfectly respectable study. Nevertheless, there is a way of approaching 'lost' wisdom which is much more fanciful and related to the old idea that the past was a 'golden age' compared to the present. Sometimes this is linked to ideas that do have a fundamental meaning, but perhaps not in the way that the 'prisca sapientia' enthusiasts would claim. We know, for example, that Platonic solids and Golden ratios were of interest to the Greek geometers and architects because of their innate aesthetic appeal, and the Platonic solids may well have been around in prehistoric times as well. They almost certainly have a part to play in nature at a fundamental level, and serious scientists use them today in studying such things as biological growth patterns and quasicrystals. They are also used, however, by purveyors of 'New Age' wisdom, for ideas with no scientific basis. So the fundamental nature of such concepts by no means guarantees that they are connected with correct ideas, only that their generic nature encourages recognition.

It is in this borderland area between serious science and fanciful myth-making that we find Newton's proposed Scholium to the *Principia*, on the 'Pipes of Pan', in which he claimed that Pythagoras's legendary discovery of a connection between string length and pitch, which provided the harmony of the music of the crystalline spheres carrying the planets round the Earth in the old geocentric cosmology, was really a disguised form of Newton's own inverse-square law of gravitation, which he kept secret, leading Ptolemy to create the Earth-centred system of epicycles.[10] Like all 'conspiracy theories', it is fabricated in a way that is almost impossible to disprove. Because a 'conspiracy' is secret, and involves a necessary difference between appearance and 'reality', the believer is always able to adjust the difference to make up any 'reality' required. Perhaps it could be considered a result of Newton's inability to believe that knowledge is an evolutionary process and that rationality is the result of accumulated wisdom rather than an innate ability possessed by humans in all previous ages. The same attitude would also

---

[10] McGuire and Rattansi (1966): proposed Scholium for *Principia*.

produce bizarre results in areas like history and prophecy though, even here, Newton's tendency to rationalise things that were in origin nonrational led to some conclusions of at least semi-rational interest.

7.2. Theology

Before everything else Newton was a theologian, and a monistic one at that. A universal, all-powerful God was, for him, the source of everything in nature. There could be no compromise on this. He writes in a draft Query:

> An Atheist will allow that there is a Being absolutely perfect, necessarily existing & the author of mankind & call it Nature: & if you talk of infinite wisdom or of any perfection more then he allows to {say}in {natur}heel reccon at a chemæra & tell you that you have the notion of finite or limited wisdom from what you find in your self & are able of your self to {prefin}the word no{t} or more then to any verb or adjective & without the existence of wisdome not limited or wisdome more then finite to understand the meaning of the phrase as easily as Mathematicians understand what is meant by an infinite line or an infinite area.[11]

In all things he was a monist, obsessed with the idea of oneness or unity, though his obsession revealed itself in many different forms. His real ancestor was the late mediaeval philosopher, William of Ockham. Ockham, like his immediate Oxford predecessor John Duns Scotus, has been associated with a 'voluntarist' tradition, in which everything in nature is determined by the will of God acting directly, a metaphysical position which resonates with that of Newton.[12] In fact, in his determination to represent God as absolute and noncontingent, in opposition to the contingencies of nature, Ockham set about reducing the multifarious Aristotelian categories used, by mediaeval predecessors such as Thomas Aquinas, to describe natural things, declaring that 'entities must not be multiplied unnecessarily' (though the phrase itself dates from a much later period). Believing that theological knowledge was based on faith, and not on reason, meant separating faith and reason and excluding natural knowledge from faith. Ockham's God was not to be discovered through the laws of nature. Removing theology from nature also meant becoming an empiricist in natural things, with knowledge acquired from experience.

Newton's theological conceptions were by no means kept separate from his work in physics but were, indeed, the very reason for its existence.

---

[11] Draft Q, CUL, MS Add. 3970.3, 619^r, NP.
[12] Spade and Panaccio (2011).

To see this, we have only to look at the strongly theological tone of *De gravitatione et aequipondio fluidorum*, possibly his first extended essay on a physical subject.[13] And in the General Scholium he writes: 'And thus much concerning God; to discourse of whom from the appearances of things, does certainly belong to Natural Philosophy'. Here also, we find his views on God in words which reveal a man searching for a monist conception of the universe, a secret uncompromising Arian whose ultimate driving force was a theological conception of oneness or unity:

> He is eternal and infinite, omnipresent and omniscient; that is, his duration reaches from eternity to eternity; his presence from infinity to infinity; he governs all things, and knows all things that are or can be done. He is not eternity and infinity, but eternal and infinite; he is not duration or space, but he endures and is present. He endures forever, and is every where present; and, by existing always and every where, he constitutes space.

God is a 'governor' ($\pi\alpha\nu\tau o\kappa\rho\alpha\tau o\rho$), the direct source of physical attractions for which mechanical sources need not be sought, able to move all things 'within his boundless uniform Sensorium'.[14] With this there was the undercurrent of Neoplatonism in the belief in a unifying 'vital spirit', the cause of all physical processes, merging with that of the alchemists, which had originated in Hellenistic Egypt at about the same time, probably from the same source. With Newton, these ideas blended into a very individualistic, but distinctly monistic theology, which denied the Trinity and anything which limited the absolute perfection of the One God.

At Cambridge Newton had been influenced by the Platonist philosopher Henry More, another native of Lincolnshire, who thought that Descartes' views were unnecessarily materialist and left no room for the concept of *spirit*. At the same time, through means unknown but for reasons that are obvious in respect of his restless drive for intellectual stimulus of every kind, he had seemingly become linked to a clandestine group of alchemists operating at the university[15] and had begun a deep study of work whose original conception may have been based on a semi-religious conception of the unity of the physical world and the existence of the Neoplatonists' unifying 'vital spirit'. It was, to some extent, the impression of such philosophies on Newton's naturally analytic mind that led to the extreme unitarianism in

---

[13] *De gravitatione*, Unp., 89–156.
[14] Q 31.
[15] The existence of such a group has been proposed on reasonable grounds but not definitely established.

his theological beliefs which he secretly maintained, at great danger to his person and position in society, for over 50 years.

Unlike his contemporaries in natural philosophy, Newton was a complete Ockhamist, determined never to multiply entities without necessity. It was the mark of all his work as a scientist; as we have seen, he detested the kind of science in which complicated hypotheses were put forward, and then made to fit existing facts by adding further assumptions. His belief in universal abstract laws undoubtedly owed a great deal to his theological inspiration. Underlying causes, he thought, ought to be simple and abstract. It was natural to Newton to push everything to the limits, and it led him at one stage to develop an extremely singular view of religion. He lived in an age when strong religious opinions were relatively normal, at least in the early part of his life, and his whole manner of life was determined by intense religious beliefs, of a kind quite alien to those of the twenty-first century. In this, he reflected his time and was only different from other seventeenth century thinkers in his unwillingness to compromise.

Newton's work in theology, prophecy, history, chronology and archaeology was an intensely private study, not intended for publication. This is a significant fact which should be taken into account when the work is assessed against the science which he did choose to publish. It would certainly have been dangerous to be open about the theology. For Richard S. Westfall, probably no man has been a more assiduous scholar in whatever field he studied; he is reputed to have left over 6 million words in the manuscripts that survive. In theology, he made himself familiar, not only with the works of Athanasius, but also those of Augustine and many other early Church theologians.[16] Once deeply involved in something he became obsessed.

Gradually as he read these authors, he became convinced that 'a massive fraud' had taken place in the vital fourth and fifth centuries.[17] Several passages of Scripture, he was sure, had been deliberately corrupted. In the process he performed remarkable work as a textual scholar, going to the earliest authorities to identify the corrupted passages, and must be regarded as a pioneer of the Biblical textual criticism, which became a major movement a century and a half later. Two particularly significant ones were 1 John 5: 7 and 1 Timothy 3: 16, identified in *An Historical Account of Two Notable Corruptions of Scripture*, a work written in 1690, and sent in manuscript

---

[16] Keynes MS 2, Westfall (1980, 312–313). Westfall lists 22 of these, but states that there were also many others.
[17] Westfall (1980, 313).

to John Locke, but there were more than twenty others.[18] In the orthodox version of 1 John 5: 7, he read: 'For there are those that bear record in heaven, the Father, the Word, and the Holy Ghost: and these three are one'. But, according to Newton's annotation: 'It is not read thus in the Syrian Bible. Not by Ignatius, Justin, Irenaeus, Tertullian, Origen, Athanas, Nazianzen Didymus Chrysostom Hilarius, Augustine, Beda, and others. Perhaps Jerome is the first who reads it thus'.[19] As for 1 Timothy 3: 16, orthodoxy insisted on: 'And without controversy great is the mystery of godliness: God was made manifest in the flesh'. But, according to his reading, the passage should have been: 'And without controversy great is the mystery of godliness, which was manifested in the flesh'. Arius, he came to believe, had been right after all in denying the Trinity, and Athanasius and the people who came after him, who suppressed Arius's beliefs, had perpetrated a terrible deception.[20]

A key word in Newton's argument was the Greek adjective *homoousios*, referring to a Son who was consubstantial with the Father. In a draft history of the Church in the fourth century, he claimed that the opponents of Arius at the Council of Nicea in 325 had intended to show that Scripture justified their conviction that the Son was the eternal *logos*, but the debate had forced them into a corner, and had obliged them to make the Son *homoousios* with the Father, even though the word could not be found in Scripture. 'That is, when the Fathers were not able to assert the position of Alexander [Bishop of Alexandria] from the Scriptures, they preferred to desert the Scriptures than not to condemn Arius'. Athanasius had then distorted the words of the third century theologian, Dionysius of Alexandria, to make it appear that he had been in favour of a term which he had made a point of describing as heretical.[21] For Newton: 'It was ye Son of God wch he sent into ye world & not a humane soul y^t suffered for us. If there had been such a human soul in o^r Saviour, it would have been a thing of too great consequence to have been wholly omitted by the Apostles'.[22]

The extraordinary *Paradoxical Questions concerning the Morals and Actions of Athanasius and his followers*, probably from the early 1690s,[23] is, according to Derek Gjertsen, 'in many ways the strangest of all Newton's

---

[18] *Two Notable Corruptions* (1690/1754), *Corr.* III, 83–129.
[19] Yahuda MS 14, f. 57^v. Westfall (1980, 313).
[20] Keynes MS 2, f. XIII^v. Westfall (1980, 313–314).
[21] Yahuda MS 2.5b, ff. 40v–41. Westfall (1980, 314).
[22] Yahuda MS 14, f. 25. Westfall (1980, 316).
[23] *Paradoxical Questions*, probably early 1690s, NP.

works. A passionate document, it is also instantly readable. Sixteen questions are posed and discussed'. 'In the manner of a modern investigative journalist', Newton sets out 'the charges laid against Athanasius 1,200 years before', charges upon which Athanasius had been tried and acquitted by the imperial court. Newton, however, has 'no difficulty finding him guilty on all counts'. Unusually for Newton, the work is described by McLachlan as 'marked by a vein of irony'.[24]

In the course of his studies of these questions, Newton began intensive reading of the earlier Platonist theologians on which Arius had drawn for his view of God as an unapproachable Lord of Creation, for whom Christ was the necessary mediator with mankind, taking note of passages from such as Clement of Alexandria: 'God is not divided, not disjointed, not moving from place to place, any way circumscribed, but existing everywhere always...'; and Novitian: 'immense and without limit, not one who is enclosed in a place but one who encloses every place...'.[25]

The words 'in heaven, the Father, the Word, and the Holy Ghost: and these three are one', in 1 John 5:7, which gave support to the doctrine of the Trinity which he opposed, did not appear in the original Greek text. He showed in *Two Notable Corruptions* that the words, which appeared for the first time in the third edition of Erasmus's New Testament, crept into the text of Latin versions after being added as marginal: 'the Æthiopic, Syriac, Arabic, Armenian, and Slavonic versions, still in use in the several Eastern nations, Ethiopia, Egypt, Syria, Mesopotamia, Armenia, Muscovy, and some others, are strangers to this reading', he said. 1 Timothy 3:16, had 'God was manifest in the flesh' in the King James version, as opposed to 'He was manifest in the flesh'.[26] Newton did not publish the work in his lifetime for fear of being accused of heresy, and it was not until the nineteenth century that editions of the Bible began to correct the passages. It is now thought that the first corruption originated in a late fourth century gloss by the Spanish theologian Priscillian which found its way into Jerome's Latin Vulgate, while the second appeared first in an early sixth century Greek version as an interpolation added by Macedonius, patriarch of Constantinople. Newton's textual scholarship, like much of his work in science, has apparently stood the test of time.

---

[24] Gjertsen (1986, 423). McLachlan (1950, 60).
[25] Keynes MS 4, ff. 14 and 41. Westfall (1980, 318).
[26] *Two Notable Corruptions* (1690/1754, 25).

We may well be intrigued by what kind of religion was supported by this intense study of biblical sources, and we might well be surprised to find out how much Newton applied his abstract rationalising even in this area, at least in his earlier theological writings, though the results may be hard to imagine as 'religion' in any sense that would be familiar to us, and certainly not as regular Christianity. His religious views were undoubtedly a strangely abstracted exercise rather than a matter of emotional conviction, though he did set about to practise what he professed in terms of moral behaviour. Passionately committed as he was, he was an intellectual not a fundamentalist, and he had a notable lack of interest in 'enthusiasm' or the evangelical aspect of Christianity.[27] Passion only came into his religious writing, it has been supposed, at the odd moments when everyone was stirred up, especially against Roman Catholics. One of the most famous instances in his lifetime was the so-called 'Popish plot' of 1679, when mass hysteria, aroused by the claims of a charlatan to have discovered a large-scale anti-state conspiracy, led to the deaths of many innocent Catholics. There is, however, no known reference by Newton to the Popish plot in the whole of his voluminous works.[28]

It may be impossible to get inside the mind of a man whose cultural background on such issues was totally different from our own, or to develop imaginative sympathy for views which seem at odds with genuinely rational thought, but, if not rational in our sense, there seems to have been a rationale behind Newton's viewpoint. In the same way as he sought to make science universal, he seems to have been driven also towards seeking out a universal type of religion. He came up with a creed in which Christ was the 'divine mediator between God and man ... subordinate to the Father Who created him'. He was not the union of divine with human, but 'the created *logos* incarnate in a human body', a human form 'animated by a divine or semidivine spirit', exalted by his suffering to sit at God's right hand: 'to say there are two Gods would suppose them collaterall and univocall'.[29] Jesus, however, 'was not an ordinary man but incarnate by the almighty power of God & born of a Virgin without any other father than God himself'.[30] 'Though created by God in time, Christ existed before the world began.' He was the 'spirit of prophecy ... the angel of God who

---

[27] Iliffe (2016, 488).
[28] Westfall (1980, 345). Newton, in fact, hardly ever discusses contemporary political or religious affairs in any of his writings, published or unpublished (Iliffe and Smith, 2016, 29).
[29] Yahuda MS 14, f. 173, Westfall (1980, 317).
[30] Westfall (1980, 315, 324). Yahuda MS 15.3, f. 146.

had appeared to Abraham, Jacob and Moses, and who had guided Israel in the days of the judges.'[31]

It is conceivable, as was asserted by Betty Jo Teeter Dobbs, in *The Janus Faces of Genius*, that Newton associated Jesus as *logos* or mediator between divine and human with the universal vital agent described in *Of Natures Obvious Laws*. God, acting in the world perceived by humans, but unapproachable by them, would always act through an intermediary who had access to Him: 'God does nothing by himself wch he can do by another'.[32] But mental association of one idea with another does not necessarily imply a single coherent picture incorporating both. Newton was generally an unconscious synthesiser. His conflation or syncretism of ideas from various sources was mostly unprogrammed. Ideas developed by association into a recursive pattern but the creation of a unified vision came on a more selective basis, and, though theology may have been the ultimate driving force behind Newton's many intellectual quests, this does not necessarily mean that he believed he had created a complete system based on it. His theology was an end to be achieved, rather than a system to be imposed on other sources of knowledge. Its all powerful creator God provided a final, but not immediate, explanation of the universe and its processes.

It is, therefore, an error, in my view, to portray Newton as a monolithic figure driven only by bizarre theological concerns, and only inventing modern science by accident in the process, even though the theology was clearly extremely important to him. He spent more than thirty years of his life, after all, working for the totally secular and worldly institution of the Mint, when he was independently wealthy and not in financial need of such employment. His various manuscripts, written in the totally different styles and terminologies appropriate to widely different areas of study suggest that he was easily able to compartmentalise his activities,[33] and his work in all areas of science suggests a strong degree of rationalism and even an extreme degree of rationalising in his accommodation of religious texts to a scientific explanation. If anything, the rationalism tended to increase as he grew more confident in the scientific basis of his work.

The linking of science and religion was not in any way unusual for the age in which Newton lived or even for a long time afterwards, and heterodox religious opinions were not particularly rare in the seventeenth century. Newton pushed his investigations in these areas to extremes, but he did

---

[31] Westfall (1980, 824).
[32] Dobbs (1991); Yahuda MS Var. 1, Newton MS 15.4, f. 67^r, quoted on p. 36.
[33] See Iliffe and Smith (2016, 28–30) on this.

the same in areas such as mathematics and natural philosophy, which we now regard as purely rational. The majority of people who met Newton in daily life clearly didn't see him as acting like a Messianic prophet. He didn't behave like a man possessed except in the two-year period when he was writing the *Principia*. He carried on as a public servant and relatively normal member of society who spent his spare time occupied with studying the things that interested him.

We don't, in fact, have the evidence to suggest that Newton's scientific work was simply a component in a grandiose plan with theology, or, in an even more extreme version, theological alchemy, at the centre. He naturally made associations between different areas of activity because this was the way his mind and his creative thinking tended to work, but this is not necessarily evidence that one activity was driven by another. Relating how an individual of such restless intellect would push his investigations in all areas of interest to him to the limit is a totally different thing to proposing that anyone who acts in this way should be portrayed as a man with a mission whose every action is determined by this end. Authors who imagine that the career of anyone as complex as Newton can be reduced to a single explanatory hypothesis are likely to end up generating a spectacular thesis which turns out to be just another self-fulfilling prophecy.

In about 1680, he began his most important theological work, *The Philosophical Origins of Gentile Theology* (*Theologiae gentilis origines philosophiae*),[34] which he later extended in a work called the *Irenicum*.[35] These expressed his theological views at the time of the *Principia*, but he seems to have moved away from them at a later date.[36] The most ancient type of religion, he asserted in these works, was that associated with the Vestal cult; 'the rationale of this institution was that the God of Nature should be worshipped in a temple which imitates nature, in a temple which is, as it were, a reflection of God'.[37] Religion, in this sense, was scientific or 'philosophical' in origin. Newton went on to great lengths to reconstruct the plan and dimensions of the Temple of Jerusalem, learning Hebrew to read the relevant passages in Ezekiel. He devised an argument to determine the exact length of the sacred cubit. He thought that understanding Jewish ceremony and worship would produce a greater understanding of the prophecies.

In his *Irenicum*, he claimed that: 'All nations were originally of one religion & this religion consisted in the Precepts of the sons of Noah...'

---

[34] *Origins of Gentile Theology*, Yahuda MSS, NP.
[35] *Irenicum*, NP.
[36] Iliffe (2016, 510); Iliffe and Smith (2016, 21).
[37] Yahuda MS 17.3, ff. 9–11. Westfall (1980, 354).

Based on love of God and love of neighbour, this religion was passed from Abraham, Isaac and Jacob, and brought to Israel by Moses. It was the one that Pythagoras learned during his travels and passed on to his disciples. It was the same one that was taught by Socrates, Cicero, Confucius and other philosophers. 'This religion', he proclaimed, 'may be therefore called the Moral Law of all nations'. When Jesus answered that the two great commandments were to love God and to love your neighbour: 'This was the religion of the sons of Noah established by Moses & Christ & still in force'. These two commandments 'always have & always will be the duty of nations & The coming of Jesus Christ has made no alteration in them'.[38] 'Thus you see there is but one law for all nations the law of righteousness & charity dictated to the Christians by Christ to the Jews by Moses & to all mankind by the light of reason & by this law all men are to be judged at the last day'.[39] To persecute people now for anything added beyond the beginning was making 'war on Christ'.[40]

'The true religion', he wrote in the first chapter of a history of the church,

> was propagated by Noah to his posterity, & when they revolted to ye worship of their dead Kings & Heros & thereby denied their God & ceased to be his people, it continued in Abraham & his posterity who revolted not. And when they began to worship the Gods of Egypt & Syria, Moses & the Prophets reclaimed them from time to time till they rejected the Messiah from being their Lord, & he rejected them from being his people & called the Gentiles, & thenceforward the beleivers both Jews & Gentiles became his people.[41]

That was his view of early Church history. The true religion had been corrupted in Egypt after the time of Noah by ancestor cults. All later nations had adopted the same set of 'divinised ancestors', whom they worshipped under different names, but who were twelve in number. The chief among these was the 'old man' considered as the founding ancestor, variously called Noah, Saturn, Janus, and other names.[42] It was the origin of his belief that other ancient peoples had legends of a flood similar to Noah's,[43] a belief which would, in a sense, be vindicated by George Smith's discovery in 1872, of the description of a flood very like that in Genesis inscribed on cuneiform

---

[38] *Irenicum*, Keynes MS 3. Westfall (1980, 820–821).
[39] *A short Schem of the true Religion*, Keynes MS 7, 2–3. Westfall (1980, 822).
[40] Westfall (1980, 821).
[41] Yahuda MS 15.3, f. 57. Westfall (1980, 820).
[42] Westfall (1980, 352).
[43] Westfall (1980, 355).

tablets from Nineveh containing parts of the Babylonian epic of Gilgamesh.[44] It was also the origin of an otherwise inexplicable passage which Newton placed at the end of his book on *Opticks*.

Newton was more interested in abstract notions than in ethics, but he thought that the perfection of natural philosophy would lead to an improvement in moral philosophy — not a bad argument — and said so here in his scientific work. He wrote in Query 31:

> And if natural Philosophy in all its Parts, by pursuing this Method, shall at length be perfected, the Bounds of Moral Philosophy will also be enlarged. For so far as we can know by natural Philosophy what is the first Cause, what Power he has over us, and what Benefits we receive from him, so far our Duty towards him, as well as that towards one another, will appear to us by the Light of Nature. And no doubt, if the Worship of false Gods had not blinded the Heathen, their moral Philosophy would have gone farther to the four Cardinal Virtues; and instead of teaching the Transmigration of Souls, and to worship the Sun and Moon, and dead Heroes, they would have taught us to worship our true Author and Benefactor, as their Ancestors did under the Government of *Noah* and his Sons before they corrupted themselves.[45]

He implied that idolatry had not only produced false theology, but had also led to false natural philosophy. Ancient peoples, following the most ancient form of religion, had represented the structure of the heavens by what Newton called a 'prytaneum', a sacred circular space with a fire at the centre representing the Sun, a temple *representing* the whole universe as 'ye true & real temple of God'[46], an idea stated in many ways by his predecessors both ancient and modern, and with some justification in historical fact.[47] However, men of many nations were 'led by degrees to pay a veneration to their sensible objects', rather than the true God that they *represented*,

> & began at length to worship them as the visible seats of divinity. <ffor tis agreed that Idolatry began in ye worship of the heavenly bodies (*deleted*)> And because ye sacred fire was a type of ye Sun & all ye elements are part of that universe wch is ye temple of God they soon began to have these

---

[44]George Smith's findings were first presented to the Society of Biblical Archaeology on 3 December 1872, and immediately drew attention to the close similarities between the Babylonian account and Genesis. Smith later found Babylonian origins for 'the traditional Old Testament stories of the creation, the fall of man, and the Tower of Babel', summarizing his findings in Smith (1876). (See Dean (2004)).
[45]Q 31.
[46]Yahuda MS Var. 1, Newton MS 41, Dobbs (1991, 151).
[47]This issue is discussed extensively in Dobbs (1991, 150–166).

also in veneration. For tis agreed that Idolatry began in ye worship of ye heavenly bodies & elements.[48]

Though idolatry would progress at a later date to the worship of dead ancestors, of animals, and even of their inanimate representations, the first stage involved twelve gods which 'were the seven Planets with the four elements and the quintessence of the Earth'.[49] In Dobbs's interpretation, alchemy, which was permeated with these associations, had come down only in a corrupted form, and Newton's search for the deepest truths in the alchemical authors should be seen as an attempt at transcending the accumulated idolatrous associations to 'focus on 'the fire at the heart of the world' in micromatter',[50] and recover its lost secrets.

The same applied to cosmology. Preaching in Trinity College Chapel on 2 Kings 17.15, 16, the text dealing with the idolatry of the Golden Calf, Newton asserted that 'the study of the frame of the world as the true Temple of ye great God they worshipped.... For there is no way (wthout revelation) to come to ye knowledge of a Deity but by ye frame of nature'.[51] The Earth-centred astronomy of the Egyptian, Ptolemy, he held, was a product of the imposition of a false religion — an interesting contrast, indeed, to the Roman Church's view on Galileo's attempt at replacing it with a Sun-centred system. The Christian religion was, for Newton, no more true than that which had been held by the sons of Noah, and he saw the same pattern of corruption repeating itself with Athanasius, again, of course, in Egypt. For him, the second coming of Christ was more important than the first. This would not be so much the ending of the physical world in a 'cataclysmic destruction', as 'ye establishing of true religion' and 'the preaching of ye everlasting gospel to every nation & tongue & kindred & people'.[52]

While Newton mostly abandoned scientific work after 1717, he continued to work on his studies of theology, prophecy and chronology to the end. We have been conditioned by several centuries of advancing secularism, and it is difficult to imagine a time when people could see science and religion as so closely connected. Of course, people are mainly anxious to

---

[48]Yahuda MS Var. 1, Newton MS 42, f. 8^r, quoted in Dobbs (1991, 161).
[49]Yahuda MS Var. 1, Newton MS 16.2, f. 1, quoted in Dobbs (1991, 161).
[50]Dobbs (1991, 165). Dobbs may overstate the link between alchemy and theology for Newton, but, in his later years Newton produced detailed accounts of the corruption of early Christianity by the gnostics, whom he called 'Chymical Cabbalists' (Iliffe, 2016, 514–517).
[51]Yahuda MS 41, ff. 6–7. Westfall (1980, 355).
[52]Yahuda MS 1.3, ff. 35–36. Westfall (1980, 330).

show God's providence when they live in a sceptical or materialist culture. Westfall thought that Newton's rationalising of religion was an attempt to preserve it by cutting out the parts that looked increasingly indefensible in a rapidly secularising culture.[53] By 1700, the general attitude of society to religious matters was very different to what it had been in 1650. The increasing urgency of the arguments from design by such as Boyle, Ray, Newton, Whiston and Derham would not have happened in an age like the mediaeval period when faith was unquestioned; the very existence of such arguments suggest that they were part of a receding tide.

7.3. Chronology and Prophecy

In his old age Newton worked hard on his prophecy and chronology, the chronological research largely taking over from that on the true primitive religion.[54] His *Chronology of Ancient Kingdoms Amended* had to be squared with the account of the development of idolatry and paganism among the descendants of Noah, so he had to take 400 years somehow out of Greek history, using arguments involving astronomy and the probable regnal lengths of kings.[55] Newton wanted to show the antiquity of Jewish culture with respect to those of other ancient peoples, and he placed the construction of Solomon's Temple in 1015 BC, a little before the conventional (but by no means established) date of the mid-tenth century. The Jews (who had been allowed to return to England in the 1650s) were the people chosen to carry God's message, and it was not unusual to see them as having the greatest historical antiquity. The Israelite civilisation had to be seen to precede all pagan ones, or, at least, such pagan ones had to be seen to descend from the sons of Noah.[56] Because Newton concealed the theological truth behind his calculations, or no longer saw it as central, the work as published had no obvious point;[57] and was quickly attacked by rivals anxious to push their own chronological schemes. Ironically, though, some of his views on Greek chronology have now been resurrected, though not convincingly established, by serious, though not mainstream, scholarship.

---

[53] See, for example, Westfall (1980, 826).
[54] Iliffe (2016, 510); Iliffe and Smith (2016, 21).
[55] *Chronology* (1728).
[56] Newton did concede that the Egyptians had attained civilisation before the Jews, at 1125 BC compared to 1059 BC for King David, but he claimed that the Israelites had writing first.
[57] Westfall (1980, 805).

As Frank Manuel has written: 'The habits of the Master of the Mint were not absent in the Bible commentator and the chronologist'.[58] On probability grounds, Newton argued that the assignment of reign lengths for kings of 35–40 years by Greek chronologers was 'much beyond the course of nature'; 18–20 years was much more probable on statistical grounds.[59] According to O. B. Sheynin: 'In his reasoning Newton used the same idea which underlies the law of large numbers: the random quantity (the time interval) should be approximately equal to its mathematical expectation'.[60] Though he had no general statistical rule for his assignment, his work was quoted by later statisticians such as Condorcet, Ellis and Pearson.[61] Newton also argued against the possibility of any ancient peoples using a calendar of exactly 360 days without correcting it from time to time by lunar or solar observations. With typical knowledge of the ancient sources, he showed that none of the calendars he examined had this structure. If they had, 'The beginning of such a year would have run around the four seasons in 70 years, and such a notable revolution would have been mentioned in history, and is not to be asserted without proving it'.[62]

He was sceptical, in another context, about the authenticity of the dates at which Christian festivals were celebrated, arguing, correctly, that the Nativity had been fixed at December 25 to coincide with the winter solstice, and that other festivals had been arranged to take the place of the cardinal points of the pagan year. He says in the *Observations on the Prophecy of Daniel and the Apocalypse of St. John*:

> The times of the Birth and Passion of Christ, with such like niceties, being not material to religion, were little regarded by the *Christians* of the first age. They who began first to celebrate them, placed them in the cardinal periods of the year; as the annunciation of the Virgin *Mary*, on the 25th of *March*, which when Julius *Cæsar* corrected the Calendar was the vernal Equinox; the feast of *John* Baptist on the 24th of *June*, which was the summer Solstice; the feast of St. *Michael* on *Sept.* 29, which was the autumnal Equinox; and the birth of *Christ* on the winter Solstice, *Decemb.* 25, with the feasts of St. *Stephen*, St. *John* and the *Innocents*, as near it

---

[58] Manuel (1963, 10).
[59] *The Original of Monarchies* (MS 1702); *Chronology* (1728, 52).
[60] Sheynin (1972).
[61] Todhunter (1865), §746; Ellis (1844), see *Works*, 13. 179; Pearson (1928, 424). These citations are by Sheynin (1972, 221).
[62] 'A Letter from Sir Isaac Newton to a Person of Distinction, who had desired his opinion of the learned Bishop Lloyd's Hypothesis concerning the Form of the Most Ancient Year', *Gentleman's Magazine*, 25, 3–5, 1755 (written in 1713 to Humphrey Prideaux, Dean of Norwich).

as they could place them. And because the Solstice in time removed from the 25th of *December* to the 24th, the 23d, the 22d, and so on backwards, hence some in the following centuries placed the birth of *Christ* on *Decemb.* 23, and at length on *Decemb.* 20: and for the same reason they seem to have set the feast of St. *Thomas* on *Decemb.* 21, and that of St. *Matthew* on *Sept.* 21. So also at the entrance of the Sun into all the signs in the *Julian* Calendar, they placed the days of other Saints; as the conversion of *Paul* on *Jan.* 25, when the Sun entred ♒ [Aquarius]; St. *Matthias* on *Feb.* 25, when he entred ♓ [Pisces]; St. *Mark* on *Apr.* 25, when he entred ♉ [Taurus]; *Corpus Christi* on *May* 26, when he entred ♊ [Gemini]; St. *James* on *July* 25, when he entred ♋ [Leo]; St. *Bartholomew* on *Aug.* 24, when he entred ♍ [Virgo]; *Simon* and *Jude* on *Octob.* 28, when he entred ♏ [Scorpio]: and if there were any other remarkable days in the *Julian* Calendar, they placed the Saints upon them, as St. *Barnabas* on *June* 11, where *Ovid* seems to place the feast of *Vesta* and *Fortuna*, and the goddess *Matuta*; and St. *Philip* and *James* on the first of *May*, a day dedicated both to the *Bona Dea*, or *Magna Mater*, and to the goddess *Flora*, and still celebrated with her rites. All which shews that these days were fixed in the first *Christian* Calendars by Mathematicians at pleasure, without any ground in tradition; and that the *Christians* afterwards took up with what they found in the Calendars.[63]

Newton also introduced advanced astronomical methods into chronology, attempting to base ancient chronology on the precession of the equinoxes. Frank Manuel writes:

> ... there was no sharp distinction in Newton's mind between the physical history of the universe and the history of nations, since both histories could be learned by man and there were continual correspondences between them. In his monist system a chronological event in the history of kingdoms could be translated into an astronomical event in the physical history of the universe and vice versa, for there were parallel histories in the heavens and on earth.[64]

The methods could also be used for much more recent history. Newton's colleague, Edmond Halley, for instance, sought, in 1691, to establish the exact date and place of Julius Caesar's first landing in Britain, based partly on an eclipse of the Moon. Newton also used his methods to fix the date of a

---

[63] *Observations* (1733), opening paragraph of Chapter XI. The zodiac names are symbols in the text. The sign for Cancer (♋) mistakenly replaces that for Leo (♌).

[64] Manuel (1968, 377). Though astronomical events had played a part in previous work in cosmology, Manuel (1963, 191), considered that Newton's dating of events 'through computations of the equinoctial precession' was 'an act of sheer genius', which could even have an effective modern application if relevant ancient astronomical observations should 'come to light'.

significant recent event.[65] In his *Observations on the Prophecy of Daniel and the Apocalypse of St. John*, he set the date for Christ's crucifixion as 7 April 34 AD, with 3 April 33 AD as second choice.[66] The most serious available scholarship on this issue seems to make the latter date the preferred one.[67]

Newton fixed a new date for the Trojan war, which would significantly affect Egyptian chronology. The date was fixed by its close proximity to the mythical voyage of Jason and the Argonauts which could be dated to c 936 BC by a change in precession of seven degrees from the 'fifteenth degree' for the solstices and equinoxes in the time of the mythical Chiron to the 'eighth degree' for the summer solstice of the historical Meton. Hipparchus, writing in c 135 BC, had found that the positioning of the constellations on the celestial sphere given by Eudoxus two centuries earlier were no longer correct. Assuming that this was because of the precession of the equinoxes, Newton calculated (though not without further assumptions that were quickly attacked by his opponents) that the celestial globe described was made in 939 BC. If he supposed that it was constructed by the astronomer Chiron, who had been contemporary with the Argonauts, he could date the Argonauts' expedition. With the fixing of the date of the Argonauts, the fall of Troy was placed around 904 BC, considerably later than generally supposed, with no Greek Dark Age following it. Newton thought that the antiquity of Egyptian history had been exaggerated by several centuries, as had those of all nations other than the Jews. He claimed that 'the Sesostris of Herodotus, whose conquests were the same as those of Tuthmoses III, was the Biblical Shishak'.[68] The telescoping of the chronology meant that he also had to provide a revised account of the growth of the world population from the eight people who survived Noah's Flood. Clearly, many earlier estimates of antiquity were absurdly high, in any case, as Newton stated at the beginning of his work.[69] While the majority of scholars support the conventional chronology, and seem to have the evidence on their side, there is a respectable minority of revisionist archaeologists who favour a reduction in length of the sort proposed by Newton, though for a variety of different reasons.

---

[65] Halley (1691).
[66] *Observations* (1733), Part I, Chapter XI; 1922 ed., 235–250, especially 250. Newton's argument for AD 34 was based on the fact that the Gospels described the Passion as having occurred two years after Jesus and his disciples had walked through fields of ripe corn at the time of the Passover.
[67] Pratt (1991).
[68] *Chronology* (1728).
[69] Buchwald and Feingold (2012), Chapter 5. See also Feingold (2016).

The Trojan war may well have been a real event. While the voyage of Jason and the Argonauts almost certainly never happened, the fall of Troy probably did — at least one of the cities at the site now securely identified as Troy (VIIa) shows evidence of war destruction (dated to the 1180s BC) — so this is of some interest. Whether such an argument could ever be thought to have any validity is another matter. The kind of intuitive connections which had served Newton well when they were based on genuinely objective scientific data and a method of observation which eliminated prior prejudice had exactly the opposite effect when the only available data was a series of unreliable texts with a multiplicity of interpretations, and a prior prejudice which he couldn't eradicate.[70] The actual chronology Newton arrived at, with its telescoping of Egyptian history, is, as we have said, a minority view, based on a large number of arguments that we would now consider arbitrary and on sources that are, by modern standards, totally unhistorical, but it has not been totally rejected. Such contemporary scholars as Peter James and David Rohl, for example, have pursued similar adjustments to Egyptian history, though from a rather different perspective.[71]

Although Newton's actual chronology may be massively defective in many respects, and irreparably flawed in its basic assumptions, it represents one of the earliest serious attempts to synchronise biblical and gentile history on scientific principles, and it would have been almost impossible for *any* scholar to have devised a correct chronology with the information available at the time. In fact, even after archaeology was established as a scientific subject, prior prejudice determined the relative chronologies of different civilisations until well into the twentieth century. Only when radiocarbon and dendrochronology were able to apply a completely objective method of dating were the prejudices that all civilisations diffused from a single centre with a 'superior' culture finally eradicated.[72]

---

[70] One of the major flaws of his work was his reliance on ehemerism, or the belief that mythological deities had really existed as humans at an early era.

[71] James *et al.* (1991); Rohl (1995), (2002). Rohl agrees with Newton's identification of Sesostris with Shishak, but claims that Sesostris was actually Ramesses II (1995, 400). Majority opinion does not support Newton's prejudice, based on his theological views and historical account of the corruption of religion, in favour of the relative antiquity of the Hebrews with respect to the Mycenean Greeks.

[72] Buchwald and Feingold (2012), 1, cite a case from as late as 2006 in which objective information of this kind has invalidated a consensus view that the Minoan civilisation in Crete could have overlapped and maintained contact with the New Kingdom in Egypt. (It is an interesting irony that scientific methods in modern chronology depend so much on the modern form of Newton's 'alchemy'!)

The point of the *Chronology* was partially lost because of its exclusion of the now discarded earlier views on theology which seemingly had been its original instigation. Westfall called it: 'A work of colossal tedium... read today only by the tiniest remnant who for their sins must pass through its purgatory'.[73] Rupert Hall has a different, but equally striking, take: 'A mind as remarkable for its clarity as its fullness is at work on every page of the *Chronology*. Only hindsight makes the book read like an immensely clever fairy story, as if it were the work of a seventeenth century Tolkien'.[74]

Buchwald and Feingold, who considered the work 'stupefyingly tedious', believed, nevertheless, that it 'developed a novel theory of the origin of civilization'. 'Human populations, [Newton] decided, had to expand according to certain rules that implicate particular stages in the development of civilization'. In the words of the reviewer, Jonathan Rée, Buchwald and Feingold argued 'that it may be the source of one of the formative ideas of our time: that every society must pass through the same stages — savagery, pastoralism and agriculture — on the way to civilised maturity', leading to the claim that 'Newton's mature vision of the early history of mankind heralded the stadial method of societal progress, which would become the hallmark of such Enlightenment figures as Turgot and Adam Smith'.[75] In their view Newton had applied 'scientific' reasoning to chronology in the same way as he had to natural philosophy, especially in his scepticism of written sources (which had been enhanced by his examination of unreliable witness statements during his years at the Mint) and his use of the novel averaging technique that he had employed in optics and astronomy to convert multiple unreliable observations into a single reliable one.[76]

For Newton and other contemporary thinkers, chronology went hand-in-hand with prophecy. Works such as Revelation were of great interest at the time, the Roman Church often representing the Beast as a result of the great apostasy of c 400 (which happened to coincide, more or less, with the Athanasian incident); the vision of the 1260 days was sometimes taken as 1260 years, leading, rather conveniently to the Restoration of the Church of England in 1660. But not for Newton; he didn't see the Church of England in this perspective at all, and he never took prophecy as being relevant to

---

[73] Westfall (1980, 815).
[74] Hall (1992, 344).
[75] Buchwald and Feingold (2012, 5), 428, 5; Rée (2013); Feingold (2016, 528–529).
[76] Manuel (1963) says that 'Newton's work as a historian, seen amid the compilations and 'synchronisms' of his own epoch, makes a decent showing'. (191) 'He was on the road to positivistic historiography'. (192).

contemporary events. As he believed that the great apostasy had reached its peak only in 607, the great event for him was pitched about 200 years into the future, and maybe well into the third millennium.[77] Ruthless and systematic pursuer of the truth as he was, Newton set about organising information on historical events like the barbarian invasions of Europe in a way that no subsequent historian has ever equalled.[78] His scholarship and textual analysis were impeccable, his approach as rational as could be accommodated within his programme, and based on historical sources as good as could be found. His methodology was not significantly different from that of modern biblical scholars.

Prophecy for him was not intended to foretell the future, but was intended to be a record which could be used retrospectively to show the workings of God's providence in human history. Writing of the Apocalypse in *Observations on the Prophecy of Daniel and the Apocalypse of St. John*, he says:

> The folly of Interpreters has been, to foretel times and things by this Prophecy, as if God designed to make them Prophets. By this rashness they have not only exposed themselves, but brought the Prophecy also into contempt. The design of God was much otherwise. He gave this and the Prophecies of the Old Testament, not to gratify men's curiosities by enabling them to foreshadow things, but that after they were fulfilled they might be interpreted by the event, and his own Providence, not with the Interpreters, he then manifested thereby to the world. For the event of things predicted many ages before, will then be a convincing argument that the world is governed by providence.[79]

Most of the book is concerned with matching up the prophetic utterance with events that took place in earlier times, mainly during the period of the Roman Empire. There is comparatively little about what might happen in the future.

Newton was reluctant to give dates for future events — most of the dates have to be worked out by the reader, and are not so much intended as a prediction as evidence that they would happen at some time in the distant future, rather than in his own lifetime, as many of his contemporaries believed. They were minimal rather than actual. Earlier prophecies were interpreted as referring to real events in figurative language. Prophets were members of an 'Elect', people who had the responsibility of revealing the

---

[77] Yahuda MS 7.3g, f. 13^r; cf. MS 7.3i, f. 54^r. Snobelen (1999, 391–392).
[78] Westfall (1980, 329).
[79] *Observations* (1733), Part II, Chapter I, 1922 ed., 305–306.

workings of God's providence; but this was a relatively normal occurrence. To be a prophet did not confer special status, as it did in antiquity. Newton thought of himself as one of this group of people, who might, in modern terms, be compared to those people in a purely secular society who, realising they have special talents, see themselves as having some mission in some such field as politics, science or the arts.

While Newton was concerned to show that the apocalyptic language of prophetary texts could be interpreted via the relatively normal events of human history, he at the same time believed that the Holy Temple in Jerusalem contained the structure of events beyond history, after the Second Coming. This was why it was important for him to try to discover every detail of the Temple's construction. He made sketches of the Temple, drawing heavily on Ezekiel and subsequent commentators, and tried to find its exact dimensions, believing that it contained a key to the universe revealed by his physical investigations. He also thought that Stonehenge, which he never visited, held similar secrets.

> In England near Salisbury there is a piece of antiquity called Stonehenge which seems to be an ancient Prytanæum. For it is an area compassed circularly with two rows of very great stones with passages on all sides for people to go in and out at. Tis said that there are some pieces of antiquity of the same form & structure in Denmark. For its to be conceived that the Vestal Temples of all nations as well as of the Medes & Persians were at first nothing more then open round areas with a fire in the middle, till towns & cities united under common councils & built them more sumptuously. In Ireland one of these fires was conserved till of late years by the Moncks of Kildare under the name of Briget's fire & the Cænobium was called the house of fire.[80]

Inigo Jones had previously suggested that Stonehenge was an ancient temple in *The Most Notable Antiquity of Great Britain Vulgarly called Stone-Heng on Salisbury Plain Restored* (1655), as had John Webb in *A Vindication of Stone-Heng Restored* (1665).[81] It is interesting that Newton's friend, William Stukeley, later visited and wrote on Stonehenge, discovering some of its actual secrets.[82]

Newton thought that temples showed that ancient peoples believed that the Sun was the centre of the universe, and he found the size of the royal cubit by which they were constructed. An 'ingenious set of hypotheses by Newton' determined both 'the ancient cubit of Memphis' and 'the sacred

---

[80]'The Original of Religions', Yahuda MS 41, fol. 3r.; quoted Morrison (2011, 56).
[81]Jones (1655); Webb (1665).
[82]Stukeley (1724, 1740).

cubit of the Jews', the scale of the latter being used in plate III, p. 346, of Newton's *Chronology of Ancient Kingdoms Amended*.[83] According to George Sarton, his calculation of the royal cubit contained three fundamental errors, but still came amazingly near to the correct value, at 527.9 mm (or 20.79 inches) by comparison with the 'true' length of about 523–525 mm: 'In spite of the fact that his argument was based upon three wrong premises, he found a length for the royal cubit which was incredibly close to the correct one!'.[84] This was the first investigation of the standards of length in the ancient world. Newton's only predecessor in publishing on the 'mathematical side of Egyptian archaeology' was Hans Gram, author of *De origine geometriae apud Aegyptios*, published at Copenhagen in 1706, though Newton's work, beginning in the 1680s, certainly predated this publication and had no connection with it.[85]

Newton's remarkable work as an architect and architectural historian is recovered in Tessa Morrison's reconstruction of his two- and three-dimensional plans for Solomon's temple from Babson MS 0434.[86] For Kim Williams, who has written the Foreword to Morrison's *Isaac Newton's Temple of Solomon and his Reconstruction of Sacred Architecture*, it is 'even more fascinating' that 'Newton studied Vitruvius and derived his own Vitruvian man' than that he tried to discover the size of the sacred cubit. 'Thus Newton is given his rightful place in the tradition of architectural thinkers such as Alberti and Leonardo. Further, his criticism and revision of [Juan Battista] Villalpanda's reconstruction of Solomon's temple shows him to be an acute architectural analyst'.[87]

Like others of his time, Newton saw the Scriptures as concealing a secret code. The prophets — and that included other writers from the Near East, such as the Chaldeans, and even the pagan Greeks, as well as the prophets of the Bible — all wrote in the same language, based on 'the analogy between the world natural, and an empire or kingdom'.[88] The language was

---

[83] *A Dissertation upon the Sacred Cubit* (1737); *Chronology* (1728), plate III, 346; Manuel (1968), 375 and 464, n. 30.
[84] Sarton (1936). This was in the *Dissertation*, where his value for the *sacred* cubit was 24.6 inches. In the *Chronology*, Chapter V, the sacred cubit is 'about $21\frac{1}{2}$, or almost 22 inches'. In the *Prolegomena ad Lexici Prophetici* (Babson MS 434) it is between 31.2 and 33.6 inches. (Gjertsen, 1980, 551).
[85] Gram (1706).
[86] Morrison (2011); Babson MS 0434.
[87] Kim Williams, Foreword to Morrison (2011).
[88] *Observations* (1733), Chapter II 'Of the Prophetic Language', 16. Also, *The First Book, Concerning the Language of the Prophets*, c 1680, KCC, Keynes MS 5, I–VI, 1–6, quoted McLachlan (1950, 119).

not specific to individual prophets, such as Daniel or John. According to Newton's *Rules for Interpreting the Words and Language in Scripture*, these analogies followed specific rules, similar to those outlined in the *Principia* for explaining physics. According to these rules, all available texts should be considered, each symbol should have a unique meaning, prophecies should relate only to important events of the age, and the simplest and most literal interpretations should be sought wherever possible.[89]

Newton had remarkably few books of prophecy in his library, and, except for the works of Joseph Mede and Henry More, relied mostly on his own interpretations. His *Observations* have been spoken of with respect in certain quarters, though some of the more radical aspects of his prophetic writing were left out by his posthumous editors. According to a contemporary 'historicist' source, in helping to establish the Protestant historicist interpretation of Bible prophecy, Newton, 'the heir to Mede's methodology', 'extended Mede's syllogistic logic into a completed interpretive system which would stand the test of time'.[90] Gjertsen considers it more analogous to Claude Lévi-Strauss's attempts to understand the mythology of the Amerindians in his *Mythologiques* (1964–1971) than to contemporary attempts to decipher the hidden meanings in a writer like Nostradamus.[91] It is certainly less of a pseudoscience than some of the Freudian analysis that has been used to explain why Newton showed so much interest in the subject.

The Newtonian interpretations have generated a few odd coincidences. Newton, for example, is wrongly supposed to have predicted the end of the world (actually the Second Coming) in 2060, but this is only a lower limit.[92]

> And the days of short lived Beasts being put for the years of lived [*sic*] for 'long lived' kingdoms, the period of 1260 days, if dated from the complete conquest of the three kings A.C. 800, will end A.C. 2060. It may end later, but I see no reason for its ending sooner. This I mention not to assert when the time of the end shall be, but to put a stop to the rash conjectures of

---

[89] Manuel (1974, 116–122). Gjertsen (1986, 537), says that, application of the 15 rules today would eliminate texts like Nostradamus and much other 'fanciful nonsense', and that, apart from the last rule, they could be used with profit today by any scholar facing 'an obscure and difficult text'. They would, in fact, be unlikely 'to provoke any opposition from a modern secular critic'.

[90] 'The Protestant Interpretation of Bible Prophecy — The Biblical Alternative' (http://historicist.info/ newton/title.htm).

[91] Gjertsen (1986, 275).

[92] Yahuda MS 7.3o, f. 8^r; Yahuda MS 7.3g, f. 13^v; Westfall (1980, 816–817); Snobelen (1999, 391–393); Snobelen (2003). An alternative date of 2694 is cited in Chapter 4 of Buchwald and Feingold (2012).

fanciful men who are frequently predicting the time of the end, and by doing so bring the sacred prophesies into discredit as often as their predictions fail. Christ comes as a thief in the night, & it is not for us to know the times & seasons wch God hath put into his own breast.[93]

It also seems that he was not referring so much to a destruction of the world as the creation of a new world order. It may have been the Holy Roman Empire, founded by Charlemagne in 800, whose end was being predicted 1260 years after its formation. In fact, the conquests of Napoleon would make the Empire redundant by 1806.[94]

Again, in manuscripts written at different times, Newton considers the commencement of the 1260-year cycle at 607, 609, 774, 788 and 841, leading to respective conclusions of the papal ascendancy in 1867, 1869, 2034, 2048 and 2101. The earlier dates are interesting as coinciding with the loss of temporal power on the annexation of the Papal States by the Kingdom of Italy in 1870, after the capture of Rome on 20 September, but, as we know from other examples, coincidence is common in such prophetic occurrences.[95] A story that was popular in the nineteenth century was that Newton had interpreted the 'time of trouble' in Daniel 12:4, when 'many shall run to and fro, and knowledge shall be increased', as meaning that the increase in knowledge would lead mankind to being able to travel at a speed of 50 miles per hour. While Voltaire had considered this patently absurd idea to be merely a sign of dotage, the speed predicted coincided exactly with that which railways had by then made possible.[96] For most modern commentators, of course, such prophesied occurrences, whether 'true' or not, have about as much meaning as relating the events of daily life to the 'predictions' of popular horoscopes in newspapers and magazines. The main interest that the prophetic work has for us today is the grasp shown by Newton of the sources of ancient history and his occasional applications of the scientific method to incidental details.

Something that is not so much of a coincidence occurs with the prophesied return of the Jews to Israel at a particular time in history

---

[93] Yahuda MS 7.3g, f. 13^v; Snobelen (2003).
[94] Snobelen (2003).
[95] Hall (1992, 372).
[96] The story is found in *The Friend*, Volume 22, Number 11, 1 November 1865, which quotes the book *Astral Wonders*, by the Rev. Mr. Craig, Vicar of Leamington, a friend of David Brewster. It probably has about as much credibility as the story of Diamond, the dog supposedly responsible for burning some of Newton's papers, but it is remarkable that it brings a *third* person named John Craig into the Newton story, after the contemporary mathematician and the twentieth century economic historian.

after nearly 2000 years of captivity.[97] Such a prophecy was essential to the millennial thinking that dominated Newton's theology. Newton was one of a seventeenth century school of commentators who used texts like Romans 11 and Matthew 24. He also made references to statements in the Old Testament prophets. One manuscript has a call for the return in 1895–1896 and a conclusion in 1944 (1290 and 1335 years after 609), much later than the suggestions by Mede (1715), Whiston (1766), and others.[98] *Observations* (113–114) saw the end as between 2000 and 2050. *Observations* (122) has the cleansing of the sanctuary between 2370 and 2436. Newton was always reluctant to set exact dates.[99]

The return was, in fact, a decisive event, and Newton may actually have contributed to it. Responding to systematic persecution, Jews began to leave Eastern Europe for the Holy Land in large numbers from 1881. Theodor Herzl published his vision of a future Jewish state in 1896 in *Der Judenstaat* (*The State of the Jews*), the year before he presided over the first World Zionist Congress.[100] According to Newton, the Turkish Empire would survive until then; its fall would signify the Restoration.[101] Newton's influence may not have been particularly strong, but his published *Observations* was a significant source for fundamentalist interpretations of the prophecies in later centuries. According to Stephen Snobelen, 'Newton's exegesis merged with a prophetic tradition that helped create during the nineteenth and twentieth centuries the religious and political climates that paved the way for the resettlement of Jews in Palestine–the longed-for vision of the Restoration'.[102] Perhaps fittingly, Professor Abraham Shalom Yehuda, an expert on affairs in the Middle East, acquired a large number of Newton's manuscripts concerning these matters at the Sotheby sale in 1936, and they are now held at the National Library of Israel in Jerusalem.

7.4. Psychology

As the early issues of the Royal Society's *Philosophical Transactions* indicate, at the time when Newton began his scientific career, the clear-cut distinction between what was properly scientific and what was not still had to be

---

[97] Yahuda MS, 10.2, 1^r; Snobelen (2001, 106–107).
[98] Two incomplete treatises on prophecy, c mid-1680s and c 1705–1710, KCC, Keynes MS 5, 138^v, NP; Snobelen (2001, 110).
[99] *Observations* (1733, 113–114, 122).
[100] Herzl (1896).
[101] Yahuda MS, 7.1k, ff. 8^r–9^r, Snobelen (2001, 105).
[102] Snobelen (2001, 110–111).

made. The distinction, in fact, owes a great deal to Newton's own example. Although he pursued researches that we now regard as nonscientific in the areas of theology, prophecy and chronology, in his specifically scientific work, there is nothing that lies outside the boundaries of science as we now understand it. Even borderline areas that, in other hands, might lead to dubious conclusions are, in his writings, subjected to the processes of careful observation and scientific analysis. Psychology never amounted in his work to more than a few youthful speculations and experiments, but they were scientific in a way that cannot always be applied to analyses of his personality and neuroses by some twentieth century scholars.

It is clear, however, that here, as in some other areas, a properly scientific language had yet to be constructed, and so, in *Quaestiones quaedam philosophicae*, the young Newton talks about the detrimental effects of ageing in the vocabulary of his time: 'The boyling blood of youth puts ye spirits upon too much motion or else causet too many spirits. But cold age makes ye brain either two dry to move roundly through or else is defective of Spirits'.[103] Newton understood how his own creativity depended on his physical state, and that it could be improved by particular health regimens. He also suffered psychological problems at various times in his life, including minor breakdowns, which he attributed to physical causes. In the *Quaestiones*, he says: 'Phantasie is helped by good aire moderate wine. but spoiled by drunkenesse, Gluttony, too much study (whence & from extreame passion cometh madnesse), dizzinesse commotions of the spirits. Meditation heates the braine in some to distraction in others to an akeing & dizzinesse'.[104]

He was aware that his habits of concentration on difficult matters for long periods of time affected both his physical and mental health. He was interested in dreams, a problem which we have yet to solve in any convincing way. In 'Of Sleepe & Dreams', from the *Quaestiones*, he asks: 'How is it that the Soule so often remembers her dreams by chance otherwise not knowing shee had dreamed, & thence whither she be perpetually employed in Sleepe. whether dreams are of the body or soule. Why are they patched up of many fragments & incoherent passages'.[105]

From an early date Newton set out to investigate the soul- or mind-body problem using experimental techniques, both psychological and physiological.[106] His main psychological experiment was a very risky attempt, inspired

---

[103] QQP, f. 109^r, NP.
[104] *ibid.*
[105] QQP, Of Sleepe & Dreams', f. 132^v, NP.
[106] See Iliffe (1995).

by a report in Boyle's *Experiments and Considerations Touching Colours* (1663),[107] to determine how far imagination could be used to conjure up strong images, and he decided to test this using the image of the Sun, which he observed with one eye in a looking-glass before turning to a 'dark corner', after 'ye motion of ye spirits in my eye were almost decayed', to look for the after-images, or, as he put it in a later account, 'to observe the impression made, & the circles of colours which encompassed it & how they decayed by degrees & at last vanished'.[108] He saw spots in front his eye when closed, and, upon opening it, saw pale objects as red and dark ones as blue. He did this three times.

> At the third time, when the phantasm of light & colours about it were almost vanished, intending my phansy upon them to see their last appearance, I found to my amazemt, that they began to return & by little & little to become as lively & vivid as when I had newly looked upon ye sun. But when I ceased to intende my phansy upon them, they vanished again. After this, I found that as often as I went into ye dark & intended my mind upon them, as when a man looks earnestly to see any thing wch is difficult to be seen, I could make ye phantasm return without looking any more upon the sun. And the oftener I made it return, the more easily I could make it return again.[109]

Without even looking at the Sun, he found that, looking on any bright object, he 'saw upon it a round bright spot of light like ye sun'; and, even though he had only looked at the Sun with his right eye, he could see the same images almost as plainly with the left. Eventually, he could no longer look at any bright object with either eye without seeing the Sun, and had to shut himself in a darkened room for three full days together, and use every possible means to divert his imagination from thinking about the Sun. For, whenever he did so, he would see the Sun's image even though he was in the dark. Eventually, by avoiding looking at bright objects, he recovered most of his sight, though, for some months afterwards, just thinking about the Sun brought the image before him 'even tho I lay in bed at midnight with my curtains drawn'. In 1691, he reported that, though he had been recovered for many years, he still felt that if he thought hard enough he could make 'ye phantasm return by the power of my fansy'. However, as he told John Locke, the explanation was 'too hard a knot' for him 'to untye'. 'It seems ... to

---

[107] Boyle (1663).
[108] QQP, 'Immagination. & Phantasie & invention', f. 43, NP; Newton to Locke, 30 June 1691, *Corr.* III, 152–154.
[109] Newton to Locke, 30 June 1691, *Corr.* III, 152–154, quoting 153.

consist in a disposition of ye sensorium to move ye imagination strongly & to be easily moved both by ye imagination and by ye light as often as bright objects are looked upon'.[110]

Ultimately, this psychological work merged into Newton's deep and penetrating investigations into metaphysics and his speculations on the mind-matter problem, which were discussed in *Newton and Modern Physics* (7.4). He no doubt remembered his psychological experiences when he was constructing a theory of how the brain was able to create motor function in the human body by sending electrical impulses through the nerves, and how electrical signals of the same kind were responsible for the transmission of visual stimuli to the brain. Though the details were still lacking, he seems to have finally established that the mind was as much part of the natural world and governed by the same fundamental forces as inanimate nature.

7.5. Economics

Economics should certainly be regarded as a scientific study, though it is normative because dependent upon human behaviour, rather than absolute, and no one has found economic laws that can be used to confidently predict outcomes on the basis of well-established equations, as physical laws are expected to do. Policies depend on the conditions prevailing at the time, and outcomes cannot be predicted with the accuracy of physical theories. Newton held what would be considered minority opinions on economics at the time, which were not rated highly by earlier commentators. Modern discussion, however, seems to hold a much better opinion of Newton's prowess as an economist as more fashionable attitudes to the subject have become less dominant than previously. Introducing a collection of Newton's Mint papers in 1896, W. A. Shaw wrote:

> They will speak for themselves, with their masterly brevity and clearness. And if the train of thought which runs through this volume has been grasped, it will be at once apparent how different, and more expert and true, was his attitude of mind towards the monetary difficulty of his time

---

[110] Newton to Locke, 30 June 1691, *Corr.* III, 152–154, quoting 153–154. Buchwald and Feingold (2012) stress the novelty of this experiment and that with the bodkin, described in *Newton and Modern Physics*, 2.1, in relation to an examination of the reliability of the senses: 'Hooke never did anything like this, and neither Boyle nor Huygens were interested in examining perception itself. Newton alone among his contemporaries decided to undertake what can only be described as intervening experiments in human visual perception'. (42). QQP, 'Immagination. & Phantasie & invention', f. 43, NP; Newton to Locke, 30 June 1691, *Corr.* III, 152–154.

than that of John Locke in 1696. It would be difficult to express sufficient admiration of the skill and modesty with which Newton pierced the secrets of the exchanger's craft.[111]

Newton, however, was an economist for pragmatic reasons, not because he was driven to it by inner urges, as he was almost everything else. Essentially, it was part of his job, an aspect of his work at the Mint. So we should not expect to find general ideas and wide-ranging theories, but something more practically-based and limited in perspective. This is an interesting contrast to the grand theoretical schemes of classical economics, which are based on the style of Newtonian physics, and it is perhaps significant, as we have seen, that the great twentieth century classical economist, John Maynard Keynes helped to preserve many of Newton's manuscripts and wrote the often-quoted essay on Newton as an alchemist which we discussed earlier. Typically, Newton pursued his task with the customary vigour, making, for example, quite extensive studies of the coinages of many parts of Europe and of North and South America, which led to his first suggestion that the price of gold should be lowered.[112]

His great feat as an administrator consisted in taking the major role in directing the great recoinage of 1696–1699, which helped to save the economy of England from collapse at a time of war and national crisis. In fact, only an administrator of extraordinary ability could have solved the immensely difficult logistical problems involved, and rescued the process from the incompetence and corruption of the then Master of the Mint, Thomas Neale. Nothing on this scale had ever been done before in any European state.[113] Unprecedentedly and uniquely, after Neale's death in 1699, Newton, as Warden, succeeded to the Mastership, iconically on 25 December, his fifty-seventh birthday. He could now leave the prosecution of coiners to others and concentrate on the production of coins of the realm and on advising on economic policy.

Of course, even in this purely vocational field, Newton couldn't help being an innovator, and he shows a rather modern understanding of money in seeing its value as grounded in confidence. He is quite radical when he makes statements such as: "'Tis mere opinion that sets a value upon [metal] money.

---

[111] Shaw (1896), Introduction, reprint, 132–134.
[112] Hall (1992, 330); Craig (1946, 44); *The Values of Several Foreign Coyns* (1702), *Corr.* IV, 388–390.
[113] The recoinage followed immediately after the creation of the Bank of England and the National Debt by Newton's patron Lord Halifax in 1694, making this 'the most extraordinary financial revolution' in England's history (Iliffe and Smith, 2016, 24).

We value it because we can purchase all sorts of commodities and the same opinion sets a like value upon paper security'.[114] Accordingly, he supports the use of paper money, which, had only been introduced for practical purposes in June 1685, saying that governments could borrow at interest to make up for problems with supply in relation to money defined in terms of metallic currency.

He also gave support for an inflationary policy, observing that a shortage in the money supply would generate high interest rates, which had damaging effects on commerce and led to higher unemployment, while a low interest rate would make imports too expensive and lead to the export of bullion, arguing, in 'a strikingly modern sounding passage', that: 'If interest be not yet low enough for the advantage of trade and designs of setting the poor on work...the only proper way to lower it is more paper credit till by trading and business we can get more money'.[115] Even before taking up his position at the Mint, Newton was among those who had been consulted by the Secretary to the Treasury, William Lowndes, concerning the problem of the coinage. Like Lowndes, he had urged that the recoinage be accompanied by devaluation, though, unlike Lowndes, he thought this should be done by weight of the coin, rather than by reducing the nominal value and producing impractical fractional units of currency.[116] Unfortunately, this sensible advice was ignored by the Government, who went for recoinage without devaluation.

He recognised the difficulties inherent in a bimetallic currency, which in the case of England meant the simultaneous use of coins made of silver and gold. In his view, the relative cheapness of gold in China and India with respect to Europe drained silver out of Europe, and he strove to prevent the continual loss of silver from Britain created by competition with an overvalued golden guinea, and by export to the East through the commercial interests of the East India Company.[117] In 1717, Newton sought to overcome the problem by reducing the price of the guinea by sixpence and setting a fixed value for the currency at twenty-one shillings in silver, effectively shifting the country's economy to a Gold Standard, which lasted

---

[114] Shirras and Craig (1945); Hall (1992); Levenson (2009, 242). Levenson credits Newton here with a close approximation to 'the modern conception of money'.
[115] Shirras and Craig (1945); Hall (1992, 330); Levenson (2009, 242).
[116] *Concerning the Amendment of English Coyns*, autumn 1695, University of Goldsmiths' Library, MS 62, ff. 34–36; Craig (1946, 8–9); Westfall (1980, 554–555); Levenson (2009, 117–119).
[117] Shirras and Craig (1945); Hall (1992, 329–332); Levenson (2009, 242).

until 1931.[118] This had been an inevitable consequence of the Government not accepting Newton's recommendation of devaluation in 1695. In addition, long after the guinea coin had disappeared, up to two and half centuries after this action, retailers set prices at a nominal value in guineas to add a five per cent disguised surcharge to their wares.

In practical terms, Newton, the most conscientious of civil servants, saved the country massive sums of money by eliminating the practice of culling by which the Master and his associates were able to melt down heavier coins and recycle them as bullion. He effectively reduced the standard deviation on the mass of minted gold guineas from 1.3 grains or 85 mg to 0.74 grains or 49 mg, by methods which may have involved using his law of cooling. He was also responsible for at least formalising the introduction of using small samples of coins to reduce errors.[119]

Overall, Newton's pragmatic and totally practical treatment of economic matters was at the opposite end of the spectrum to his approach to world affairs through prophecies issued at a remote time in a highly-figurative and almost indecipherable language, though there is a strange similarity between them. Curiously, both economic and historical events emerge from processes which are far too complex and unrepeatable to be predictable in advance to any significant degree and can only be subjected to a retrospective analysis. Even with prophecies available as a guide, Newton saw that future events could not be successfully forecast on the basis of limited human knowledge. His refusal to indulge in the grand economic theorising that would be tried out by some of his successors was perhaps an instinctive realisation that, as a particular aspect of the human historical process, economic history could not be determined on the basis of simple fundamental laws like those of physics. In this, again, he seems to have opted for the most appropriate methodology, and taken a position far in advance of that of many of his successors.

Newton's work on economic theory is scientific as far as it goes. It differs from his other scientific work in being resistant to large-scale theorising, but it fulfils his criterion of being derived from generalisations based on observation. It is also completely objective in a way that his work on chronology and prophecy could never have been. Nevertheless, even in these

---

[118] *State of the Gold and Silver Coin* (1717); Royal Proclamation (22 December 1717); Fay (1935).

[119] Belenkiy (2013). Belenkiy speculates that the melting points of silver, gold and copper could have been left out of the 1701 paper on the law of cooling as 'state secrets'. As an administrator, Newton also improved the efficiency of the Mint by 'pioneering 'time and motion' studies' (Iliffe and Smith, 2016).

esoteric fields, we can detect the use of aspects of the scientific method, certainly on Popper's criterion of falsifiability, and, in this context, Newton's approach works well in comparison with the nineteenth- and twentieth century pseudoscientific theories that Popper rightly rejected. A theory, like Newton's chronology, in which good scientific principles are applied to dubious data, will eventually use the principles to show up the flaws in the data. However, a theory, like Freudian psychoanalysis, in which dubious principles are applied to reliable data, has no mechanism from within for showing its false starting-point, as the data will always end up supporting the conclusion. It didn't take long for the Newtonian chronology to be rejected by his contemporaries — it happened, in fact, even before the work was actually published. However, it has taken many decades for some of the pseudoscientific theories applied to history and biography since Newton's time to be eradicated, and, in some cases, it hasn't happened yet.

Chapter 8

Securing the Legacy

8.1. Science Becomes Politics

It is, of course, absurd to see the history of science as representing the unimpeded development of pure knowledge. Special interests, personalities, nationalities have all played important parts in what has become the accepted knowledge at any particular time, though in the long-term perspective they become insignificant. Science in the seventeenth century was highly competitive and often strongly political in flavour, with bitter clashes between warring factions. Newton found himself drawn into this scene, at first against his will, but, in later life, he became a master at it and manipulated it to his own ends. Hooke, Flamsteed and, most especially, Leibniz found themselves up against a ruthless, determined and invincible opponent. Historians, perhaps still influenced by the ideal of the scientist as selfless seeker of truth, have tended to side with the losers, but, in truth, they deserve little sympathy; if seen in some purely 'moral' terms (impossible in the circumstances), none of the parties would emerge with credit intact. Certainly, people who have used underhand methods to try to outwit their opponents can hardly complain if those opponents turn out to be superior practitioners of the same kind of tactics.

Newton's actions in the late disputes, especially with Leibniz, must be seen in the perspective of the earlier six-year controversy over his optical discoveries. Reasoned argument had not then won other scientists over to his point of view; his new position of political authority now would. Much more was at stake than personal reputation. Regardless of the supposed 'rights' or 'wrongs' in a very complex dispute, the battle with Leibniz was an important event in the history of science, for, if Leibniz had outlived or outmanoeuvred Newton, the whole development of Newtonian science would have been put in jeopardy and this would have had serious consequences. Leibniz's natural philosophy was too strongly linked to that of Descartes to be considered as a serious alternative to that of Newton. While Leibniz was brilliant in many areas, and sought unity in nature, Newton actually delivered it, finding a method by which all knowledge could be integrated into a single system, and he was unique in this regard. The loss of his authoritative voice and clear-cut programme for development would have been catastrophic for science.

The two men differed significantly even in their attitudes to the new mathematics. Newton developed his analytical techniques as a tool to solve the problems of dynamics. The discovery of calculus for him was an integral part of the discovery of the laws of nature. In first applying calculus to physical problems, Newton had realised that physics required a mathematics that was intrinsically continuous, this was a moment of exceptional significance in the history of physics. Leibniz developed his mathematics as a separate intellectual exercise which he applied occasionally by illustration to the philosophy of nature. The Continental mathematicians who became the first major exploiters of Leibniz's techniques, Euler, Lagrange and Laplace, all succeeded by applying them to Newtonian physics. It should not be thought, then, that the science that Newton produced could have been successfully different in style, that it could somehow have been 'Hookean' or 'Leibnizian'. For all the arguments of the sociologists on the influence of historical conditions on its practitioners, science itself is actually a search for ultimate truths and, after a time, becomes independent of the manner in which it is produced.

What Newton did, which probably no other man could have done, was to create a personal style capable of being applied with confidence by his successors to a complete explanation of the universe. The sense of certainty and authority, which is an unmistakable characteristic of this style, derived from the realisation that it represented truths of a higher and more abstract order than could be found from the haphazard results of experimental evidence alone. By developing a unified view of nature based on the simplest possible principles, Newton showed that an individual approach to science could be universal. From then on, science would become the only form of human knowledge in which individual insights could be seen as components of a single progressive enterprise leading to ultimate truth. The effects of this style on the progress of science are incalculable. If Newton had never existed many of the scientific truths he established would still have been discovered, but we can only guess how many centuries it would have taken to develop his strong unifying vision.

In the last resort, Newton needed to establish his priority over the calculus because of the concerted attacks on his world-system which had been maintained by the Cartesian-inspired philosophers ever since the publication of the *Principia*. His analytically-derived system had only achieved its present credibility because of the power of the synthetic mathematical developments associated with it, but if his opponents had managed to prove that he was a mathematical charlatan, then his whole position would have become untenable. Brilliant political manipulator as he was, Newton made sure of the victory of his system in England, using his position as President of the Royal Society and his influence in Government circles to appoint his own nominees to important academic positions. The victory which had begun with his mastery of the art of mathematics in the *Principia* was consolidated through his mastery of the very different art of politics.

The importance of this cannot be overstated. The complete conversion of England to the Newtonian system provided the ultimate basis for its success on the Continent, and subsequently world-wide. It still had to be established after the *Principia* even in Cambridge. The great classical scholar, Richard Bentley was a particularly dynamic Master of Trinity College, determined to establish lectures in the most up-to-date physics and mathematics at Cambridge. Early in his career he had delivered the Boyle lectures of 1692 and 1693 on natural theology, choosing to discuss the Newtonian system as evidence for the existence of a deity.[1] His appointees, Cotes and Whiston, made a particularly outstanding team. Whiston introduced the formal teaching of Newtonianism into Cambridge which, true to form, had continued to teach natural philosophy of a wholly Cartesian kind from the textbook of Rohault! Whiston spoke enviously of the work of Gregory at Edinburgh: 'He caused several of his Scholars to keep Acts, as we call them, upon several Branches of the Newtonian Philosophy; while we at Cambridge, poor Wretches, were ignominiously studying the fictitious Hypotheses of the Cartesian'.[2] It was unfortunate that Bentley's plans were wrecked by Whiston's public declaration of his unitarian beliefs in October 1710.

Securing a London position for Newton proved to be a key move in the process, leading to his election as President of the Royal Society in 1703 once he knew there would be no further trouble with Hooke. The Royal Society, which had given its imprimatur to the *Principia*, was the centre of support for the dissemination of Newtonian ideas. Newtonianism was established in England both through Newton's Presidency and the placing of prominent Newtonians in academic positions: Cotes, Whiston and Smith at Cambridge;

---

[1] Bentley (1693). See *Newton and the Great World System*.
[2] Whiston (1749, 36).

Gregory and Keill at Oxford; Halley at Oxford and Greenwich Observatory; Maclaurin at Aberdeen first and then at Edinburgh. A group of disciples surrounding Newton began to publish his long-buried mathematical treatises and their own explicatory tracts on his system.[3]

In Great Britain, the adoption of the Newtonian system was, to a certain degree, instantaneous among men of science, if not among the more reactionary academics, and swiftly became part of the popular culture of the age. Whereas the Royal Society had been a subject for satire during the early period, the images of science in literary works now became increasingly positive. This was reflected in Pope's famous couplet, written in 1727: 'Nature and nature's laws lay hid in night/God said, let Newton be, and all was light'. Poets such as James Thomson saw a whole new range of subject matter, appropriate to the didactic style then in vogue, and developed new types of imagery based largely on the optics.[4]

Partly because of the personal battles with Leibniz and Bernoulli, and because of the earlier misrepresentation of the optics, the system took longer to take hold on the Continent, where Cartesianism was still strong. 'A Frenchman who arrives in London', wrote Voltaire, 'will find Philosophy, like everything else, very much changed there. He had left the world a *plenum*, and now he finds it a *vacuum*'.[5] (He was referring here to the contrast between the Cartesian aether and Newtonian action-at-a-distance.) There were, however, early followers in the Netherlands, where the prominent Newtonian, Archibald Pitcairne, had a considerable impact in the 1690s. A significant improvement occurred after the most important optical experiments were demonstrated successfully to French visitors by the Huguenot, John T. Desaguliers on two occasions, in 1715 and 1722, and a sumptuous French edition of the text was published in the latter year.[6] Wherever it was accepted, Newtonianism was applied well outside physics,

---

[3] Wallis published the *Epistola prior* and *Epistola posterior* in 1699; Clarke the *De quadratura curvarum* and *Enumeratio linearum* in 1704; Whiston the *Arithmetica universalis* in 1707; Jones the *De analysi* and *Methodus differentialis* in 1711; and Colson the *Methodus fluxionum* in 1736.

[4] James Thomson's *The Seasons* (1730) is the classic example. Poetic tributes to Newton from the 1720s include John Hughes, *The Ecstasie. An Ode*, 1720; Alexander Pope, *Epitaph. Intended for Sir Isaac Newton, In Westminster-Abbey* (*Epitaph XII*, 1727, p. 1730, quoted in Voltaire's *Cambridge Notebook*); James Thomson, *A Poem Sacred to the Memory of Sir Isaac Newton*, published June 1727; David Mallet, *The Excursion*, 1728; John T. Desaguliers, *The Newtonian System of the World the Best Model of Government*, 1728; Richard Glover, *On Sir Isaac Newton*, 1728.

[5] Voltaire, *Lettres philosophiques*, quatorzieme letter, 1733.

[6] *Optiques*, second French edition of *Opticks*, 1722.

in medicine, for example, with Pitcairne, Cheyne and Boerhaave.[7] Hume in 1739 claimed to have done for philosophy what Newton did for natural science.[8]

The key figure in the transition to Newtonianism on the Continent was probably the brilliant writer Voltaire, who, forced to flee France in his youth through his too effective wit, lived in England from 1726 to 1729, and became a considerable Anglophile. Converted during this period to the Newtonian system, he popularised it very effectively in his *Lettres Philosophiques* (1733) and his *View of Newtonian Philosophy* (1737), written under the tutorship of Maupertuis and Émilie du Châtelet.[9] Voltaire's opinion, according to Brewster, was 'that though Newton survived the publication of the *Principia* more than forty years, yet at the time of his death he had not above twenty followers out of England'.[10] Through Voltaire especially, however, Newtonianism began to replace Cartesianism in France. Thereafter, practically everyone became a 'Newtonian' — Daniel Bernoulli and Clairaut were early converts — but the post-Voltairean type of Newtonianism had an 'enlightenment' aspect, as Newton's own work had not. That most famous monument of the Enlightenment, the French *Encyclopédie*, edited by Diderot and d'Alembert between 1751 and 1772, acclaimed the work of Newton and his successors in astronomy and mathematical physics as the greatest human intellectual achievement so far accomplished and fully accepted the role of the unknown forces that Newton had struggled so hard to persuade his contemporaries to adopt. However, the 'Newtonianism' that the Enlightenment accepted was a rationalised version based on the clear-cut application of mathematical laws to well-defined physical systems of material particles, paying no heed to the inductive methodology that had created it, or to the concerns that Newton still had with many fundamental issues. It is still the most popular view of the meaning of Newtonianism today.[11]

The Cartesian philosophy was still in place on the Continent up to about 1740, and still influencing Leonhard Euler, who around that time developed a theory of magnetism based on vortices. However, Euler had, from

---

[7] Pitcairne (1701); Cheyne (1702); Boerhaave (1708, 1724, 1727, 1732).
[8] Hume (1739).
[9] Voltaire (1733, 1737).
[10] Brewster (1831; 1855, I, 291).
[11] It is no doubt a consequence of this Enlightenment perspective that many people approach the *Principia* expecting it to be a well-proportioned masterpiece of 'classical architecture', like Laplace's *Traité de Mécanique Céleste*, and are astonished to find it more like a gothic cathedral.

1736, been translating the key propositions of the *Principia* into algebraic form, and, from 1740, he effectively became the most significant Continental Newtonian — although he probably wouldn't have described himself in such terms, as he seems to have generally regarded his own work as starting a new line of investigation.[12] In addition, experimental results began gradually to tell in favour of the Newtonian system. One came from the simultaneous expeditions of Maupertuis to Lapland and La Condamine to Peru which, in 1736, measured the curvature of the Earth, and proved Newton to be correct about it being an oblate spheroid.[13] Later on there was the return of Halley's comet in December 1758, as predicted by Halley in 1705, on the basis of Newton's theory, with the corrections for perturbations by the larger planets by Clairaut.[14] Herschel's astronomical proof of 1802 that binary stars revolved round each other demonstrated the effects of Newtonian gravity outside the Solar System.[15]

In general, scientific developments in the eighteenth century followed very different paths in England and on the Continent. In England, where the Baconian option had always been preferred, the Royal Society, had declined from its peak of activity in the 1660s through each successive decade until the mid-eighteenth century, when it reached its nadir under the Presidency of Martin Folkes, as little more than a gentleman's social club. This was partly due to the using up of the easier options for new lines of enquiry, and partly due to the loss of the early vigorous generation of its first decade. It was all too easy for purely empirical research of the Baconian kind, with no significant underpinning of fundamental theory, to develop into triviality. Under the personal influence of Newton, from 1703 to about 1722 there was a temporary revival but it was centred on a single powerful personality. However, valuable experiments were performed by a succession of capable demonstrators, Hauksbee, Taylor and Desaguliers; and the belated publication of Newton's mathematical tracts generated an interest in the Newtonian version of the calculus, which was pursued with considerable

---

[12] Euler (1736–1742). Euler was well-versed in Newtonianism from an early age, as he obtained a master's degree at the University of Basel in June 1724, when just seventeen years of age, with a thesis that compared the philosophy of Newton with that of Descartes. After Clairaut's success in deriving the lunar apogee, he proclaimed the validity of the 'Newtonian hypothesis' of inverse-square law gravity as the foundation of all celestial mechanics (1752, 4f, quoted Smith, 2012, 379).
[13] Todhunter (1873).
[14] Halley (1705); Clairaut (1760); Wilson (1993).
[15] Herschel (1803–1804).

success by Cotes, Taylor, De Moivre, and several others.[16] Despite the appointment of mathematicians of this calibre to university chairs in the early part of the period, the English universities were in severe decline by the mid eighteenth century; and no longer made any pretence at providing an academic education; Henry Cavendish, probably the most remarkable British scientist of the second half of the century, found Cambridge totally useless in 1749 and left without a degree. Scottish universities were still important and helped to save the day, but more in medicine and philosophy than in physics and mathematics. Colin Maclaurin of Edinburgh was the last British mathematician who contributed significantly to the development of mathematical dynamics, and he died in 1746.

On the Continent a highly mathematical style developed, with emphasis on algebraic or analytical approaches. Continental calculus was ostensibly largely non-Newtonian, stemming directly from the work of Leibniz and Bernoulli. A major theme was the analytical development of dynamics, sometimes directly translated from the geometricised theorems of the *Principia*. It was the Continental mathematicians, in particular Euler, who largely put Newtonian mechanics into the form which is familiar to us today. Euler developed a more algebraic presentation of the mechanics, including the familiar modern form of Newton's force equation. He also worked on problems like the theory of the Moon, which Newton and his English followers had abandoned.[17] It was also left to Continental mathematicians, like Clairaut, to perfect Newton's work on the figure of the Earth.[18] Nonetheless, British mathematicians especially Maclaurin, were influential on Continental ones, and Clairaut, in particular, expressed many debts to Maclaurin in *La Figure de la Terre*.

If new developments in Britain were relatively few, the Newtonian tradition persisted, and Maclaurin's *Treatise on Fluxions*, whose second book was largely algorithmic and algebraic, especially 'served to transmit Newtonian ideas in calculus, improved and expanded, to the Continent'. The standard story of the complete decline of the Newtonian mathematical tradition, promoted by the Cambridge Analytical Society in the early nineteenth century, certainly needs modification. Apart from the direct Newtonian influence in dynamics, the Newtonian definition of limits, emphasis on power series, and basing of foundations on inequalities became significant in the works of d'Alembert, Euler, Lagrange, Cauchy and Weierstrass,

---

[16] Guicciardini (2003a).
[17] Euler (1736–1742).
[18] Clairaut (1743).

and many others, often through the mediation of Maclaurin, who supplied extra rigour to Newton's often intuitive concepts.[19] Judith V. Grabiner considers it 'both amusing and symptomatic of the misunderstanding of Maclaurin's influence that Lacroix's one-volume treatise on the calculus', *Traité élémentaire de calcul différentiel et de calcul intégral*, of 1802, which was 'translated into English by the Cambridge Analytical Society with added notes on the method of series of Lagrange, was treated by them, and has been considered since, as a purely 'Continental' work. But Lacroix's short treatise was based on the concept of limit, which was Newtonian, elaborated by Maclaurin, adapted by d'Alembert and L'Huilier, and finally systematised by Lacroix'.[20]

Euler was probably the century's most remarkable pure mathematician, and, like Newton, he was intensely religious. Like Galileo and the nineteenth century Belgian physicist, Plateau, he went blind through making risky experiments on the Sun; Newton, who had stared at the Sun's reflection as a young man to record the after-images, got away with a few days' indisposition in a darkened room. Blindness, however, did not affect Euler's rate of production, for he had a phenomenal memory, and he wrote well over a thousand mathematical papers during his career. Euler was followed by Lagrange and Laplace, who by, the end of the eighteenth century, had taken Newtonian physics to unparalleled heights. Though there were many significant developments in science in Britain during this period, they were not mainly concerned with mathematical physics or celestial mechanics on the Newtonian pattern.

8.2. Completing the Picture

Newton's system became powerful because it was extreme, and it is, of course, extreme because it stemmed from the uncompromising metaphysics that Newton had derived from his theology. His successors — Euler, Maupertuis, Lagrange, Hamilton — produced improved versions of his dynamics, but each used an extremum principle of some kind, defining a quantity involving mass and the differentials of space and time, and showing its extreme behaviour. In virtually exact repetition of the Newtonian pattern, there was always a law which defined a quantity (force, momentum, energy, action, Lagrangian, Hamiltonian) and another which set this quantity to zero, or a constant value, or a maximum or minimum. At the same time, a

---

[19] Grabiner (1997).
[20] Grabiner (1997); Lacroix (1802).

further ('physical') law specified the operation of the source of the defined quantity, as Newton's law of gravitation had defined the operation in his system of the quantity force. Emphasizing the Newtonian distinction between the abstract system and the process of measurement, the new dynamics eventually required a revision in the mathematical style, though not the substance, of Newtonian mechanics, the generalised coordinates of Lagrangian dynamics removing the need to trace the particular histories of individual particles.[21]

By the end of the eighteenth century, however, extremum principles were so familiar and so mathematically structured that their ultimately metaphysical and even theological origin in Newton's work could be disregarded. Even the indeterminacy in measurement apparent since Newton first tackled the three-body problem, and essential to Newton's separation of the abstract system and the measurement process, could be contradicted. Laplace, for example, writes:

> If an intellect were to know, for a given instant, all the forces that animate nature and the condition of all the objects that compose her, and were also capable of subjecting these data to analysis, then this intellect would encompass in a single formula the motions of the largest bodies in the universe as well as those of the smallest atom; nothing would be uncertain for this intellect, and the future as well as the past would be present before its eyes.[22]

And again:

> All events, even those which on account of their insignificance do not seem to follow the great laws of nature, are a result of it just as necessary as the revolutions of the sun. In ignorance of the ties which unite such events to the entire system of the universe, they have been made to depend on final causes or upon hazard, according as they occur and are repeated with regularity, or appear without regard to order.[23]

According to a well-known legend, Laplace joked about having 'no need' for the 'hypothesis' of God (double blasphemy for Newton!), but that was

---

[21] It should be pointed out, however, that, though methods such as least action (in the form of Hamilton's principle) are in many ways more powerful and efficient than Newton's laws, they cannot be applied to systems in which forces are dissipative and so not expressible in terms of potentials, whereas Newton's laws can be applied to nonconservative, as well as conservative systems.

[22] Laplace (1825/1995). This imaginary intellect has been described as the 'Laplace demon'.

[23] Laplace (1814/1902).

because his entire work was a synthetic development from already-existing Newtonian principles and not an analytical development of new ones.[24]

The eighteenth century struggled to establish the electrostatic force as inverse-square, like gravity, while Kant showed that this was a natural result of 3-dimensional space. Eventually, Priestley used an argument by direct analogy with one of Newton's gravitational theorems,[25] before Coulomb established the point conclusively by direct experiment.[26] Out of this came a new mass-like quantity, charge, and this was probably the last truly fundamental addition to the physical parameters, because, even though two new forces, the strong and weak nuclear interactions, were discovered in the twentieth century, they could be conceived of as being produced by charge-like source quantities, which, under ideal conditions, would behave in the same way as electric charge.

Current electricity and electromagnetism complicated the picture, but eventually, Maxwell set down, in mathematical form, all the laws that were known to be valid for electric and magnetic fields — in particular, those of Coulomb, Ampère and Faraday — and in doing so noted that there was an asymmetry which could be corrected by the addition of another term to the law of Ampère.[27] This term was the so-called displacement current — a current which he supposed must exist when charge was supplied to a parallel plate capacitor, where it could be assumed to remain static. Even though there is no physical justification for such a current except by analogy, the assumption had a remarkable effect, for he was immediately able to generate wave equations whose velocity was exactly that of light. Maxwell was able to show that light was purely electromagnetic and to generate a purely abstract theory which, in a few equations, incorporated the whole of electricity, magnetism and optics.

---

[24] Rouse Ball (1908): 'Laplace went in state to Napoleon to present a copy of his work, and the following account of the interview is well authenticated, and so characteristic of all the parties concerned that I quote it in full. Someone had told Napoleon that the book contained no mention of the name of God; Napoleon, who was fond of putting embarrassing questions, received it with the remark, 'M. Laplace, they tell me you have written this large book on the system of the universe, and have never even mentioned its Creator'. Laplace, who, though the most supple of politicians, was as stiff as a martyr on every point of his philosophy, drew himself up and answered bluntly, *Je n'avais pas besoin de cette hypothèse-là.* ('I had no need of that hypothesis'.) Napoleon, greatly amused, told this reply to Lagrange, who exclaimed, *Ah! c'est une belle hypothèse; ça explique beaucoup de choses.* ('Ah, it is a fine hypothesis; it explains many things'.)'.
[25] Priestley (1767).
[26] Coulomb (1785a, 1785b).
[27] Maxwell (1861–1862, 1865).

Sir William Hamilton, who was professor of metaphysics at Edinburgh from 1836 to 1856, is reputed to have influenced Maxwell into believing that true knowledge existed in relations between things, rather than in things themselves, and that all knowledge was subject in some form to a general 'principle of relativity' — this may well have contributed to Maxwell's curiously detached attitude to the competing wave and particle theories of light, which he regarded purely as analogies.[28] Maxwell's great contemporary, Ludwig Boltzmann was also prophetic of later developments in his philosophy of the 'complementarity of contradictory hypotheses', reminiscent as it is of the complementarity principle of Niels Bohr — himself supposedly influenced by the philosopher Søren Kierkegaard.[29]

Characteristically, Maxwell's theory of the electromagnetic field was for many years disregarded, on account of its intrinsically abstract nature, William Thomson, soon to be Lord Kelvin, famously declaring: 'I never satisfy myself until I can make a mechanical model of a thing. If I can make a mechanical model I can understand it. As long as I cannot make a mechanical model all the way through I cannot understand; and that is why I cannot understand the electromagnetic theory'.[30] Though Maxwell's theory explained all the facts in a simple way without needing a mechanical model, Kelvin persisted in a fruitless search for just such a model, delaying the acceptance of Maxwell's theory for more than 20 years. However, all the physical explanation and ingenious mechanistic hypothesizing once used to explain electromagnetic and optical effects has now been swept away and all that remains is Maxwell's purely abstract system of equations. The very fact that such a distinguished physicist could even contemplate the development of physics on mechanistic lines shows that the true nature of the Newtonian revolution in physics had never been fully understood.

There are many similar cases in the history of science, and that history has been distorted by the public positions that individual scientists have been forced to adopt. Thus the public image of Michael Faraday is of a brilliant experimentalist, a sort of magician conjuring up results on electricity and magnetism from nowhere as the result of superb experimental technique. This picture could not be further from the truth; Faraday was above all a theorist developing a unified view of the forces of nature through what we

---

[28] Maxwell (1856), cited Everitt (1974, 206); Olson (1975); Keswani and Kilmister (1983, 343).
[29] Cited in Coveney and P. Highfield (1990, 331), from McGuinness (1974).
[30] Thomson (1904).

now call field theory, but compelled by the opposition he knew he would face to suppress the theoretical basis from which he generated his brilliant experimental results.

8.3. Newton and the Enlightenment

There was, in fact, a very considerable contrast everywhere between the eighteenth century and the seventeenth. The rational tone of the Enlightenment began to replace the religious enthusiasm of the earlier period. Newtonianism had come just in time to feed off the work of Newton's mediaeval predecessors. The eighteenth century religious establishment was largely Latitudinarian — Newton's 'placemen' contemporaries at Cambridge had finally found their places (!) — while deists and atheists began to come forward in increasing numbers. One aspect of this transformation was an emphasis on the material particles rather than the 'active powers'. By a process of popular vulgarisation, the Newtonian mathematical structure became associated with a universe that was materialist and determinist, a considerable deviation from Newton's own views. Later attacks on 'Newtonianism' by literary and other nonscientific writers like William Blake gave further impetus to this notion, and it was extended by implication to the whole structure of physical science itself as well as its effect on philosophy, society, and so forth. As a result of this, it is frequently made out that modern physics represents a fundamentally different world-view from that of Newton. Newton, we are told, was an Apostle of the Age of Reason who believed in a 'clockwork universe'; in the modern period, however, we have learned through quantum mechanics that physics is probabilistic and nondeterministic and, by implication, non-Newtonian.

It is astonishing that the inadequate appreciation of Newton's work by his contemporaries led to him being labelled by their characteristic appellation of mechanist. His whole approach had superseded mechanism but the lack of understanding ended up with him being labelled as their arch-apostle. Blake might have been astounded to realise that he had so much in common with Newton. If Newton had never existed, a different, and truly mechanistic, kind of physics would very likely have been created; we would certainly have had an inverse square law of force acting on the planets, but it is impossible to imagine how long we would have had to wait for the Newtonian concept of an abstract and universal law of gravitation.

The conventional view sees Newton as emerging ultimately triumphant in the clashes with his opponents, perhaps even too triumphant for the overall good of science. But the clashes did not result in a total victory for Newton. They were bought at the expense of the suppression or downgrading of some of his most important ideas; the main casualty was the 'theory of matter' and the forces acting upon it, in particular the electrical force, which only appeared in print in a confused mass of Queries at the end of the *Opticks*. The manuscripts for the *Opticks* show that vast areas of Newtonian thought — about forces and the nature of matter, in particular — were suppressed in the process of composition. The Queries, especially, contain many statements, presented as conjectures, which the manuscripts reveal to be based on sound analytical, and sometimes quantitative, reasoning.[31] Though they contain some of the most remarkably prescient guesses ever made by a practising scientist, the Queries, taken at face value, were the last speculative efforts of an ageing natural philosopher on subjects which were not yet amenable to the more rigorous methods previously employed in his optics and dynamics. But the Queries were not founded on speculation. They were founded on extended and precise chains of physical and mathematical reasoning, and were all that he dared to publish from a vast amount of manuscript material accumulated over a long period of years. The meaning of the Queries cannot be found in the published texts alone but must be sought also in the draft material from which they were composed. Ideas that could have had a massive influence on later developments were omitted altogether.

And there was a very good reason for their omission. Drafting the suppressed fourth Book of the *Opticks* in the early 1690s, shortly after publishing the *Principia*, Newton wrote: 'This principle of nature being very remote from the Conceptions of Philosophers I forebore to describe it in that Book leas[t it] should be accounted an extravagant freak & so prejudice my Readers against all those things wch were ye main designe of the Book...'[32] At an earlier date he had written to Boyle: '...what I am not satisfied in I can scarce esteem fit to be communicated to others, especially in natural Philosophy where there is no end of fansying'.[33] Oldenburg was informed that: 'I had formerly purposed never to write any Hypothesis of light & colours, fearing it would be a means to ingage me in vain disputes...';[34] for, as he told Halley, when he was in the middle of his work in mechanics, 'Philosophy is such an impertinently litigious Lady that a man had as good

---

[31] For example, the draft Queries in UL Cambridge, MS Add. 3970.3, ff. 231^r–301^r, 359^r, 477^v–478^r, 610^r–612^r, 618^r–623^r, NP.
[32] Draft conclusion for *Opticks*, Book IV, c 1690, Hypothesis IV; Westfall (1980, 521).
[33] Letter to Boyle, 28 February 1679; *Corr*. II, 288.
[34] Newton to Oldenburg, 7 December 1675; *Corr*. I, 361.

be engaged in Law suits as have to do with her. I found her so formerly & now I no sooner come near her again but she gives me warning'.[35]

Newton was not mistaken in the reactions of his contemporaries. His principles were indeed 'remote from the Conceptions of Philosophers', who continued to promote mechanistic explanations of both colours and gravity long after he had shown them to be unnecessary. It seems at all times that the ideas of even advanced and successful thinkers can only be accepted when packaged in a way which lessens their revolutionary impact. Though I don't personally subscribe to the view, some scholars, for example, believe that Queries 17–23 represent concessions on a mechanical cause of gravity finally wrung out of an ageing man by the relentless criticisms of mechanistic philosophers.[36]

The suppression, though necessary in the context of the social structure of science at the time (and even in our time), was undoubtedly to the disadvantage of science, and there is a sense in which reading Newton's suppressed drafts becomes a disturbing, as well as fascinating, exercise. It makes us realise what price we have had to pay in scientific terms for maintaining scientific social structures. The General Scholium, as we have seen, contained tantalizing hints about an 'electric and elastic spirit' operating in nature, but no more on this was published. No attempt was made to get at the manuscript material behind this statement until relatively recent times. Faraday, we have been told, was fond of quoting the passage from Newton's correspondence with Bentley which begins: 'Tis unconceivable that inanimate brute matter should (without ye mediation of something else wch is not material) operate & affect other matter without mutual contact ...'.[37] It is interesting to imagine how he would have reacted to the draft statements concerning the universal actions of the electric spirit if these had appeared in the published Queries. David Brewster wrote a famous biography of Newton during Faraday's lifetime, publishing many manuscript documents;[38] if he had published some of the draft Queries, or *De vi electrica*, or the rejected versions of the General Scholium, these

---

[35] Newton to Halley, 20 June 1686; *Corr.* II, 437.
[36] Hall (1992, 355–357), provides a classic example of this approach. It is, in my view, an interpretation that should be totally rejected, and as wrong a view of the significance of aether theories and their place in scientific history, as was the view of the nineteenth and most of the twentieth century that Newton's corpuscular theory of light had been completely superseded by the theories of Young and Fresnel and the experiments of Foucault.
[37] Tyndall (1870, 81), cited in Cohen (1952, liii).
[38] Brewster (1855).

would have been available to Faraday. The reception of Faraday's and Maxwell's work on the electromagnetic field might well have been very different.

It is important to ask why scientists, even one of the calibre of Newton, are put into a position where they have to suppress some of their work to get other parts accepted, and how this process affects the development of science; or how often revolutionary thinkers in the history of science have been forced into presenting their work in a way which makes it acceptable to their more conservative contemporaries, but makes it lose its cutting edge; or, again, how much is lost by forcing modern scientists to write in an impersonal style which observes proprieties but lessens the impact they would like to have. Perhaps we are being optimistic in imagining that the Newtonian revolution has actually succeeded. It should certainly cause some reflection that there are ideas of startling originality which would greatly advance the progress of science but which are so far from the accepted, and extremely limited, canons of the day that even a Newton is afraid of bringing them out into the open. If historical evidence is anything to go by, it is likely that some such process is still going on today.

8.4. The Power of Abstraction

If anything can explain the universality and lastingness of Newton's physics it must surely be his great abstracting power. He alone, among all his contemporaries, was able to build a system of the world on a purely abstract concept of gravitation, without any concession to a mechanistic explanation. All the great unifying ideas in physics — the electromagnetic theory of light, quantum mechanics and quantum field theory, the gauge theories of particle interactions — have been abstract in this way, though this trend towards abstraction has been strongly resisted at every stage in the process, and even today there are many who seek a unified theory of physics at the mechanistic level and believe that it is possible to explain fundamental truths, such as the unidirectionality of time and the conservation of charge, in a mechanistic way. Ultimately Newton's ideas were based on a faith not shared by his contemporaries, one that has only slowly filtered through into science and has always been resisted, that nature is based on simple and abstract principles. Modern physics seems to represent a gradual coming to terms with the abstract Newtonian legacy.

A great deal of the most successful aspects of fundamental physics in the last two centuries has been developed using such abstract metaphysical principles as analogy and symmetry. Following the special theory of relativity, for example, the whole of electromagnetic theory became explicable as an extension of Coulomb's inverse-square law by the addition of a fourth dimension onto that of space in Minkowski's space-time, which did away with the need even for Einstein's simplified kinematics.[39] The simple parallel between electromagnetism and gravity (or between mass and charge) had at last been established, and it was natural to assume, in general relativity, that the 4-dimensional space-time connection applied to the gravitational, as well as the electromagnetic, whether or not this was actually true.

Four-dimensional space-time has a remarkable analogy, which has been little explored, with the mathematical structure known as quaternions, discovered by Sir William Rowan Hamilton in 1843.[40] Hamilton refers to 'a certain synthesis of the notions of time and space, or in the greatest abstraction of Uno-dimensional and Tri-dimensional Progression, the result being a Quaterno-dimensional Progression, or what I call a Quaternion'.[41] This was one of many pre-relativity attempts to define a fourth dimension, usually referring to time, another being by d'Alembert, discoverer of the wave equation, which is the four-dimensional analogue of Laplace's analytical equation for Coulomb's law.[42] Earlier references to a 'fourth dimension' (including one by Newton's friend, Henry More) had been specifically spiritual or theological; and some later references, including one by Oliver Lodge, were spiritualist, though Lodge also had a concept equivalent to the worldlines of special relativity.[43]

The method of analogy ultimately has no justification except that of 'the uniformity of nature', which is, in essence, a purely metaphysical concept, recognisable to such as Ockham or Newton, and theological to them. However, the method is widely pursued even today, for the modern programme of uniting all the forces and all the particles is essentially that of Faraday and others in the nineteenth century, and ultimately of the programme originally laid down by Newton. The method of analogy has been widely used purely because it has been successful, and not because of any ultimate origin it may once have had. But is there any more fundamental principle that explains its success, and can we use this fundamental

---

[39] Einstein (1905); Minkowski (1909).
[40] Hamilton (1843/1848, 1844). These are cyclic and dimensional square roots of -1, with the real unit 1 needed for closure. They are similar in structure to 4-vectors with the real and imaginary parts exchanged.
[41] Quoted in Graves (1882–1891, 3), 253–354 and 2:478.
[42] d'Alembert (1747).
[43] More (1671); Rucker (1985); Lodge (1891).

principle (or 'magic ingredient') to explain the special success of physics itself? Were Newton and others inadvertently using this ingredient? Did Newton's ruthless analytical techniques, based on his theological persuasions, lead him to discover solutions which incorporated it without his direct knowledge?

The method of analogy presupposes the more fundamental concept of symmetry, and this seems ultimately to be the magic ingredient that makes physics 'work' as the ultimate explanation of nature. Newton had an interest in symmetry, using it in his optics and in many aspects of his dynamics,[44] and there is a flicker of its importance, as he considers, at the end of Query 31, why animals are all formed on the same plan, a long time before this question would become really meaningful:

> Such a wonderful Uniformity in the Planetary System must be allowed the Effect of Choice. And so must the Uniformity in the Bodies of Animals, they having a right and a left side shaped alike, and on either side of their Bodies two Legs behind, and either two Arms, or two Legs, or two Wings before upon their Shoulders, and between their Shoulders a Neck running down into a Back-bone, and a Head upon it; and in the Head two Ears, two Eyes, a Nose, a Mouth, and a Tongue, alike situated. Also the first Contrivance of those very artificial Parts of Animals, the Eyes, Ears, Brain, Muscles, Heart, Lungs, Midriff, Glands, Larynx, Hands, Wings, swimming Bladders, natural Spectacles, and other Organs of Sense and Motion; and the Instinct of Brutes and Insects, can be the effect of nothing else than the Wisdom and Skill of a powerful ever-living Agent, who being in all Places, is more able by his Will to move the Bodies within his boundless uniform Sensorium, and thereby to form and reform the Parts of the Universe, than we are by our Will to move the Parts of our own Bodies.

A Short Scheme of the True Religion extends the argument:

> Whence is it that all the eyes of all sorts of living creatures are transparent to the very bottom and the only transparent members in the body, having on the outside a hard transparent skin and within transparent juyces with a crystalline lens in the middle and a pupil before the lens, all of them so truly shaped and fitted for vision that no Artist can mend them? Did blind chance know that there was light and what was its refraction, and fit the eyes of all creatures after the most curious manner to make use of it?[45]

This is, of course, an example of the argument from design (which we have already seen elsewhere in Newton's writings), and we are now all aware

---

[44] Symmetry principles figure significantly in the optical work in the minimum deviation principle, the explanation of the rainbow and in the Newton's rings experiment.
[45] *A Short Scheme*, KCC, Keynes MS 7, Cohen and Westfall (1995, 344).

that all the features previously explained by design by a Creator would eventually be used to support evolution by natural selection. But it is a mistake to suppose that this means that features have evolved by 'blind chance' in a random arrangement or rearrangement of particles of matter in motion, as imagined, for example, by Lucretius or Descartes; convergent evolution tells us otherwise; they are based on structural processes which are self-similar and based on abstract laws; despite the inherent degree of randomness and uniqueness in the process, there is 'purpose' directing evolutionary development though it may not be the continual working of a designer-Creator.

Symmetries also occur in the paired opposites of Newton's alchemical writings, as in his *Commentarium on the Emerald Tablet of Hermes Trisnegistus*, of the early 1680s, where he says: 'Inferior and superior, fixed and volatile, sulphur and quicksilver have a similar nature and are one thing, like man and wife. For they differ one from another only by the degree of digestion and maturity'.[46] Other examples are collected from the alchemical literature in the first chapter, 'De materiis spermaticis', of his *Praxis* of 1693.[47] Betty Jo Dobbs points to similar dual pairings in modern chemistry, including acid and base, anion and cation, electron and proton, and says that 'He continued to argue for many years for a matched pair of opposites as basic constituents of nature, speaking of them in many of his alchemical papers'.[48] Ultimately the chemical examples, and also some biological ones, all stem from the duality of the electrical force in its positive and negative manifestations.

Symmetry allows us to do what Newton wished to do: to define an abstract, unknowable reality, combined with a process of observation or measurement of its parts. Symmetry between two concepts means absolute identity in most respects, combined with absolute opposition in one. So symmetry allows us to characterise a part of reality without characterising the whole. Only through symmetry can unity result in diversity. And physics works in such a way that when you characterise a part of reality in a certain way, you are necessarily characterising the rest as different (that is, opposite). This is what explains Newton's success in introducing mass as a conserved quantity opposed to the variables space and time, and also the success of his opposition of matter and 'spirit' (ultimately resolving itself as charge

---

[46] *Commentarium*, Keynes MS 28, f. 6ʳ,v. Dobbs (1991, 276).
[47] *Praxis*, 'Cap. 1. De materiis spermaticis', in Dobbs (1991, 296).
[48] Dobbs (1991, 71).

and mass). But he didn't consciously set out to do this; he developed by analytical means the only procedures that would work.

The idea of a union of opposites, although probably of very ancient origin, was very prominent in the nineteenth century idealist philosophy inspired by Kant and Hegel, and often influenced mathematicians and physicists, as symmetry principles do today. William Rowan Hamilton, for instance, was particularly struck by a passage in the poet, Coleridge's publication, *The Friend*:

> Every power in nature and in spirit, must evolve an opposite, as the sole means and condition of its manifestation: and all opposition is a tendency to re-union. This is the universal Law of Polarity or essential Duality. ... The Principle may be thus expressed. The Identity of Thesis and Antithesis is the substance of all Being; their Opposition the condition of all Existence, or Being manifested: and every Thing or Phaenomenon is the Exponent of a Synthesis as long as the opposite energies are retained in that Synthesis.[49]

Coleridge implied here that observed phenomena were the result of unobservable opposed powers. As Hamilton added in his own annotations:

> Power can be manifested only by its effects, that is, by overcoming Resistance, which is Contrary Power. ... The thought of *Being* or *Existence general*..., as distinguished from phenomena, that is, from manifestations of existence, arises in us along with and as a realization or externalization of our belief in a common ground, a hidden principle of unity, belonging to the two opposite tendencies.[50]

Symmetry makes it possible to avoid characterising reality, and this seems to be the unconscious aim of the modern physicist, just as it was the conscious aim of Newton. If to every concept there is an exactly symmetrical opposite, then we never have to specify reality if we use both at the same time. Symmetry can also be exact, in a way that no specific idea can. It is important, however, that symmetries are not pushed into being *identities*. Minkowski, for example, went too far when, referring to his discovery of the mathematical combination of space and time, he stated that: 'From now on, space by itself, and time by itself, are destined to sink into shadows, and only a kind of union of both to retain an independent existence'.[51] Space and time are not identical, but *typically symmetrical* concepts: they have some points of identity and others which are different, and indeed symmetrically opposite — facts which may well go a long way towards

---

[49] Coleridge (1969, 158).
[50] Graves (1882–1891), I: 439. Hankins (1980, 285).
[51] Minkowski (1909), 1923 translation, 104.

explaining wave-particle duality and Zeno's paradoxes. The combination of space and time leads to a *broken*, not a perfect, symmetry, and these are especially characteristic of physics at the fundamental level.[52]

It follows also that the programme to reduce everything to an aspect of space in multiple dimensions — the modern version of Descartes' privileging of extension — is unlikely to achieve its objectives because physics needs its symmetrical opposites for its success. Symmetrical opposites are, again, characteristic aspects of *broken* symmetries, and the multi-dimensionalities which result are not purely spatial, and introduce incompatibilities with a spatial description, just as time does in quantum mechanics. The fundamental status of the general theory of relativity looks problematic in this context, unless it is supplemented by Newtonian gravity, to overcome its otherwise obvious incompatibility with the highly successful quantum theory.[53]

It is clear that the modern search for a *unified* theory, based on symmetries and analogies, is essentially at one with the originally theologically-inspired project of the fourteenth and seventeenth centuries, in its metaphysical basis, and especially with its culmination in the work of Newton, and we now have a better understanding of what such a theory would actually look like. It would certainly be characterised by abstraction, simplicity and symmetry. It would also be, in principle, extreme, no compromise being allowed in the best theological tradition. There would be no mathematics, other than that derived through symmetry principles, no model-dependent structures of any kind, and no arbitrary cosmology. It would certainly look different from any theory yet devised for a particular aspect of physics, yet these would all be ultimately deducible from it. Though purely secular in itself, such a theory would derive much of its power, in the same way as Newton's physics, from the fact that its origin was in a metaphysics ultimately closely connected with theology, which required an uncompromising logic. Searching for a unified theory which will explain everything in physics, we are confronted with a multiplicity of approaches, many of fearful complexity and 'mechanistic' in principle, being built on particular models of the universe. If Newton has anything to teach us here, and I believe that he has, the final result will be not be complex, but of an

---

[52] Rowlands (2007, 92–93).

[53] General relativity is a generic theory based on truths in pure mathematics. It has no intrinsic physical application but incorporates aspects of Newtonian gravity to create physical meaning. The proper relationship between general relativity and Newtonian gravity is a subject that requires proper discussion (see *Newton and the Great World System*). It is certainly not a matter of alternative hypotheses or competing paradigms as Eddington presented it in 1919.

extraordinary simplicity, and, unlike all other physical theories, be expressible in terms of complete abstraction, without any mechanistic element whatsoever.

In science Newton's contribution is surely unsurpassed and probably unequalled. He remains a unique figure in the history of science because he emerged with the startling and unexpected discovery that the basic truths of physics were fundamentally abstract. Physics could not be based on a 'story-book' picture which privileged modelling based on concepts related to everyday experience. This was so extraordinary a conception that his contemporaries could not accept it and *it still remains unaccepted today*. The Newtonian revolution was in this sense only partially successful. Even the reason for the special nature of Newton's contribution is not yet fully recognised. Though the Newtonian revolution in physics was never completely established, permanent success has eluded physics based on non-Newtonian approaches; and though the synthetic aspect of his work has long been established, the analytic is at least as important and has a powerful message for contemporary physicists.

Newton was a much more powerful and original thinker than could be imagined from his popular image as a great mathematician; his pre-eminence among scientists was not an accident of the particular moment when he lived, but of the unique nature of his particular thought-processes. So many of his ideas penetrate to such deep levels that they continue to stand up well even when compared to modern developments in subject areas that did not even exist in his own time. There must have been something in the very nature of his thought which makes it the perfect exemplar of scientific method even today, and there can be no doubt that this was intimately bound up with his extraordinary powers of abstraction and simplification. It was his unique achievement to create a programme for the whole future development of physics. No lasting success has ever been achieved outside this programme, though the attempt has been made many times. It is possible for us, in a way which was not possible to our predecessors, to read Newton as though he were our contemporary, and to find fresh inspiration for our own researches in the sheer authority of his style and the modernity of his thought.

Newton was extreme, pushing everything to the utmost, disregarding utterly the objections of his contemporaries; but modern developments have shown that he was right in taking this uncompromising attitude, for it is the one that has now been universally adopted in modern theories, such as quantum mechanics and quantum field theory. The quasi-theological nature of the proposition that nature obeys universal and abstract rules and will be found to be ultimately simple has often been noticed with respect to modern science; it is not always realised that it does actually stem from a metaphysics that was always closely allied to theology.

The extremely abstract nature of Newton's qualitative reasoning seems to point directly to the similarly abstract development of twentieth- and twenty-first century physics. Fundamental physics has never been more abstract than it is at the present, and we have learned to live with theories which have no conceivable 'mechanistic' explanation. It appears that there is a sense in which we are only now beginning to respond to the most important element in the Newtonian revolution. If this is the case, then something subsequently went wrong. The revolution only partially succeeded; the reason for the special nature of Newton's contribution was not yet fully recognised. The Newtonian programme was partly subverted by a return to the mechanistic theory of a deterministic universe described by 'billiard-ball' atoms in various states of motion; and this 'clockwork universe' has remained the popular image of Newtonianism down to this day. Newton's mathematical theorems were accepted but his qualitative analyses never caught on. However, if the true Newtonian style was sufficiently powerful to short-cut three hundred years of experimental work and theory on so many aspects of the structure of matter, on the fundamental forces, and the fundamental physical laws, and with only the crudest empirical data as primary evidence, then it must have been based on some very powerful patterns of analytical thought, perhaps ones which are superior to our own.

Appendix

Some Results of Recent Scholarship

Newton is, in some respects, well understood, but in other respects he is an enigma. Later research has proved many earlier positions wrong, but the inertia in the system tends to lead to the same overall structure, especially in the more broad-brush and popular approaches. Sometimes the most important research on the scientific side (which is my own area of expertise) is by those who understand the more technical details, and this often takes a long time to filter through. In history of science, technical detail is very important and there has been a great deal in this account. Readers can find out here what both I and other scholars have added in this direction, but I think it worthwhile, for future reference, to emphasize some points where specific details have been corrected or modified, without necessarily being fully incorporated into the overall picture.

Many biographers have contributed to a reading of Newton's life and character, although I think the influence of Frank E. Manuel's Freudian interpretation in *A Portrait of Isaac Newton* (1968) still weighs too heavily in the balance.[1] Newton's personal life was certainly subjected to the usual twentieth century biases in the works written in that period. The sensationalisers are always with us, but, fortunately, in the scholarly field, most of this is recognised as nonsense. However, some of the more sensational claims have crept into popular culture, and need to be eradicated by firm scholarship. A few people have been unrelentingly hostile, based on emotional prejudice rather than rational argument.

Westfall, whose magnificent work will remain the basis of many aspects of Newton and his work for a very long time, was I think subliminally influenced by Manuel into too harsh a reading of Newton's character. Hall's work of

---

[1] Manuel (1968).

1992 offered a more balanced and contextual approach, though it is not as comprehensive in technical detail.[2] Many things said about Newton's quarrels with contemporary scientists are based on a lack of understanding generally of seventeenth century controversy, and sometimes his behaviour is taken in isolation rather than in relation to the behaviour of his opponents. Many early biographers were perhaps too generous to Leibniz, and later ones perhaps to Hooke. Individual authors will have their prejudices, but opinions and speculations are not facts. It is a massive speculation to assume that Newton, in quoting a mediaeval author about 'Standing on the shoulders of giants', was making a perverse reference to Hooke's small stature. It is also beyond the evidence to assert as 'fact' that Hooke's portrait is no longer extant because Newton had it destroyed.[3]

The exact facts are indeed crucial if we can ever establish them. If, for example, Brown is correct in his conclusion that Leibniz's algorithmic calculus postdated his second visit to London, and this can be documented, then we can no longer be sure that it was independent of his reading of Newton's *De analysi*, and relevant letters. If it was not, then Newton's actions in the whole controversy look very different to the way they have been viewed for the last two centuries. Rather than being an example of 'paranoia', they become consequences of his particular skill at analysing documentary evidence for historical and legal purposes. Seemingly 'unreasonable' behaviour would become much more reasonable, or at least understandable, in the new context. Again, if there was any influence from Newton, direct or subliminal, we would have to determine to what degree and in what way it contributed to what Brown believes was a Kuhnian paradigm shift in human knowledge. It would certainly be impossible to maintain that Newton's refusal to have his work printed in journal or book form meant that it played no significant part in the history of the subject. It is astonishing that such major shifts in interpretation could depend on the finest of judgments of a few disputed facts.

The extraordinary extent of Newton's mathematical work was uncovered by the great work of Whiteside, though, inevitably, it needs correcting on a few points, one of the most significant being his evaluation of the great Lemma 28 of Book I of the *Principia*, on transcendental curves. Fortunately, there have been a considerable number of analyses of the validity and

---

[2] Hall (1992).
[3] Hooke's biographer, Allan Chapman (2004), makes a strong case for the portrait never having existed.

significance of this theorem.[4] On the historical side, correction is needed on the use of calculus by Newton and after Newton. Guicciardini and a number of other scholars have shown that, contrary to the story in many accounts, Continental mathematicians were influenced by Newton and Newtonians, and Newtonian mathematics maintained a standard at least comparable with that on the Continent almost to the middle of the eighteenth century.[5] The old story that it was a question of notation seems to have very little validity, even though Leibniz's notation is particularly powerful, and, in relation to the dispute with Leibniz, Guicciardini has extended our understanding of what was meant by 'publication'.

In philosophical terms, Newton structured his physics on metaphysical ideas such as duality, symmetry, self-similarity, simplicity, abstraction. He was an effective proponent of Ockham's razor, though it would be interesting to know how he acquired this way of thinking. He gained no credit for his philosophy, even though it was probably the most significant ever produced in changing our understanding of the world, because he wrote no systematic treatise on the subject. Modern scholars, however, have begun to understand how profound it was, and how superficial previous discussions were on his understanding of such things as space and time.

Because he privileged the concept of mass, Newton was really the first physicist to emphasize and correctly enunciate the fundamental conservation principles — of mass, momentum, energy and angular momentum — and the first extremum principle — of zero force — and to structure physics upon them. This universalized physics at least as much as the law of gravitation. It also needed calculus, to distinguish the conserved from the nonconserved, especially in the form of differential equations, and has remained the basis for all physical theories ever since. Newton's mathematics was largely created on the basis of physical meaning, and the disputes on the foundations of calculus which occupied mathematicians for many decades after Newton occurred mainly because this wasn't understood. Some of the foundational issues were only resolved in the twentieth century, and then only amid further controversy. Newton also made qualitative physics so abstract that it became quasi-mathematical, making it possible to create generic pictures easily amenable to generalising mathematical laws.

Some of Newton's work on dynamics was misrepresented at earlier times. It was not until the 1960s, for example, that it was realised that Newton

---

[4] Arnol'd (1989); Pourciau (2001a); Pesic (2001).
[5] Guicciardini (2003a, 2003b); Grabiner (1997).

actually did know about conservation of energy, and wrote about it in the *Principia*, along with the conservation of momentum and of angular momentum. Even his early work on collisions makes no sense without all three principles. He also related them to the third law of motion through the conservation of mass. Currently, some authors have proclaimed that he had no knowledge of the conservation of angular momentum, and of the behaviour of rigid bodies. This is clearly not true, either in the published work or in the manuscripts, although it is very likely true that his knowledge of the *applications* was limited.

The Moon test of gravity undertaken in the 1660s has been misrepresented for a variety of reasons. One was a desire to avoid the 'annus mirabilis' 'myth', which seems to have led to something of an over-correction. The crucial paper of the 1660s describing the test of gravity wasn't printed until 1957.[6] Before that it was possible for authors to claim that Newton made up his earlier exploration of an inverse-square law. After that, the earlier position was defended by an insistence on the work's limited significance. At present, I think, there is not enough understanding of the extent to which Newton was able to think abstractly from the beginning, and how he brought together material from other sources to correspond increasingly with abstract positions he had already reached. It has not been fully realised how quickly he was able to derive abstract mathematical theories of physics which were only tangentially related to hypotheses. They were separate streams in his development. So, I don't believe his mathematical treatment of gravity was influenced significantly by ideas about Cartesian aethers or by alchemy,[7] although I do believe that he tried to bring these ideas into line with the position he had reached in the mathematics.

I have not found many points where Newton was significantly untruthful, even in his quarrels with contemporaries, where it might have been to his advantage.[8] It was apparently not one of his major character flaws, and we should be very careful before we say that a so-far uncorroborated statement must be untrue. Nearly all the claims that can be checked stand

---

[6]Hall (1957).
[7]Westfall (1980); Dobbs (1975, 1991).
[8]This is certainly the case with statements made in his own name. It could be argued that the 'anonymous' claim that he used analytical methods in discovering most of the propositions in the *Principia* (*Account*, 1715) is at least a massive overstatement of the position, and this may well be the case, but the claim is not what many have interpreted it to mean, the composition of a version of the *Principia* in analytical form. It is a statement about a method of *discovery*, and may even refer to mental, rather than scribal, composition.

up to scrutiny. Therefore, it seems rather astonishing that some people still think that he made up the apple story, or that someone else did, even though several accounts exist, stemming back from Newton himself. In view of the recursive way of thinking which we have identified in *Newton and Modern Physics*, it is not at all astonishing, and even highly likely that the incident happened exactly as he described it. Discoveries in this mode are often triggered by the most inconsequential of events. Also, looking at the manuscript of the Moon test, it does not seem at all that the idea generated was a long way from universal gravitation, for it is impossible to connect falling bodies and all the planets and satellites in the Solar System by the same force without it being 'universal' in some sense of the term. It is not yet a law connecting all particles of matter, but it does seem to be a law connecting astronomical bodies.

All scholarly work stands to be corrected by the discoveries of later authors. So Westfall, who produced much inciteful work on Newton's dynamics, wrote largely before Brackenridge and Nauenberg realised the significance of Newton's curvature method, and before Nauenberg's penetrating work on the lunar theory.[9] So Westfall, and also Hall, thought that Newton's dynamics of orbits was influenced by Hooke. Hooke has been wrongly interpreted as the originator of the dynamics he expounded to Newton, *but never claimed as his discovery*, when Bennett had already shown that it was actually originated by Wren.[10] Contrary to a whole host of authorities, including Whiteside, Westfall, and Kollerstrom, Nauenberg[11] has shown that the lunar theory was *successful* and Wepster[12] has shown that it was used by Tobias Mayer in the main purpose for which it was intended, as an aid to navigation. This is crucial because much of the negative comment about Newton's attempts to understand the lunar problem has been based on its supposed 'failure', and the story has been repeated by authority after authority, despite the evidence to the contrary. One of the main problems was that Newton couldn't establish an accurate value for the lunar mass and density, but he achieved remarkable results in spite of this. In addition, he was unable to obtain more than half the motion of the apsides, though he understood the source of the discrepancy in qualitative terms. In such cases, success was eventually achieved, not by introducing entirely new

---

[9] Westfall (1980); Brackenridge (1992, 1995); Brackenridge and Nauenberg (2002); Nauenberg (1994, 1999).
[10] Bennett (1982, 60–62).
[11] Whiteside (1976); Nauenberg (2000, 2001a, 2001b).
[12] Wepster (2010, 91).

principles, but by taking the problems to a higher order of calculation in the perturbation theory. Newton knew the correct method even in those cases where he was unable to achieve close numerical agreement with the observations. In general, I believe, many authors have given insufficient credit to Newton for his work on the System of the World (though Nauenberg is an exception), and there is still prejudice against doing so. Westfall was, I think, also wrong on Newton using 'fudge factors', partly perhaps because of his relatively harsh analysis of Newton's character.[13] I don't believe that any of Westfall's 'fudges' stands up to scrutiny, although for different reasons in each case, and only one of the calculations, the revised version of the precession of the equinoxes, falls below the standards of his best practice.

Some of Newton's more obscure discoveries are only slowly coming to light, even in the dynamics of the *Principia*. Chandrasekhar opened up corners of the work previously unnoticed.[14] He made errors in some of his judgements, and sometimes didn't make clear the distinction between Newton's original work and his own reconstructions.[15] He was, nevertheless, a great mathematical physicist, and could offer special insights from his own unique experience.[16] The relevance of some ideas is sometimes only realised when new discoveries are made. Newton's work on dual laws is now considered very important, partly because of the discovery of a cosmological constant. I have a personal interest in the virial theorem and was intrigued to find it discussed as a Corollary to Proposition 40 in Book I, immediately after a demonstration that potential energy acquired did not depend on the path.

The least understood part of the *Principia* has been Book II. Truesdell and others were negative partly because, as George E. Smith has shown, it provided something different to what they expected.[17] It offered a theory of

---

[13] Westfall (1971, 1973, 1980).
[14] Chandrasekhar (1995). With something like 800 Propositions, Corollaries, Cases, Examples, Lemmas, Scholia, Hypotheses, Definitions, Laws, Rules, Phenomena and occasional additional comments, it is by no means impossible that the *Principia* still has secrets waiting to be uncovered.
[15] See, for example, Nauenberg (1995); Dobson (1991, 1998, 2001).
[16] Guicciardini (2005), who considered Chandrasekhar's work an 'exceptional book', nevertheless criticised its lack of doubt in Newton's infallibility. Chandrasekhar, however, seems to have left out of his discussion most of the aspects of the *Principia* which might be considered contentious, and Guicciardini himself is one of a number of authors who have shown that some earlier criticisms of Newton's work were incorrect. He refers to some areas that *contemporaries* might have thought 'problematic', but the only one that seems genuinely problematic today is the rotational dynamics of rigid bodies.
[17] Truesdell (1960, 1967, 1968a, 1968b); Smith (1999, 2005).

resistance, rather than a theory of fluid mechanics. It is often stated that Book II contains mostly incorrect results, but most of the results are, in fact, correct, some only recently established as such. Some authors, for instance, have been wrongly negative about Proposition 34 and the Scholium which follows it; later analyses have shown that both are correct within their terms of reference. Proposition 36 has been seen as a disaster by almost every authority, when it is actually a spectacular success, with at least four major innovations. Many people have also had completely the wrong idea about the Boyle's law derivation. This is classic generic thinking by Newton, and is totally correct in these terms. It doesn't involve a physical model at all. That would only come with new evidence, essentially Brownian motion and it doesn't falsify Newton's mathematical analysis. Also, the study of Newton's disproof of Cartesian vortices using Propositions 51 and 52 has concentrated on the Propositions rather than the Corollaries, where Newton shows that he clearly understands all the main types of vortex motion and sees that none of the mathematical relations between velocity and distance can offer a fit to Kepler's laws.

No one, in my view, has fully understood Newton's physics in a modern context. I don't claim that I have, but I think I have given it more emphasis than most other authors. The electrical work has to be seen as an extraordinary breakthrough and a major instance of Newton's recursive method, but it has never really been presented in that light. It is not always stressed how much Newton contributed to all aspects of wave theory, one of the most important aspects of contemporary physics. His anticipation of wave-particle duality has often been pointed out as a remarkable coincidence, which is, in my view, a philosophical rather than historical misinterpretation. It has been claimed that its similarity to modern ideas was merely fortuitous and that Newton's ideas had no influence, neither of which is true.[18] The same has been said of Newton's particulate theory of light in relation to the modern quantum/photon theory. Every effort has been made to separate the two theories, by physicists as well as historians, but, as I previously explained in detail in *Waves Versus Corpuscles*, this is neither physically true nor historically true.[19] Some authors have noticed the parallels to quantum theory in many aspects of Newton's work, which are neither accidental nor fortuitous. They are there because Newton thought generically whenever possible, and this is where one ends up if one thinks

---

[18]Thomas Young had already recognised that Newton's theory was dualistic as early as 1817 (1817b, 325), well before the twentieth century revival of wave-particle duality.
[19]Rowlands (1992).

this way. The intimations of quantum theory and wave-particle duality are also especially important in showing the highly significant degree of continuity and absence of dislocation between Newtonian ideas and quantum mechanics.

Nearly everyone has misunderstood the significance of aether theories, mainly because of the emphasis on their supposed rejection in Einstein's relativity and their supposed disproof in the Michelson-Morley experiment. However, the aether-theory background is very important to the quantum theory, both historically and at the present day.[20] If we want to penetrate deeper into the nature of fundamental physics, we will have to understand why this kind of thinking emerged at regular intervals in scientific history, and remains with us today in the concept of the quantum mechanical vacuum. Newton's investigation of aether theories was no more a blunder or retrograde step than his continued interest in a particle theory of light. The relationship of Newtonian theory to relativity, as discussed extensively in *Newton and the Great World System*, is another subject which requires a deeper understanding. The historical picture is still largely coloured by the spin introduced by Eddington in 1919. Again, most authors, who are concerned with using the word 'Newtonian' as the antithesis of 'modern' or 'contemporary', have taken little notice of the extensive use of Newtonian theory in modern cosmology, and the demonstration of its validity in the most accurate modern experiments.

The theory of matter, opened up by authors such as Hall and Thackray, is a huge potential field for work in Newton studies because the Newtonian development is still so little understood.[21] Newton has been given little credit for eighteenth century chemistry because of the concentration in the later part of the century on systematics, as opposed to his search for physical solutions. His influence, however, has surely been underestimated, or wrongly described in negative terms, along with his significance, with Robert Boyle, in translating alchemical practice into recognisably chemical processes. The idea that alchemy was a fundamentally religious practice for Newton, and the sensationalist and negative interpretations, beginning with Brewster and

---

[20] Frank Close has, for example, explained (2014) that the reason why Michelson and Morley failed to detect an aether was because the gauge boson of the electromagnetic interaction they were using was massless and so its speed in vacuum couldn't be altered, whereas the gauge bosons for the weak interaction (W and Z) are massive and allow the detection of both the continuous aether-like Higgs field and its excitation as the Higgs boson.

[21] Hall and Hall (1962); Hall (1992); Thackray (1970).

Keynes, seem to be giving way through the work of William R. Newman, to an emphasis on its scientific character and relation to chemistry and the structure of matter.[22] In addition to chemistry, alchemy also led in Newton's work to cross-over developments in optics and thermometry. Heat was an important study for Newton and there is a long article by D. L. Simms which collects many of Newton's thoughts on the subject, although I think some of its conclusions have been superseded by later scholarship.[23] Even more important, however, were the glimmerings of an understanding of thermodynamics, which I think is unique to Newton in this period. Of special importance was his realisation, as his *own* contemporaries protested, that the universe was not run on clockwork and was not deterministic, even though this has become part of the popular understanding of the Newtonian world picture.

Always aware of the tentative nature of his conclusions, and always attempting to provide a generic theory rather than one that was hypothesis-specific, Newton made no errors in science with major consequences. The denial of achromatic lenses, on which he may have changed his mind more than once, and which began to be corrected soon after his time, hardly counts in this category, nor do the unsatisfactory hypotheses of whose failure he was fully aware.[24] All his really 'dubious' work — by anachronistic modern 'scientific' standards — is related to non-scientific areas such as religion and theology. Even here, there is much of interest from a scientific point of view, especially a tendency towards abstraction and the general. The study of Newton's work in theology, prophecy and chronology has been revolutionised by in-depth scholarship which has shown that it deserves the same detailed attention as the science in understanding the extraordinary mind of its creator. Even for a study like mine, which concentrates mostly on the science, some consideration of this rich and larger context is essential, and the work of the scholars quoted and referenced is beginning to create a coherent picture of Newton's complex activities within these fields, and his equally complex relationships with his predecessors.

It is clear now that Newton's theologically-inspired metaphysics was the foundation of and driving force behind all his extraordinary achievements in science. Though he wrote very little directly on the subject, Newton was undoubtedly one of the greatest of all metaphysical thinkers in that

---

[22] Newman (2002).
[23] Simms (2004).
[24] These included two successive 'laws' of optical dispersion and the attempt at a rule for double refraction, which was stated only in a query.

he produced a system that penetrated to some of the most profound depths of abstract thought ever imagined. It has seldom been stated until recently that Newton's work on textual analysis is sound, as are some of the principles he applied to chronology. He also made shrewd guesses concerning cultural history. While his work on Solomon's Temple has only recently been understood and favourably estimated, his work on economics remains to be properly evaluated, and freed especially from the mid-twentieth century prejudices of Sir John Craig.[25] Newtonian studies are flourishing at present, after a long period of neglect before the middle of the twentieth century. However, the fruits of all the dedicated scholarship need to be fully incorporated into the overall picture.

---

[25] Westfall (1980, 616–617), believed that Craig's judgment of Newton's time at the Mint (Craig, 1946) was seriously flawed because he criticized Newton for failing to use powers that he never possessed. Newton could give advice but he could not dictate policy. Shirras and Craig (1945) were no more positive about Newton as an economist, but views on economic theory have undergone many changes since their time, and a proper assessment of Newton's contribution awaits further research. I think it is very unlikely that there was *any* subject on which, as Westfall claimed with regard to economics, Newton's 'views simply repeated accepted opinions' (*op. cit.*, 619).

Bibliography

Abbreviations

Corr.   H. W. Turnbull, J. F. Scott, A. R. Hall and L. Tilling, *The Correspondence of Isaac Newton*, 7 vols., Cambridge University Press, 1959–1977.
CUL   Cambridge University Library.
Hyp.   *An Hypothesis Explaining the Properties of Light*, 1675, *Correspondence*, I, 361–86.
KCC   King's College, Cambridge.
MP   D. T. Whiteside, *Mathematical Papers*, 8 volumes, Cambridge University Press, 1967–1981.
NP   Rob Iliffe and Scott Mandelbrote, The Newton Project, http://www.newtonproject.ox.ac.uk/
NPA   William R. Newman, The Chymistry of Isaac Newton Project, http://webapp1.dlib.indiana.edu/newton/.
NPC   Stephen D. Snobelen, The Newton Project Canada, http://www.isaacnewton.ca/Isaac Newton.
OP   Alan E. Shapiro, *The Optical Papers of Isaiac Newton*, vol. 1, Cambridge University Press, 1984.
QQP   *Quaestiones quaedam philosophicae*, NP, *Certain Philosophical Questions: Newton's Trinity Notebook*, c 1664, ed. J. E. McGuire and M. Tamny, Cambridge, 1983, Q1-31 *Opticks*, Queries 1–31.
Unp.   A. Rupert Hall and Marie Boas Hall, *The Unpublished Scientific Papers of Isaac Newton*, Cambridge University Press, 1962.

Works Cited

Aiton, E. J., *The Vortex Theory of Planetary Motion*, Science History Publications, New York, 1972.

Alexander, H. G., *The Leibniz-Clarke Correspondence*, Manchester University Press, Manchester, 1956.

Anderson, John D., Jr, *A History of Aerodynamics*, Cambridge University Press, Cambridge, 1997.

Arnol'd, V. I., *Huygens and Barrow, Newton and Hooke*, Birkhäuser Verlag, Basel, 1990.

Bacon, Francis, *Novum Organum*, 1620.

Bacon, Roger, *The 'Opus maius' of Roger Bacon*, ed. J. H. Bridges, 3 vols., 1897–1900.

Badash, Lawrence, 'Radium, Radioactivity and the Popularity of Scientific Discovery', *Proceedings of the American Philosophical Society*, 122, 145–154, 1978.

Barkla, H. M. and Auchterlonie, L. J., 'The Magnus or Robins effect on rotating spheres', *J. Fluid Mech.*, 47, part 3, 437–447, 1971.

Barrow, Isaac, *The Usefulness of Mathematical Learning Explained and Demonstrated: Being Mathematical Lectures Read in the Publick Schools at the University of Cambridge*, translated John Kirkby, London, 1734.

Bechler, Zev, 'Newton's Search for a Mechanistic Model of Colour Dispersion: A Suggested Interpretation', *Archive for History of Exact Sciences*, 11, 1–37, 1973.

Bechler, Zev, 'Newton's law of forces which are inversely as the mass: a suggested interpretation of his later efforts to normalise a mechanistic model of optical dispersion', *Centaurus*, 18, 184–222, 1974.

Bechler, Zev, "A less agreeable matter": The Disagreeable Case of Newton and Achromatic Refraction', *British Journal for the History of Science*, 8, 101–126, 1975.

Becquerel, Henri, *Comptes Rendus*, 122, 501–503, 559–564 (read 9 March), 689–694, 762–767, 1896.

Belenkiy, Ari, 'The Master of the Royal Mint: how much money did Isaac Newton save Britain?', *J.R. Statist. Soc. A*, 176, Part 2, 481–498, 2013.

Bentley, Richard, *A Confutation of Atheism from the Origin and Frame of the World*, London, 1693 (the Boyle lectures for 1692–1693).

Bernal, J. D., *The Extension of Man: A History of Physics before 1900*, M.I.T. Press, Cambridge, Massachusetts and London, 1972.

Berry, A. J., *From Classical to Modern Chemistry: Some Historical Sketches*, Cambridge University Press, 1954.

Blank, Brian E., 'T'he Calculus Wars', Review of Jason Socrates Bardi, *The Calculus Wars: Newton, Leibniz, and the Greatest Mathematical Clash of All Time*, Basic Books, 2007, *Notices of the AMS*, 56, no. 5, 602–610, May 2009.

Blitzer, L., 'Satellite Orbit Paradox: A General View', *American Journal of Physics*, 39, 882–886, 1971.

Boerhaave, H., *Institutiones medicae*, Leiden, 1708.

Boerhaave, H., *Institutiones et Experimenta chemiae*, Paris, 1724 (unauthorised).

Boerhaave, H., *A New Method of Chemistry; Including the Theory and Practice of That Art*, translated by P. Shaw and E. Chambers, 2 vols., London: J. Osborn & T. Longman, 1727.

Boerhaave, H., *Elementa chemiae*, Leiden, 1732.

Boltzmann, Ludwig, 'Über das Wärmegleichgewicht zwischen mehratomigen Gasmolekülen', *Wiener Berichte*, 63, 397–418, 1871.

Boltzmann, Ludwig, 'Einige allgemeine Sätze über Wärmegleichgewicht', *Wiener Berichte*, 63, 679–711, 1871.

Boltzmann, Ludwig, 'Analytischer Beweis des zweiten Hauptsatzes der mechanischen Wärmetheorie aus den Sätzen über das Gleichgewicht der lebendigen Kraft', *Wiener Berichte*, 63, 712–732, 1871.

Boltzmann, Ludwig, 'Weitere Studien über das Wärmegleichgewicht unter Gasmolekülen', *Wiener Berichte*, 66, 275–370, 1872.

Boltzmann, Ludwig, 'Uber die Beziehung eines Allgemeine Mechanischen Satzes zum zweiten Hauptsatze der Warmetheorie' ('On the Relation of a General Mechanical Theorem to the Second Law of Thermodynamics'), *Sitzungsberichte Akad. Wiss.*, Vienna, Part II, 75, 67–73, 1877.

Boltzmann, Ludwig, *Sitzungsberichte der Kaiserlichen Akademie der Wissenschaften (Wien)*, 373–435, 1877 (submitted 11 October).

Bošković, R. J., *De cometis*, 1746.

Bošković, R. J., *Theoria Philosophiae Naturalis*, 1748.

Bošković, R. J., *De Lumine*, 1749; *A Treatise of Natural Philosophy*, translated by J. M. Child, Cambridge, Mass., 1966.

Boyle, Robert, *New Experiments Physico-Mechanical, Touching the Spring of the Air*, 1660.

Boyle, Robert, *Certain Physiological* Essays, 1661.

Boyle, Robert, *Experiments and Considerations Touching Colours*, December 1663.

Boyle, Robert, *New Experiments and Observations Touching Cold, or An Experimental History of Cold Begun*, 1665.

Boyle, Robert, *The Origine of Formes and Qualities*, printed by H. Hall for Ric. Davies, Oxford, 1668.

Boyle, Robert, *Certain Physiological Essays*, second edition, Henry Herringman, London, 1669.

Boyle, Robert, 'Of the Incalescence of *Quicksilver* with *Gold*, generously imparted by B.R.', *Philosophical Transactions*, 10, 515–533, 1676.

Boyle, Robert, *The Christian Virtuoso*, London, printed by E. Jones for J. Taylor, London, 1690.

Brewster, Sir David, *The Life of Sir Isaac Newton*, John Murray London, 1831.

Brewster, Sir David, *Memoirs of the Life, Writings, and Discoveries of Isaac Newton*, 2 vols., Edinburgh Thomas Constable & Co., 1855, New York: Johnson Reprint Corporation, 1865.

Brougham, H. P. and Routh, E. J., *Analytical view of Sir Isaac Newton's Principia*, Longman, London, 1855.

Brown, Richard C., *The Tangled Origins of the Leibnizian Calculus: A Case Study of a Mathematical Revolution*, World Scientific, Singapore, Hackensack, NJ and London, 2012.

Brückmann, F. E., *Magnolia Dei in Subterraneis, oder Unterirdische Schatzkammer aller Konigreiche und Lander*, Helmstadt, 1, 204, 1727.

Buchwald, Jed Z. and Feingold, Mordechai, *Newton and the Origin of Civilization*, Princeton University Press, Princeton and Oxford, 2012.

Buffon, Comte de, *Histoire naturelle*, 1749–1767.

Buttazzo, Giuseppi and Fredianao, Aldo (eds.), *Variational Analysis and Aerospace Engineering*, Springer-Verlag, New York, 2009.

Calero, Julián Simón, *The genesis of fluid mechanics*, 1640–1780, 2008.

Callendar, H. L., 'The Caloric Theory of Heat and Carnot's Principle', Presidential Address to the Physical Society of London, *Proc. Roy. Soc.*, 23, 153–189, 1911.

Cardwell, D. S. L., *From Watt to Clausius*, Heinemann Educational Publishers, London, 1971.

Carnot, Sadi, *Réflexions sur la Puissance Motrice du Feu*, 1824, translated as *Reflection on the Motive Power of Fire*, Dover, 1960.

Cascetta, Furio, 'Short history of flowmetering', *ISA Transactions*, 34, 229–243, 1995.

Chandrasekhar, S., *Newton's Principia for the Common Reader*, 1995, 201.

Chang, Hasok, *Inventing Temperature: Measurement and Scientific Progress*, Oxford Studies in the Philosophy of Science, Oxford University Press, Oxford, 2004.

Chang, Hasok, 'The Myth of the Boiling Point', online hypertext paper with video links, 18 October 2007 (www.hps.cam.ac.uk/boiling/).

Charleton, Walter, *Physiologia Epicuro-Gassendo-Charltoniana*, 1654.

Cheng, K. C., 'Some Observations on the Origins of Newton's Law of Cooling and Its Influences on Thermofluid Science', Applied Mechanics Reviews, *Transactions of the ASME*, 62, 060803–060816, 2009.

Cheyne, George, *New Theory of Fevers*, London: printed, and sold by H. Newman, and J. Nutt, 1701, reissued 1702.

Cheyne, George, *Fluxionum Methodus Inversa*, London: J. Matthews, sold by R. Smith, 1703.

Clairaut, A.-C., *Théorie de la figure de la Terre* Paris: Drand, 1743.

Clairaut, A.-C., *Théorie du movement des Cométes*, M. Lambert, Paris, 1760.

Clausius, R., 'Über die bewegende Kraft der Wärme, Part I, Part II', *Annalen der Physik*, 79, 368–397, 500–524, 1850, translated as 'On the Moving Force of Heat, and the Laws regarding the Nature of Heat itself which are deducible therefrom', *Philosophical Magazine*, 2, 1–21, 102–11, 1851.

Clausius, R., 'Über die Wärmeleitung gasförmiger Körper', *Annalen der Physik*, 125, 353–400, 1865; *The Mechanical Theory of Heat — with its Applications to the Steam Engine and to Physical Properties of Bodies*, John van Voorst, London, 1867.

Cockcroft, J. D. and Walton, E. T. S., 'Disintegration of Lithium by Swift Protons', *Nature*, 129, 649, 1932.

Cohen, I. Bernard, Preface to I. Newton, *Opticks*, 1931, reprinted Dover Publications, 1952 (based on the fourth edition, 1730).

Cohen, I. Bernard, *Isaac Newton's Papers and Letters of Natural Philosophy*, Massachussetts, Cambridge, Harvard University Press, 1958.

Cohen, I. Bernard, 'Hypotheses in Newton's Philosophy', *Physis*, 8, 163–164, 1966.

Cohen, I. Bernard, *Introduction to Newton's 'Principia'*, Harvard University Press, Cambridge, Massachussetts, 1971.

Cohen, I. Bernard, 'A Guide to Newton's *Principia*', in I. Bernard Cohen and Anne Whitman, *Isaac Newton The Principia A New Translation*, Berkeley and Los Angeles and London: University of California Press, 1999, 1–370.

Cohen, I. Bernard and Westfall, Richard S. (eds.), *Newton A Norton Critical Reader*, W. W. Norton & Company, 1995, General Introduction, xi–xii.

Cohen, I. Bernard and Whitman, Anne, *Isaac Newton The Principia A New Translation*, Berkeley and Los Angeles and London: University of California Press, 1999.

Coleridge, S. T., *The Friend*, in *Collected Works*, ed. B. Rooke, Princeton, NJ, 1969, vol. 4, part 1.
Comte, M. and Diaz, J. I., 'On the Newton partially flat minimal resistance body type problems', 20 December 2004.
Costabel, Pierre, 'Newton's and Leibniz's dynamics', *The Texas Quarterly*, 10, 1967, 119–126.
Cotes, Roger, preface to the second edition of the *Principia*, 1713.
Coulomb, C. A., 'Premier mémoire sur l'électricité et le magnétisme', *Histoire de l'Académie Royale des Sciences*, 574, 569–577, 1785.
Coulomb, C. A., 'Second mémoire sur l'électricité et le magnétisme', *Histoire de l'Académie Royale des Sciences*, 579, 578–611, 1785.
Coveney, R. and Highfield, P., *The Arrow of Time*, London, 1990.
Craig, John, *Methodus Figurarum ... Quadraturas Determinandi (Method of Determining the Quadratures of Figures Bounded by Curves and Straight Lines)*, London, 1685.
Craig, Sir John, *Newton at the Mint*, Cambridge University Press, 1946, 44.
Curie, Pierre, Curie, Marie and Bémont, Gustave, 'Sur une nouvelle substance fortement radio-active, contenue dans la pechblende (On a new, strongly radioactive substance contained in pitchblende)', *Comptes Rendus*, 127, 1215–1217, 1898.
d'Alembert, J.-le-R., *Mémoires de l'Académie de Berlin*, 1747.
d'Alembert, J.-le-R., *Essai d'une nouvelle théorie de la résistance des fluides*, Paris, 1752.
Dalton, John, 'On the Absorption of Gases by Water and Other Liquids', *Manchester Memoirs*, i, 271–287, 1805 (read 21 October 1803).
Dalton, John, *New System of Chemical Philosophy*, 1808, 1810.
Davy, Humphry, *An Essay on Heat, Light and the Combinations of Light*, 1799.
Dean, Dennis R., 'Smith, George', *Oxford Dictionary of National Biography*, 2004.
de la Pryme, A., *Diary of Abraham de la Pryme*, ed. Charles Jackson, Durham, 1870.
Desaguliers, John T., 'Remarks on some attempts towards a perpetual motion', *Philosophical Transactions*, 31, 234–239, 1721, presented 9 November 1721.
Desaguliers, John T., 'An account of an Optical Experiment made before the Royal Society, on Thursday, Dec. 6th, and reported on 13th, 1722', *Philosophical Transactions*, 32, 206–208, 1722.
Desaguliers, John T., 'Some Thoughts and Conjectures Concerning the Cause of Electricity', *Philosophical Transactions*, 41, 175–185, 1739.
Desaguliers, John T., *A Course of Experimental Philosophy*, third edition, London, 1763.
Descartes, René, letter to Mersenne, 13 November 1629, *The Philosophical Writings of Descartes*, vol. 3 (*Correspondence*), trs. John Cottingham, Robert Stoothoff, Dugald Murdoch, and Anthony Kenny, Cambridge: Cambridge University Press, 1991, p. 9.
Dircks, H., *Perpetuum mobile: Or, A history of the search for self-motive power, from the 13th to the 19th century*, 1968.

DiSalle, Robert, 'Newton's philosophical analysis of space and time', in I. Bernard Cohen and George E. Smith (eds.), *The Cambridge Companion to Newton*, Cambridge University Press, 2002, 33–56.

Dobbs, Betty Jo Teeter, *The Foundations of Newton's Alchemy or, The Hunting of the Greene Lyon*, Cambridge University Press, 1975.

Dobbs, Betty Jo Teeter, *The Janus Faces of Genius*, Cambridge University Press, 1991.

Dobson, Geoffrey J., 'Newton's Errors with the Rotational Motion of Fluids', *Archive for History of the Exact Sciences*, 54, 243–254, 1999.

Drazin, Philip, 'Fluid mechanics', *Physics Education*, 22, 350–354, 1987.

Ducheyne, Steffen, 'Newton's training in the Aristotelian textbook tradition: from effects to causes and back', *History of Science*, 43, 215–237, 2005.

Ehrlichson, H., 'How Newton went from a mathematical model to a physical model for the problem of a first power resistive force', *Centaurus*, 34, 272–283, 1991.

Ehrlichson, H., 'Newton's first inverse solutions', *Centaurus*, 34, 345–366, 1991.

Ehrlichson, H., 'Evidence that Newton used the calculus in some of the propositions in his *Principia*', *Centaurus*, 39, 253–266, 1996.

Einstein, Albert, 'On the electrodynamics of moving bodies', *Annalen der Physik*, 17, 891–921, 26 Sep 1905 (received 30 June); translated in A. Sommerfeld (ed.), *The Principle of Relativity*, New York, 35–65.

Ellis, R. L., 'On a question in the theory of probability', 1844, in *Mathematical and other writings of R. L. Ellis*, Cambridge, 1863, 173–179.

Euler, Leonhard, *Mechanica sive molus scientia analytice exposita*, 2. vols. 1736–1742.

Euler, Leonhard, 'Recherches sur irrégularities du movement de Jupiter et de Saturne', *Recueil des pieces qui ont remporté les prix de l'Académie des Sciences*, vol. 7, Paris, 1752.

Euler, Leonhard, *The True Principles of Gunnery*, translation of two papers by H. Brown, 1777.

Everitt, C. W. F., 'Maxwell, James Clerk', *DSB*, 9: 198–230, 1974.

Falk, G., 'Entropy, a resurrection of caloric — a look at the history of thermodynamics', *Eur. J. Phys.*, 6, 108–115, 1985.

Fardin, M. A., Perge, C. and Taberlet, N., ' "The hydrogen atom of fluid dynamics" — introduction to the Taylor–Couette flow for soft matter scientists', *Soft Matter*, 10, 3523–3535, 2014.

Fatio de Duillier, Nicolas, *De la cause de la pesanteur (On the Cause of Weight)*, read to the Royal Society, 1688, published in Bernard Gagnebin, 'De la cause de la pesanteur. Mémoire de Nicholas Fatio de Duillier présenté à la Royal Society le 26 février 1690', *Notes and Records of the Royal Society*, 6, 106–160, 1949.

Fay, C. R., 'Newton and The Gold Standard', *Cambridge Historical Journal*, 5, 109–117, 1935.

Feingold, Mordechai, 'Isaac Newton, historian', in Rob Iliffe and George E. Smith (eds.), *The Cambridge Companion to Newton*, second edition, Cambridge University Press, 2016, 524–553.

Fellmann, E. A., *G. W. Leibniz — Marginalia in Newtoni Principia Mathematica (1687)*, Paris: Vrin, 1973.

Fellmann, E. A., 'The *Principia* and Continental Mathematicians', in D. C. King-Hele and A. R. Hall (eds.), *Newton's Principia and its Legacy, Notes and Records of the Royal Society*, 42, no. 1, 13–34, 1988.

Ferguson, James, 'A brief survey of the history of the calculus of variations and its applications', *arxiv.org/pdf/math/0402357*.

Figala, Karen, 'Newton as alchemist', *History of Science*, 18, 103–137, 1977.

Fontenelle, Bernard de, *Elogium*, 1728.

Freind, John,*Praelectiones chymica*, 1709.

Gagnebin, Bernard, 'De la cause de la pesanteur. Memoire de Nicholas Fatio de Duillier presente a la Royal Society le 26 fevrier 1690', *Notes and Records of the Royal Society*, 6, 106–160, 1949.

Galileo, *Dialogues Concerning Two New Sciences*, trs. Henry Crew and Alfonso de Salvio, Buffalo: Prometheus Books, 1991.

Gassendi, Pierre, *Animadversiones (Obsevations)*, 1649.

Gauld, Colin F., 'Newton's Use of the Pendulum to Investigate Fluid Resistance: A Case Study and Some Implications for Teaching about the Nature of Science', *Sci. & Educ*, 18, 383–400, 2009.

Gauld, Colin F., 'Newton's Investigation of the Resistance to Moving Bodies in Continuous Fluids and the Nature of 'Frontier Science'', *Sci. & Educ*, 19, 939–961, 2010.

Geoffroy, E. F., 'Table des differens rapports observes en chemie entre differentes substances', *Mémoires de l'Académie Royale des Sciences*, 1718, 202–212.

Giacomelli, R. (with Pistoleis, E.), 'Historical Sketch', in William Frederick Dund (ed.), *Aerodynamic Theory: A General Review of Progress*, vol. 1, Dover Publications, New York, 1963, 311 ff.

Gillmore, Gavin K., Phillips, Paul S., Pearce, Gillian and Denman, Antony, 'Two abandoned metalliferous mines in Devon and Cornwall, UK: radon hazards and geology', *2001 International Radon Symposium*, 94–105.

Gjertsen, Derek, *The Newton Handbook*, Routledge & Kegan Paul, 1986.

Gleick, James, *Isaac Newton*, HarperCollins, London, 2003.

Grabiner, Judith V., 'Was Newton's Calculus a Dead End? The Continental Influence of Maclaurin's *Treatise of Fluxions*', *American Mathematical Monthly*, 104 (5), 393–410, 1997.

Gram, Hans, *De origine geometriae apud Aegyptios*, Copenhagen, 1706.

Graves, R. P., *Life of Sir William Rowan Hamilton*, 3 vols., Dublin, 1882–1891.

Green, Robert, *The Principles of the Philosophy of the Expansive and Contractive Forces*, 1727.

Gregory, David, *Exercitatio geometrica de dimensione figurarum (Geometrical Exercise on the Measurement of Figures)*, 1684.

Gregory, David, *Astronomiae physicae et geometricae elementa*, Oxford, 1702.

Guerlac, Henry, 'Newton's Optical Aether: His Draft of a Proposed Addition to His *Opticks*', *Notes and Records of the Royal Society*, 22, No. 1/2, 45–57, September 1967.

Guicciardini, Niccolò, *The Development of Newtonian Calculus in Britain 1700–1800*, Cambridge University Press, 1989.

Guicciardini, Niccolò, 'Did Newton use his calculus in the *Principia*?', *Centaurus*, 40, 303–344, 1998.

Guicciardini, Niccolò, *Reading the Principia*, Cambridge University Press, 2003.

Guicciardini, Niccolò, 'Introduction Conceptualism and contextualism in the recent historiography of Newton's Principia', *Historia Mathematica*, 30, 407–431, November 2003.

Guicciardini, Niccolò, *Isaac Newton on Mathematical Certainty and Method*, The MIT Press, Cambridge, Mass. and London, 2009.

Guicciardini, Niccolò, "Gigantic implements of war": images of Newton as a mathematician', in Eleanor Robson and Jacqueline Stedall (eds.), *The Oxford Handbook of the History of Mathematics*, chapter 8.2, Clarendon Press, Oxford, 2009.

Hales, Stephen, *Vegetable Staticks*, London, 1727.

Hall, A. Rupert, *Philosophers at War: The Quarrel between Newton and Leibniz*, Cambridge, 1980.

Hall, A. Rupert, *Isaac Newton Adventurer in Thought*, Cambridge University Press, 1992.

Hall, A. Rupert, 'Newton versus Leibniz: from geometry to metaphysics', in I. Bernard Cohen and George E. Smith (eds.), *The Cambridge Companion to Newton*, Cambridge University Press, 2002, 431–454.

Halley, Edmond, 'A Discourse tending to prove at what time and place, Julius Caesar made his first descent upon Britain', *Philosophical Transactions*, 16, 1691, 495–501.

Halley, Edmond, *Astronomiae cometicae synopsis*, Oxford, 1705; *A Synopsis of the Astronomy of Comets*, London, 1705; *Philosophical Transactions*, 24, 1882–1889, 1704–1705.

Hamilton, Gavin, *Patterns in Fluid Flow Paradoxes Variations on a Theme*, The University of Western Ontario, London, Canada, 1980.

Hamilton, W. R., *Transactions of the Royal Irish Academy*, 21, 199–296, 1848, *Mathematical Papers*, III, 159–226, dated 13 November 1843.

Hamilton, W. R., 'On Quaternions, or a new System of Imaginaries in Algebra', *Philosophical Magazine*, 3rd series, 25, 10–13, 241–246, 489–495, 1844.

Hankins, T. L., *Sir William Rowan Hamilton*, Baltimore and London, 1980.

Harrison, David Mark, *The Influence of Non-linearities on Oscillator Noise Performance*, PhD Thesis, September 1988, Leeds University.

Hauksbee, Francis, 'Several experiments on the mercurial phosphorus, made before the Royal Society, at Gresham-College', *Philosophical Transactions*, 24, 2129–2135, 1704–1705.

Hauksbee, Francis, *Physico-Mechanical Experiments on Various Subjects*, London 1709, second edition, 1719.

Hazen, Robert M., *The Diamond Makers*, Cambridge University Press, 2003.

Heilbron, J. L., *Elements of Early Modern Physics*, University of California Press, 1982.

Helmholtz, H. von, *Über die Erhaltung der Kraft* (*On the Conservation of Force*), 1847.

Henry, John, 'Gravity and De Gravitatione: the development of Newton's ideas on action at a distance', *Studies in History and Philosophy of Science*, 42, 11–27, 2011.

Henry, William, 'Experiments on the Quantity of Gases Absorbed by Water at Different Temperatures and Under Different Pressures', *Philosophical Transactions*, 93, 29–42, 274–276 (dated 8 December 1802, read to the Royal Society on 23 December).

Herivel, John, *The Background to Newton's Principia*, Oxford: Clarendon Press, 1965, 257–289.

Herivel, John, *Joseph Fourier, The Man and the Physicist*, Clarendon, Oxford, 1975.

Herschel, John, *Philosophical Magazine* (3), 2, 401–412, 1833, 401.

Herschel, William, *Philosophical Transactions*, 1803, 339, presented 9 June.

Herschel, William, *Philosophical Transactions*, 1804, 353.

Herzl, Theodor, *Der Judenstaat (The State of the Jews)*, 1896.

Hiscock, W. G., *David Gregory, Isaac Newton and their Circle*, Oxford, 1937.

Hobden, M. K. 'John Harrison, Balthazar van der Pol and the Non-Linear Oscillator', *Horological Journal*, July 1981, 16–18, November 1981, 16–19, February 1982, 26–28, April 1982, 15–18, June 1982, 14–15.

Hobden M. K. 'As 3 is to 2', *Horological Science News*, November 2011, NAWCC Chapter. 161.

Hofmann, J. E., *Leibniz in Paris, 1672–1676*, Cambridge, 1974.

Homberg, W., *Mémoires de l'Académie des Sciences*, 1705, 88.

Hooke, Robert, *Tract on Capillary Action*, 1661, later included in *Micrographia*, 1665; reprinted in R. T. Gunther, *Early Science in Oxford*, Volume X: *The Life and Work of Robert Hooke* (Part IV), 1935.

Hooke, Robert, *Micrographia*, London, 1665.

Howorth, Muriel, *Pioneer Research on the Atom: The Life Story of Frederick Soddy*, New World, London, 1958.

Hui, Qing, 'Newton's minimal resistance problem', *Project of Math 7581*, 13 April 2006, 1–18.

Hume, David, *A Treatise of Human Nature*, 1739.

Huygens, Christiaan, *Discours de la Cause de la Pesantour (Discourse on the Cause of Gravity)*, published with *Traité de la Lumière* (Leyden, 1690), *Oeuvres complètes publiés par la Société hollandaise des Sciences*, 22 vols. (The Hague, 1888–1950), 21.

Huygens, Christiaan, *Oeuvres completes*, 22, 268, 1888–1950.

Iliffe, Rob, ' 'That puzleing Problem': Isaac Newton and the Political Physiology of Self', *Medical History*, 39, 433–458, 1995.

Iliffe, Rob, 'Abstract considerations: disciplines and the incoherence of Newton's natural philosophy', *Studies in the History and Philosophy of Science*, 35 (3), 427–454, 2004.

Iliffe, Rob, 'The religion of Isaac Newton', in Rob Iliffe and George E. Smith (eds.), *The Cambridge Companion to Newton*, second edition, Cambridge University Press, 2016, 485–523.

Iliffe, Rob and Smith, George E., 'Introduction' to *The Cambridge Companion to Newton*, second edition, Cambridge University Press, 2016, 1–33.

James, Peter *et al.*, *Centuries of Darkness*, London, 1991.

Jansson, T. R. N., Haspang, M. P, Jenson, K. H., Hersen, P. and Bohr, T., 'Polygons on a Rotating Fluid Surface', *Phys. Rev. Lett.*, 96, 174502, 2006.

Jarlskog, Cecilia, 'Lord Rutherford of Nelson, His 1908 Nobel Prize in Chemistry, and Why He Didn't Get a Second Prize', arXiv:0809.0857v1, 11.

Jones, Inigo, *The Most Notable Antiquity of Great Britain Vulgarly called Stone-Heng on Salisbury Plain Restored*, 1655.

Jones, William, *Analysis per quantitatum series, fluxiones, ac differentias: cum enumeratione linearum tertii ordinis*, London, 1711.

Joule, James Prescot, 'On Matter, Living Force, and Heat', 1847, *The Scientific Papers of James Prescott Joule*, 1884, I, 271.

Jungk, Robert, *Brighter than a Thousand Suns*, New York: Harcourt Brace, 1958.

Keill, James, *An Account of Animal Secretion, the Quantity of Blood in the Humane Body, and Muscular Motion*, 1708.

Keill, John, *Introductio ad veram physicam* (*Introduction to the New Physics*), 1702, translated, 1720.

Keill, John, *De legibus virium centripetarum* (*Concerning the Laws of Centripetal Forces*), *Phil. Trans.*, 26, 174–88, 1708, appeared 1710.

Keswani, G. H. and Kilmister, C. W., *British Journal for Philosophy of Science*, 34, 342–354, 1983, 343.

Keynes, J. M., 'Newton the Man', on Royal Society, *Newton Tercentenary Celebrations*, Cambridge University Press, 1947, 27–34.

King-Hele, D. C. and Walker, D. M. C., *Vistas in Astronomy*, vol. 30, p. 271, 1987.

Klaproth, M. H., 'Chemische Untersuchung des Uranits, einer neuentdeckten metallische Substanz', *Chem. Ann. Freunde Naturl.* 2, 387–403, 1789.

Koestler, Arthur, 'Johannes Kepler', *Encyclopedia of Philosophy*, 8 vols., New York, Macmillan, 1967, IV, 331.

Koyré, A., 'Newton's Regulae Philosophandi', *Newtonian Studies*, 1965.

Kracht, Manfred and Kreyszig, Erwin, 'E. W. von Tschirnhaus: His role in early calculus and his work and impact on algebra', *Historia Math.*, 17, 16–35, 1990.

Kuhn, Thomas S., *The Structure of Scientific Revolutions*, 1962.

Lacroix, S. F., *Traité elémentaire de calcul différentiel et de calcul intégral*, Paris, 1802; translated as *An Elementary Treatise on the Differential and Integral Calculus*, with an Appendix and Notes, by C. Babbage, J. F. W. Herschel, and G. Peacock, Cambridge: J. Deighton and Sons, 1816.

Lamb, Horace, *Dynamics*, Cambridge University Press, 1945.

Laplace, P. S., *La Théorie Analytique des Probabilités*, second edition, 1814; *A Philosophical Essay on Probabilities*, 1814, translated by F. W. Truscott and F. L. Emory, 1902, Dover, 1951.

Laplace, P. S., *Connaissance des tems pour l'an*, 1823.

Laplace, P. S., *Philosophical Essays on Probabilities*, from 5th French edition published 1825, translated A. I. Dale, New York: Springer-Verlag, 1995.

Lehn, Waldemar H., 'Isaac Newton and the astronomical refraction', *Applied Optics*, 47(34), H95–105, 1 December 2008.

Leibniz, G. W., '*Nova methodus pro maximis et minimis, itemque tangentibus, quae nec fractas, nec irrationales quantitates moratur, et singulare pro illis caclculi genus*', *Acta eruditorum*, III, C. Gunther, Leipzig, 10 October 1684, 467–73.

Leibniz, G. W., '*De geometria recondite et analysi indivisibilium atque infinitorum*', *Acta eruditorum*, VI, C. Gunther, Leipzig, June 1686, 292–300.

Leibniz, G. W., *Tentamen de motuum coelestium causis*, *Acta Eruditorum*, 2, 82–86, February 1689.

Leibniz, G. W., '*Isaaci Newtoni tractatus duo, de speciebus & magnitudine figurarum curvilinearum*', *Acta eruditorum*, January 1705, 30–35.

Leibniz, G. W., '*Analysis per quantitatum series, fluxiones ac differentias cum enumeratione linearum tertii ordinis*', *Acta eruditorum*, February 1712, 74–77.

Leicester, Henry, *The Historical Background of Chemistry*, Dover, 1956.

Levenson, Thomas, *Newton and the Counterfeiter*, Faber and Faber, 2009.

Lodge, Oliver, *Nature*, 44, 382–387, 1891 (Presidential Address to Section A of the BA at Cardiff, 20 August).

Lynden-Bell, Donald, 'The Wonderful Geometry of Dynamics Newton's Principia for the Common Reader by S. Chandrasekhar', *Notes and Records of the Royal Society*, 50, (2), 253–255, July 1996, 253–254.

Maclaurin, Colin, *Treatise on Fluxions*, 1742.

Maclaurin, Colin, *An Account of Sir Isaac Newton's Philosophical Discoveries*, 1748.

Mandelbrote, Scott, 'Newton and eighteenth-century Christianity', in Rob Iliffe and George E. Smith (eds.), *The Cambridge Companion to Newton*, second edition, Cambridge University Press, 2016, 554–585.

Manuel, Frank E., *Isaac Newton, Historian*, Cambridge, 1963.

Manuel, Frank E., *A Portrait of Isaac Newton*, Cambridge, Belknap Press of Harvard University Press, 1968, reprinted Da Capo Press, New York, 1990.

Manuel, Frank E., *The Religion of Isaac Newton*, Clarendon Press, Oxford, 1974.

Mariotte, Edmé, *De la nature des couleurs* (1681), in *Oeuvres de M. Mariotte*, 2 vols., new ed., The Hague, 1740, 1, 208.

Maxwell, James Clerk, 'Analogues in Nature', essay read to Apostles Club at Cambridge, February 1856, cited by C. W. F. Everitt, 'Maxwell, James Clerk', *DSB*, 9, 206, 198–230, 1974.

Maxwell, James Clerk, *Philosophical Magazine*, Fourth Series, 19, 19–32, January-June 1860; 10, 21–37, July-December 1860.

Maxwell, James Clerk, *Philosophical Magazine* (4), 21, 161–175, 338–342, March, April and May 1861; 23, 12–24, 85–95, January and February 1862; *Scientific Papers*, ed. W. D. Niven, 1890, 1: 451–513.

Maxwell, James Clerk, *Philosophical Transactions*, 155, 459–512, 1865 (received 27 October, read 8 December 1864).

Mayer, J. R., 'Bemerkungen ueber die Kraefte der unbelebten Natur', *Annalen der Chemie und Pharmacie* 43, 233, 1842.

McGuinness, B. (ed.), *Ludwig Boltzmann, Theoretical Physics and Philosophical Problems*, Dordrecht, 1974.

McGuire, J. E., 'Body and void in Newton's *De Mundi Systemate*: Some New Sources', *Archive for History of Exact Sciences*, 3, 206–248, 1966.

McGuire, J. E., 'Force, Active Principles, and Newton's Invisible Realm', *Ambix*, 15, 1968, 154–208.

McGuire, J. E. and Rattansi, P. M., 'Newton and the "Pipes of Pan"', *Notes and Records of the Royal Society*, 21, 108–143, 1966.

McGuire, J. E. and Tamny, M., *Certain Philosophical Questions: Newton's Trinity Notebook*, Cambridge, 1983.

McLachlan, H. (ed.), *Sir Isaac Newton Theological Manuscripts*, Liverpool University Press, 1950.

Minkowski, Hermann, *Physikalische Zeitschrift*, 10, 104–111, 1 February 1909 (read at Cologne, 21 September, received 23 December 1908); reprinted in *Das Relativitätsprinzip* (Lorentz, Einstein and Minkowski, 1913); translated into English in *The Principle of Relativity* (Lorentz, Einstein, Minkowski and Weyl, 1923), 104.

More, Henry, *Enchiridion Metaphysicum*, 1671.

Morrison, Tessa, *Isaac Newton's Temple of Solomon and his Reconstruction of Sacred Architecture*, Birkhäuser, 2011.

Motte, Andrew, *The Mathematical Principles of Natural Philosophy. By Sir Isaac Newton. Translated into English*, 2 volumes, London, 1729.

Nauenberg, Michael, 'Barrow and Leinbiz on the fundamental theorem of the calculus', arXiv:1111.6145v1.

Néményi, P. F., 'The Main Concepts and Ideas of Fluid Dynamics in their Historical Development', *Archive for History of Exact Sciences*, 2, 52–86, 1963.

Newman, William R., 'Newton's *Clavis* as Starkey's *Key*', *Isis*, 78, 564–574, 1987.

Newman, William R., 'The background to Newton's chymistry', in I. Bernard Cohen and George E. Smith (eds.), *The Cambridge Companion to Newton*, Cambridge University Press, 2002, 358–369.

Newman, William R., 'Geochemical concepts in Newton's early alchemy', in Gary D. Rosenberg (ed.), *The Revolution in Geology from the Renaissance to the Enlightenment*, Geological Society of America Memoir, 203, 41–50, 2009.

Newman, William R., 'Newton's Early Optical Theory and its Debt to Chymistry', in Danielle Jacquart and Michel Hochmann (eds.), *Lumière et vision dans les sciences et dans les arts, de l'Antiquité du XVIIe siècle*, Librairie Droz, 2010.

Newman, William R., 'A preliminary reassessment of Newton's alchemy', in Rob Iliffe and George E. Smith (eds.), *The Cambridge Companion to Newton*, second edition, Cambridge University Press, 2016, 454–484.

Newton, Isaac, *Principia*, first edition, 1687, second edition, 1713, third edition, 1726, NP; English translation by Andrew Motte, 1729 (edition cited); revised by Florian Cajori, 1934; I. Bernard Cohen and Anne Whitman, *Isaac Newton The Principia A New Translation*, Berkeley and Los Angeles and London: University of California Press, 1999.

Newton, Isaac, *Opticks*, first edition, 1704, Latin edition, 1706, second English edition, 1717, NP; fourth edition, 1730 (edition cited); *Opticks* (preface by

I. B. Cohen, foreword by A. Einstein, introduction by E. T. Whittaker), Dover Publications, 1952 (based on the fourth edition, 1730).

Newton, Isaac, *De systemate mundi (De motu corporum, liber secundus)*, 1685, published in translation as *The System of the World*, 1728, NP.

Oldenburg, Henry, *The Correspondence of Henry Oldenburg.*, ed. A. R. Hall and M. B. Hall, 13 volumes, London and Madison, WI, 1965–1986.

Olson, R., Scottish Philosophy and British Physics, Princeton, 1975.

Ord-Hume, Arthur W. J. G., *Perpetual Motion: The History of an Obsession*, St. Martin's Press, 1977.

Oresme, Nicholas, *Livre du ciel et du monde (The Book of the Heavens and the World)*, 1377, Menut, 1968.

Palmieri, Paolo, Review *of Isaac Newton's Natural Philosophy*, edited by J. Z. Buchwald and I. B. Cohen, Dibner Institute Studies in the History of Scoence and Technology, Cambridge, MA, 2004.

Pask, Colin, *Magnificent Principia Exploring Isaac Newton's Masterpiece*, Prometheus Books, 2013.

Pearson, Karl, 'Biometry and chronology', *Biometrika*, vol. A 20, pts. 3–4, 1928, 241–262, 424.

Péligot, Eugène, *Recherches sur Vurane*, 1841.

Penrose, Roger, 'Strange Seas of Thought', Review of S. Chandrasekhar, *Newton's Principia for the Common Reader*, THES, 3 July 1995.

Pitcairne, Archibald, *Dissertationes medicae*, Rotterdam, 1701.

Playfair, John, *Illustrations of the Huttonian Theory*, 1802.

Popper, Karl. R., *Conjecture and Refutations, Routledge*, 1963.

Pratt, John P., 'Newton's date for the Crucifixion', *Quarterly Journal of Royal Astronomical Society*, 32, 301–304, 1991.

Priestley, Joseph, *A History of Electricity*, 1767.

Priestley, Joseph, *The History and Present State of Discoveries Relating to Vision, Light and Colours*, 1772.

Priestley, Joseph, *Disquisitions Concerning Matter and Spirit*, 1777.

Quarles, Francis, *Job Militant*, 1624, Second Meditation.

Rankine, W. J. M., 'On the mechanical action of heat, especially in gases and vapours', *Transactions of the Royal Society of Edinburgh*, 20, Part 1, 147–190, 1850/1853 (submitted December 1849, read 4 February 1850).

Rankine, W. J. M., 'On the centrifugal theory of elasticity, as applied to gases and vapours', *Philosophical Magazine*, 2, (4), 509–542, 1851.

Rayleigh, Lord, *Messenger of Mathematics*, 7, 14, 1877.

Rée, Jonathan, 'I tooke a bodkine', review of Jed Buchwald and Mordechai Feingold, *Newton and the Origin of Civilisation, London Review of Books*, 35, no. 19, 16–18, 10 October 2013.

Rogers, A. J., 'The System of Locke and Newton', in Zev Bechler (ed.), *Contemporary Newtonian Research*, London, 1982, 223.

Rohl, D. M., *A Test of Time: The Bible — From Myth to History*, 1995.

Rohl, D. M., *The Lost Testament*, Century, 2002.

Rouse Ball, W. W., 'Sir Isaac Newton', from *A Short Account of the History of Mathematics*, 4th Edition, 1908, chapter XVI, 263–287.

Rowlands, Peter, *Waves Versus Corpuscles: The Revolution That Never Was*, PD Publications, Liverpool, 1992.

Rowlands, Peter, *Zero to Infinity The Foundations of Physics*, World Scientific, 2007.

Rowning, John, *Compendious System of Natural Philosophy*, 1735–1743.

Rucker, R., *The Fourth Dimension And How to Get There*, 1985, 51ff.

Ruffner, James A., 'Reinterpretation of the Genesis of Newton's "Law of Cooling"', *Notes and Records of the Royal Society*, 18, 138–152, 1963.

Ruffner, James A., 'Newton's *De gravitatione*: a review and reassessment', *Arch. Hist. Exact Sci.*, 66, 241–264, 2012.

Rumford, Count, 'An Experimental Inquiry concerning the Source of the Heat which is Excited by Friction', *Philosophical Transactions*, 87, 80–102, 1798, read 25 January.

Rutherford, Ernest, *Philosophical Magazine*, 47, (5), 109–103, January 1899 (dated 1 September 1898).

Rutherford, Ernest, Nobel Banquet speech, 11 December 1908, see Cecilia Jarlskog, arXiv:0809.0857v1, 11.

Rutherford, Ernest, *Manchester Literary and Philosophical Society Proceedings*, 55, 18–20, 1911 (abstract of paper read on 7 March).

Rutherford, Ernest, *Philosophical Magazine* (6), 21, 669–688, May 1911 (dated April).

Rutherford, Ernest, 'Collision of α Particles with Light Atoms IV. An Anomalous Effect in Nitrogen', *Philosophical Magazine*, 37, (222), 537–587, 1919.

Rutherford, Ernest, *The Newer Alchemy*, 1937.

Rutherford, Ernest and Soddy, Frederick, 'The Cause and Nature of Radioactivity, Part I', *Philosophical Magazine* 4, 370–396, 1902, *The Collected Papers of Lord Rutherford of Nelson*, Vol. 1, London: Allen and Unwin, Ltd., 1962, 472–494.

Sacrobosco, Johannes de, *Tractatus de Sphaera*, c 1230, translated in *The Sphere of Sacrobosco and Its Commentators*, Chicago, 1949.

Sarton, George, *Isis*, 25, 399, 1936.

Schaffer, Simon, 'Glass works', in David Gooding, Trevor Pinch and Simon Schaffer (eds.), *The Uses of Experiment: Studies in the Natural Sciences*, Cambridge University Press, 67–104, 1989, reprinted in I. Bernard Cohen and Richard S. Westfall (eds.), *Newton A Norton Critical Reader*, W. W. Norton & Company, 202–217, 1995.

Schliesser, Eric, 'Newton and Spinoza: on action and matter (and God, of course)', *The Southern Journal of Philosophy*, 50 (3), 436–458, 2012.

Scott, W. L., *The Conflict between Atomism and Conservation Theory 1644 to 1860*, Macdonald, 1970.

'sGravesande, Wilhelm, *Introductio ad philosophiam Newtoniam*, 1720.

Shank, J. B., *The Newtonian Wars*, University of Chicago Press, 2008.

Shapiro, Alan E., 'Newton's experimental philosophy', *Early Science and Medicine*, 9, 185–217, 2004.

Shaw, William A., *Select Tracts and Documents Illustrative of English Monetary History 1626–1730*, London: Wilsons & Milne, 1896 reprint, New York: Augustus Kelley Publishers, 1967, 132–134.

Sherburne, Sir Edward, *The Sphere of Marcus Manilius*, 1675.
Sheynin, O. B., 'Newton and Classical Theory of Probability', *Archive for History of the Exact Sciences*, 7, 217–243, 1972.
Shirras, G. Findlay and Craig, J. H., 'Sir Isaac Newton and the Currency', *Economic Journal*, 55, 217–241, 1945.
Simms, D. L., 'Newton's Contribution to the Science of Heat', *Annals of Science*, 61, 33–77, 2004.
Simms, D. L. and Hinkley, P. L., *Notes and Records of the Royal Society*, 43, 31–51, 1989.
Simony, Károly, *A Cultural History of Physics*, translated by David Kramer, CRC Press, 2012.
Slare, Frederick, *Philosophical Transactions*, 18, 200–18, September–October 1694.
Smale, C. V., 'South Terras. Cornwall's Premier Radium and Uranium Mine', *Journal of the Royal Institution of Cornwall*, new series, 1(3), 304–321, 1993.
Smeenk, Chris and Schliesser, Eric, 'Newton's *Principia*', in Jed Buchwald and Robert Fox (eds.), *The Oxford Handbook of the History of Physics*, Clarendon Press, Oxford, 2013, 109–165.
Smith, Crosbie and Wise, M. Norton, *Energy and Empire: A biographical study of Lord Kelvin*, Cambridge University Press, 1989.
Smith, George, *Chaldean Account of Genesis*, 1876.
Smith, George E., 'Another Way of Considering Book 2: Some Achievements of Book 2', in I. Bernard Cohen and Anne Whitman, *Isaac Newton The Principia: A New Translation*, University of California Press, London, 1999, 188–194.
Smith, George E., 'The Newtonian style in Book II of the *Principia*', in Jed Z. Buchwald and I. Bernard Cohen, *Isaac Newton's Natural Philosophy*, MIT Press, Cambridge, Massachussetts, 249–298, 2001.
Smith, George E., 'Was Wrong Newton Bad Newton?', in J. Z. Buchwald and A. Franklin (eds.), *Wrong for the Right Reasons*, Springer, Dordrecht, 2005, 127–160.
Smith, George E., 'Newton's *Philosophiae Naturalis Principia Mathematica*', *Stanford Encyclopedia of Philosophy*, 2007.
Smith, George E., 'How Newton's *Principia* Changed Physics', in Andrew Janiak and Eric Schliesser (eds.), *Interpreting Newton: Critical Essays*, Cambridge University press, 2012, 360–395.
Snobelen, Stephen D., 'Isaac Newton, heretic: the strategies of a Nicodemite', *BJHS*, 32, 381–419, 1999.
Snobelen, Stephen D., 'The mystery of this restitution of all things: Isaac Newton on the return of the Jews', in J. E. Force and R. H. Popkin (eds.), *Millenarianism and Messianism in Early Modern European Culture: The Millenarian Turn*, Kluwer Academic Publishers, 2001, 95–118.
Snobelen, Stephen D., 'Statement on the date 2060', March 2003; updated May 2003 and June 2003; http://www.isaac-newton.org/update.html.
Snobelen, Stephen D., 'Lust, Pride and Ambition, Isaac Newton and the Devil', in James E. Force and Sarah Hutton (eds.), *Newton and Newtonianism: New Studies*, Dordrecht, 2004, 155–181.

Spade, Paul Vincent and Panaccio, Claude, 'William of Ockham', *The Stanford Encyclopedia of Philosophy, Fall 2011 Edition*, Edward N. Zalta (ed.), http://plato.stanford.edu/archives/fall2011/entries/ockham/.

Starkey, George (Eirenaeus Philalethes), *Clavis*, or *The Key*, 1651, Keynes MS 18, in B. J. T. Dobbs, *The Foundations of Newton's Alchemy*, Cambridge University Press, 1975, 251–255.

Stirling, Robert, GB Patent 4081, 16 November 1816.

Stokes, G. G., *Mathematical and Physical Papers*, vol. 1, Cambridge University Press, 1880, 102–104.

Stukeley, William, *Itinerarium Curiosum*, 1724.

Stukeley, William, *Stonehenge: A Temple Restored to the British Druids*, 1740.

Tagg, R., *Nonlinear Sci. Today*, 4, 1, 1994.

Taylor, J. C., *Catalogue of the Newton Papers, sold by order of The Viscount Lymington*, Sotheby & Co., London, 13–14 July 1936.

Thackray, Arnold, 'Matter in a Nut-Shell: Newton's *Opticks* and Eighteenth-Century Chemistry', *Ambix*, 15, 29–37, 1968.

Thackray, Arnold, *Atoms and powers: an essay on Newtonian matter-theory and the development of chemistry*, Cambridge, MA: Harvard University Press, 1970.

Thomson, William (Lord Kelvin), 'On the dynamical theory of heat', *Transactions of the Royal Society of Edinburgh*, 20, Part 2, 261–268, 289–298, 1850/1853 (1851), *Philosophical Magazine*, (4), 4, 8–21, 105–117, 168–176, 1852.

Thomson, William (Lord Kelvin), 'On a Universal Tendency in Nature to the Dissipation of Mechanical Energy', *Proceedings of the Royal Society of Edinburgh*, April 19, 1852, *Philosophical Magazine*, 4th series, 4, 304–306, October 1852; *Mathematical and Physical Papers*, vol. i, art. 59, pp. 511; J. Kestin (ed.), *The Second Law of Thermodynamics*, New York, 1976, 106–132.

Thomson, William (Lord Kelvin), *Baltimore lectures on molecular dynamics and the wave theory of light*, London, 1904; originally distributed by Johns Hopkins University, 1884, reprinted MIT Press, 1987, ed. Robert H. Kargon and Peter Achinstein.

Todhunter, Isaac, *History of the mathematical theory of probability*, Cambridge, 1865.

Todhunter, Isaac, *A history of the mathematical theories of attraction and the figure of the earth from Newton to Laplace*, 1873, reprinted, Dover, New York, 1962.

Traherne, Thomas, *As in a Clock, 'tis hinder'd-Force doth bring*, from *Christian Ethicks* (1675), in H. M. Margolouth (ed.), *Centuries Poems, and Thanksgivings*, 2 vols., Clarendon Press, Oxford, 1958, 2, 186.

Trenn, Thaddeus J., *The Self-Splitting Atom: The History of the Rutherford-Soddy Collaboration*, Taylor & Francis, London, 1977.

Truesdell, Clifford, 'A Program toward Rediscovering the Rational Mechanics of the Age of Reason', *Archive for History of Exact Sciences*, 1, 3–36, 1960.

Truesdell, Clifford, 'Reactions of Late baroque Mechanics to Success, Conjecture, Error and Failure in Newton's *Principia*', in Robert Palter, *The Annus Mirabilis of Sir Isaac Newton 1666–1966*, Cambridge, Massachussetts and London: The M.I.T. Press, 1967, 192–232.

Truesdell, Clifford, *Essays in the History of Mechanics*, Berlin, 1968.

Truesdell, Clifford, 'Reactions of Late Baroque Mechanics to Success, Conjecture, Error, and Failure in Newton's *Principia*', in *Essays in the History of Mechanics*, Springer-Verlag, New York, 1968.

Truesdell, Clifford, 'History of Classical Mechanics Part I', *Die Naturwissenschaften*, 63, 53–62, 1976.

Truesdell, Clifford, *The Tragicomic History of Thermodynamics, 1822–1854*, New York, Springer, 1980.

Tyndall, John, *Faraday as a Discoverer*, new ed., London, 1870.

van Gils, D. P., Huisman, S. G., Grossmann, S. Sun, C. and Lohse, D., *J. Fluid Mech.*, 706, 118, 2012.

Vavilov, S. I., 'Newton and the Atomic Theory', in *The Royal Society Newton Tercentenary Celebrations*, Cambridge University Press, 1947, 43–55.

Verance, P. and Dircks, H., *Perpetual motion: Comprising a history of the efforts to attain self-motive mechanism, with a classified, illustrated, collection and explanation of the devices whereby it has been sought and why they failed*, 1916.

Voltaire, *Lettres philosophiques*, 1733; *Letters Concerning the English Nation*, London, 1733; *Letters on England*, Harmondsworth, Penguin, 1980, 75.

Voltaire, *View of Newtonian Philosophy*, 1737.

Webb, John, *A Vindication of Stone-Heng Restored*, 1665.

Westfall, Richard S., *Force in Newton's Physics*, Macdonald, London and New York, 1971.

Westfall, Richard S., 'The role of alchemy in Newton's career', in M. L. Righini Bonelli and William R. Shea, *Reason, Experiment and Mysticism in the Scientific Revolution*, New York: Science History Publications, 1975, 189–232.

Westfall, Richard S., *Never at Rest*, Cambridge University Press, 1980.

Whewell, William, *Astronomy and general physics considered with reference to natural theology*, 1833.

Whewell, William, *History of the Inductive Sciences, from the Earliest to the Present Time*, London, 1837.

Whiston, William, *Memoirs of the life and writings of Mr. William Whiston... Written by himself*, London, 1749; second edition, 2 vols., London, 1753.

White, Michael, *The Last Sorcerer*, Fourth Estate, 1997.

Whiteside, D. T., 'The mathematical principles underlying Newton's *Principia Muthematica*', *Journal for the History of Astronomy*, 1, 116–138, 1970.

Whiteside, D. T., *Centaurus*, 24, 288–315, 1980.

Whiteside, D. T., 'Newton the Mathematician', in Z. Bechler (ed.), *Contemporary Newtonian Research, Studies in the History of Modern Science volume 9*, D. Reidel Publishing Co, Dordrecht, Boston, London, 1982, 109–127.

Whiteside, D. T., The prehistory of the 'Principia' from 1664 to 1686', *Notes and Records of the Royal Society*, 45, No. 1, 11–61, January 1991.

Wilson, Curtis, 'Clairaut's Calculation of the Eighteenth-century Return of Halley's Comet', *Journal for the History of Astronomy*, 24, 1–15, 1993.

Young, Thomas, *Philosophical Transactions*, 95, 74–87, 1805.

Young, Thomas, *A Course of Lectures on Natural Philosophy and the Mechanical Arts*, 2 vols., London, 1807.

Young, Thomas, 'Cohesion', written 1816, in the *Supplement to the fourth, fifth and sixth editions of the Encyclopaedia Britannica*, *Works*, ed. George Peacock, 3 vols, John Murray, London, 1855, I, 454–484.

Zdravhovska, S., 'Conversation with Vladimir Igorevich Arnol'd', *Mathematical Intelligencer*, No. (4), 30, 1987.

Index

A

Arithmetica Universalis, 39, 43
A New Theory about Light and Colours, 9, 88, 105
aberration, chromatic, 194
absolute space, 44
absolute time, 44
absolute zero of temperature, 64, 68
abstract system, 257, 259
abstraction, 5, 7–10, 12, 44, 81–82, 92, 108, 112, 120–121, 136–138, 149, 152–154, 160, 184, 190, 195, 208, 220, 223, 227, 250, 258–260, 263–264, 266, 268–270, 273–274, 279–280
acceleration, 39, 108, 163
accelerators, 44, 211
Account of the Commercium Epistolicum, 29–30, 43
achromatic lens, 279
acids, 54, 103, 132, 179, 185–187, 191, 196–199, 201, 203–208, 266
action, 108, 256
action at a distance, 1, 3, 140, 186, 190, 252
active principles, 68, 70, 73, 75–76, 78, 83, 108, 164, 169, 171, 185, 195, 202
aerodynamic force, 148–149
aerodynamics, 150, 161
aether, 3, 13, 53, 56–58, 66, 70–71, 130, 139–141, 168, 170, 196, 252, 262, 274, 278
affinity, 116, 185, 195, 199, 206–207
air, 52, 54–55, 58, 117, 127–128, 135, 175, 177–178, 180, 185, 200, 202–204
air resistance, 126, 142, 144, 147, 157, 167
air, projected solidification of, 123
Alberti, Leon Battista, 237

alchemy, 3, 5, 14, 59, 65, 90, 94, 98, 100, 107–108, 119, 169, 182–186, 188, 190–191, 193, 195–196, 205, 208–210, 212–216, 225, 228, 233, 266, 274, 278
algebra, 4, 36, 39–44, 46–47, 49, 63, 179, 254–255
alkali, 197, 203–205
alpha radiation, 211
alternate spheres of attraction and repulsion, 116, 181
amber, 54
Ampère's law, 258
Ampère, A. M., 258
analysis, 1, 19, 29, 35–39, 42–44, 49
analytic tests, 185
Analytical Society, 49, 255
ancient wisdom, 216
Anderson, John D., 150
angle of attack, 148–149
angular momentum, 162, 164
angular velocity, 163
An Historical Account of Two Notable Corruptions of Scripture, 2, 220
An Hypothesis Explaining the Properties of Light, 8, 52, 69, 142, 186
antidifferentiation, 21, 23
antifermions, 91, 109
antimatter, 91
antimony, 174, 192, 197, 199, 209
aqua fortis, 132, 173–175, 197–199, 201, 207
aqua regia, 124, 173–175, 196
Aquinas, Thomas, 218
Archimedes, 29, 148
Archimedes' Principle, 127
argument from design, 2, 265
Arianism, 45

Aristotle, 45, 185
Arius, 221–222
Arnol'd, V. I., 151
arsenic, 191, 198
artificial satellites, 144
Aston, Francis, 209
astrology, 213, 215
A Short Scheme of the True Religion, 265
Athanasius, 220–221, 228
atmospheric refraction, 131, 134
atomic weight, 118
atomism, 89, 116–119, 207–208
atoms, 64, 69, 89–91, 100–101, 104, 106, 112, 117–118, 171, 181, 208, 211, 216, 257, 270
attraction by an arbitrary mass distribution, 98
Augustine, St, 220–221
Avogadro's law, 132

B

Bacon, Francis, 51
Bacon, Roger, 51
Barrow, Isaac, 22–23, 27, 30–31, 43–44
bases, 123, 185–186, 204–205, 208, 266
Basil Valentine, 192
Bayes, Thomas, 46
Bechler, Zev, 108
Becquerel, Henri, 210
Bennett, J. A., 275
Bentley, Richard, 12, 251, 262
Berkeley, George, 47
Bernoulli's theorem, 160
Bernoulli, Daniel, 42
Bernoulli, Jakob, 24, 26, 41
Bernoulli, Johann, 24–25, 27, 30–31, 36, 40–42, 72, 252
Bernoulli, Nicolaus, 30–31, 45
beta radiation, 211
binomial expansion, 37
bismuth, 62, 64, 122, 191, 193
black body, 69
Black, Joseph, 208
Blair, Robert, 46
Blake, William, 260
Bošković, R. J., 88, 116–117, 181
Boerhaave, Hermann, 253
Bohr's theory of the atom, 104, 145
Bohr, Niels, 259
boiling point of water, 60, 62, 65

Boltzmann, Ludwig, 81, 86, 259
Bonet, Frederick, 29
bosons, 108–109
Boyle lectures, 251
Boyle's law, 126, 131, 133–138, 146, 158, 277
Boyle, Robert, 3, 15, 51, 72, 90, 110, 112, 129–130, 132, 175, 183–184, 187, 191–192, 203, 206, 208, 213–214, 229, 242, 261, 278
Brückmann, F. E., 210
Brackenridge, J. Bruce, 275
Bragg reflection, 103
bremsstrahlung, 145
Brewster, Sir David, 182, 194, 253, 262, 278
Briggs, William, 10
Brougham, H. P., 153
Brown, Richard C., 19–20, 24, 45, 272
Brownian motion, 139, 277
bubbles, 106
Buchwald, Jed. Z., 234
Buffon, Comte de, 195
burning mirror, 57
butter of antimony, 174, 198, 199

C

Caesar, Julius, 230–231
calculus, 1, 11, 15, 17–20, 22, 24–26, 28, 30–32, 35–37, 39–40, 43, 45–49, 51, 63–64, 143–144, 250, 254–255, 272–273
calculus algorithm, 21, 24, 31
calculus of variations, 48, 143
calculus, differential, 19, 21, 23, 29, 43
calculus, integral, 19, 23–24
Calero, Julián Simón, 125, 128, 148–149
Callendar, H. L., 81
capillarity, 106, 110, 112–113, 171
capillary action, 88, 112, 114, 171
capillary force, 114
carbon, 117, 194
Carnot engine, 67, 80–81
Carnot, Sadi, 67, 80
Cartesian aether, 139
Cartesian plenum, 87
Cartesian vortices, 140, 159, 165–167, 253, 277
Cartesian world-view, 2
catalysis, 185, 193

Cauchy, A. L, 255
Cavalieri, Bonaventura, 30
Cavendish, Henry, 46, 255
centre of gravity, 74, 98
centres of gravity of curves, 34
centrifugal force, 128, 138, 165
centripetal force, 114, 126–127, 144–145
Châtelet, Émilie du, 253
Chalmers, Thomas, 85
Chaloner, William, 29
Chandrasekhar, S., 48–49, 59, 144, 276
change of state, 59, 121, 123
chaos, 4–5, 47, 71, 84
charge, 82, 145, 222, 258, 264, 266
Charlemagne, 239
Charleton, Walter, 89
chemical attraction, 196
Chemical Dictionary, 183
chemical elements, 115–117, 119
chemical revolution, 185, 207–208
chemistry, 3, 5, 88, 105, 115–118, 132, 183–185, 188–190, 192, 194–195, 197, 207–208, 210–214, 266, 278
Cheyne, George, 23, 45, 253
Christ, Jesus, 215, 222–224, 226, 228, 230, 232, 239
chronology, 2, 119–220, 228–229, 231–234, 241, 246, 279–280
chymistry, 189, 209
cinnabar, 123, 199
circular motion, 74, 160
Clairaut, A.-C., 41–42, 253–255
Clarke, Samuel, 2, 14, 76–78
Clausius, Rudolf, 80–81, 86
Clement of Alexandria, 222
clockwork universe, 14–15, 260, 270
Close, Frank, 278
Cockcroft, J. D., 211
coefficient of resistance, 157
coefficient of restitution, 70
coefficient of viscosity, 142, 147, 161
cohesion, 3, 55, 96–97, 110–111, 115, 121, 124, 141, 171–172, 177, 179, 181–182, 196–198, 200
Coleridge, S. T., 267
Collins, John, 17, 20–27, 29, 31–33, 35, 37, 183
colours of natural bodies, 102–106, 212
colours, theory of, 6, 8–9, 34, 51, 194–195

combustion, 202–203
cometary orbits, 76, 140, 165
comets, motions not resisted, 167
comets, tails of, 56
Commercium Epistolicum D. Johannis Collins, 29–32
complementarity principle, 259
conduction of heat, 53, 55–56, 64, 66
Conduitt, John, 215
conic sections, 21
conservation of angular momentum, 163–164, 166, 169, 273–274
conservation of charge, 263
conservation of energy, 41–42, 69, 80, 152, 157, 273–274
conservation of mass, 80, 266, 273–274
conservation of momentum, 152, 167, 169, 273–274
conspiracy theory, 33
Conti, Natale, 216
continuity equation, 166
controversy, 1–3, 6–9, 17, 45, 51, 73, 87, 121, 170, 221, 249, 272–273
convection, 56, 64, 66
convection currents, 56
convective boundary condition, 67
Copernicus, Nicholas, 216
copper, 71, 104, 122, 172, 174–175, 191, 193, 198, 202–203, 205, 207, 209
corpuscular theory of light, 4, 7, 58, 102, 105, 109, 114–115, 117, 184, 195, 262
cosmological constant, 276
cosmology, 228, 231, 268, 278
Cotes, Roger, 14, 42, 46, 70, 84, 251, 255
Couette flow, 161
Coulomb's law of electrostatics, 258, 264
Coulomb, C. A., 258
Craig, John, 22–23, 33, 48, 151
Craig, Sir John, 280
Crompton, James, 56
crystal, 124, 139, 188
crystalline spheres, heavenly, 217
cubic curves, 44
cubit, royal, 236–237
cubit, sacred, 225, 237
Curie, Marie, 210, 212
Curie, Pierre, 210
curvature method, 275
cylinder, resistance on, 155–156

D

3-dimensionality of space, 99, 258
4-dimensional space-time, 264
d'Alembert, J.-le-R., 151, 158, 253, 255–256, 264
d'Espagnet, Jean, 193
da Vinci, Leonardo, 71, 237
Dalton, John, 117–119, 208
Dary, Michael, 10
Davy, Humphry, 51, 118
De aere et aethere, 110, 134, 138
De analysi, 17–18, 20–22, 25, 27–29, 31, 272
de Broglie, Louis, 104
De computo serierum, 22
De gravitatione et aequipondio fluidorum, 92, 126, 128, 219
de la Pryme, Abraham, 191, 216
de Moivre, Abraham, 29, 40, 46, 255
de Morveau, Guyton, 117, 195
De motu corporum in gyrum, 22
de Sluse, René, 23, 30
De systemate mundi (*The System of the World*), 57
De vi electrica, 53, 56, 65, 114, 262
density, 97, 104, 117, 132, 134–135
Derham, William, 229
Desaguliers, John T., 1, 9, 57–58, 68, 72, 88, 115, 157, 182, 252, 254
Descartes, René, 8, 12, 30, 43, 74–75, 79, 90, 112, 121, 126, 158–159, 162, 165–166, 184, 215, 219, 249, 254, 266, 268
diamond, 54, 122, 194
Diderot, Denis, 253
differential equations, 38, 144, 273
differentials, 256
differentiation, 18, 20, 23
diffraction, 108
dimensional analysis, 146
dipoles, 95, 97
displacement, 78, 258
displacement current, 258
dissipation, 76, 79
dissipation of energy, 69, 78–79, 84–86
distillation, 123, 191, 196–201, 204–205
Dobbs, Betty Jo Teeter, 191, 224, 228, 266
dot notation, 32, 48
double forces, 180
double refraction, 279

double stars, 254
drag coefficient, 143, 147, 155, 157, 160
drag equation, 155
dual force laws, 276
dual orce laws, 97
dualism, 117
duality, 44, 266, 273, 278
Duns Scotus, John, 218
dynamic similarity, 141, 143, 150

E

Earth, 92, 123, 131, 133, 186–190, 192, 217, 228, 231
earth (as element), 90, 185
Earth's atmosphere, 60, 63, 111–112, 114, 127, 130–135, 149
Earth's gravitational field, 96, 173, 180
Earth's surface, pressure at, 135
Earth, internal structure of, 76
Earth, radius of, 134
economics, 243
Eddington, A. S., 278
Ehrlichson, H., 38, 143
eigenvalues and eigenfunctions, 67
Einstein, Albert, 5, 264, 278
elasticity of air, 95–96, 101
electric attraction, 94, 96
electric charge, 258
electric field, 258
electric force, 93–94, 108–109, 180, 211–212, 261, 266
electric spirit, 55, 107, 122, 181, 262
electric vibrations, 88
electrical impulses, 243
electricity, 52, 55, 88, 93–95, 108–109, 258–259, 277
electricity, conduction of, 67
electricity, current, 258
electricity, frictional, 52, 55, 115
electrified universe, 2
electro-optical force, 93, 109, 114–115
electrochemical series, 206, 208
electromagnetic field, 82, 88, 116, 259, 263
electromagnetic radiation, 211, 258
electromagnetic theory, 259, 263–264
electromagnetism, 258, 264
electron, 91, 104–105, 211, 266
electron cloud, 181
electrostatic attraction, 104

electrostatic force, 258
elliptical motion, 98
elliptical orbits, 166
Ellis, R. L., 230
Empedocles of Agrigento, 185
Encke's comet, 85–86
energy, 70, 73, 76, 81, 105, 109, 113, 145, 169, 211, 256
Enlightenment, 234, 253, 260
entropy, 80–82, 86
Enumeratio linearum tertii ordinis, in the *Acta Eruditorum*, 27
epicycles, 217
Epistola Prior, 26
Epistola posterior, 23, 26, 28
Erasmus, Desiderius, 222
ergodic hypothesis, 82
Euclid, 215, 217
Euclidean geometry, 47
Eudoxus, 232
Euler, Leonhard, 36, 41–42, 47–49, 250, 253, 255–256
eutectic alloys, 191
exothermic and endothermic reactions, 200
expansion coefficient of air, 131
experiments on resistance to falling spheres, 156–157
extremum principles, 256, 273

F

Falk, G., 81
falsifiability, 247
falsification, 6
Faraday's law of induction, 258
Faraday, Michael, 8, 88, 116, 258–259, 262–264
Fatio de Duillier, Nicolas, 26–27, 33, 45, 139–140, 152, 195
Feingold, Mordechai, 234
Fermat, Pierre de, 29–30, 43
fermentation, 52, 55, 76, 129–132, 171, 175, 177, 195, 200, 203–204
fermion number, 91
fermions, 91, 108–109, 211
Feynman diagrams, 44
Fick's first law of diffusion, 67
field theory, 109, 118–119, 207, 260
Figala, Karen, 214
figure of the Earth, 141, 255

fire, 54, 76, 96, 122–123, 133, 177, 185, 194, 198, 201–205, 227–228, 236
first and last ratios, 36–37, 43
first law of thermodynamics, 69, 79–80
fits of easy reflection and refraction, 53, 103, 106
flame, 52, 54, 96, 129, 201, 202
flame, colours of, 202
Flamsteed, John, 1, 3, 45, 56, 152, 249
fluents, 20–21, 23
fluid mechanics, 125, 128, 142, 150, 277
fluxional measure of a central force, 40
fluxions, 20–24, 27, 29, 36–37, 39–40, 43, 47, 63
Folkes, Martin, 72, 254
Fontenelle, Bernard de, 36, 79
force, 1, 14, 40, 55, 69, 74, 77–78, 84, 91, 93–94, 98, 122, 124, 127, 137, 150, 162, 169, 243, 253, 256–257, 259, 264, 270
force laws, constant force, 134–135
force laws, inverse cube, 96, 134, 145
force laws, inverse fourth power, 95–97, 134
force laws, inverse seventh power, 97
force laws, inverse square, 140, 145, 159, 217, 258, 260, 264, 274
force laws, linear, 135
forces at the centre of matter, 94, 209
forces between nucleons, 138
forces of attraction, 87, 93, 95–96, 98, 111–114, 124, 138, 169–174, 179–182, 193, 196–198, 200, 204–205, 207
forces of repulsion, 87, 93, 95–96, 112, 124, 136–138, 169, 172–173, 175–182, 185, 190, 193, 195
Fourier theory of heat, 67
Fourier, J.-B. J., 67
fourth state of matter, 139
Francis, Mr, 72
free energy, 207
freezing point of water, 59, 62, 65, 131
Freind, John, 195
French, John, 72
frequency, 109
Fresnel, A. J., 262

Freudian psychoanalysis, 4, 238, 247
fundamental theorem of the calculus, 22–23, 27

G

Galilei, Galileo, 158, 228, 256
Galileo's theorem, 152
gamma radiation, 211
Gamow, George, 155
gases, 5, 117, 123, 125, 128–133, 135–139, 161, 175, 177, 179, 187
Gassendi, Pierre, 89
Gauld, Colin F., 158
general relativity, 4, 264
generalized coordinates, 257
geocentric cosmology, 217, 228
Geoffroy, E. F., 207
geometrical calculus, 49
Gjertsen, Derek, 221, 238
Glover, Richard, 252
God, 13–15, 73, 75–76, 79, 90, 118, 180, 215, 218–219, 221–227, 229, 235, 239, 252, 257
gold, 55, 65, 99–101, 104, 107, 122–124, 132, 167, 173–174, 191–192, 196, 204, 209, 214
gold coinage, 209, 244–245
Gold Standard, 245
gold, porosity of, 107
Golden Age, 83, 217
Golden ratio, 217
Grabiner, Judith V., 256
Gram, Hans, 237
Grasseus, Johann, 186
gravitational field, 143
gravitational force, 11
gravitational radiation, 71
gravitational refraction, 71
gravity, 1, 3, 8, 11–13, 15, 17, 51, 71, 76, 90, 93–96, 99, 101, 123, 126–127, 133, 139–140, 142, 144, 170–171, 173, 180–181, 195, 258, 262, 264, 274
Green, Robert, 116
Greenwich Observatory, 252
Gregory, David, 15, 22, 24, 33, 139–141, 251–252
Gregory, James, 23, 30
Guicciardini, Niccolò, 18, 23, 36–38, 42, 45–46, 273, 276
guinea (coin), 245–246

guinea and feather experiment, 167
gunpowder, 133, 203

H

Hales, Stephen, 182
Hall, A. Rupert, 199, 215, 234, 271, 275, 278
Halley's comet, 254
Halley, Edmond, 22, 33, 47, 135, 140, 231, 252, 254, 261
Hamilton, Gavin, 162
Hamilton, Sir William, 259
Hamilton, Sir William Rowan, 256, 264, 267
Hamiltonian, 256
Hauksbee, Francis, 88, 106, 108, 110, 112–115, 157, 171, 254
heat death of the universe, 86
heat energy, 56, 59
heat radiation, 57, 64, 66
heat transfer, 55, 59–60, 64, 67, 70
heat transfer coefficient, 61
heavier-than-air-flight, 149
Hegel, G. W. F., 267
Helmholtz, H. von, 80
Henry's law, 117
Henry, William, 117
Herodotus, 232
Herschel, Sir John, 116, 118
Herschel, Sir William, 254
Herzl, Theodor, 240
Hesiod, 83
hierarchica theory of matter, 195
hierarchical structure of matter, 78, 95, 115, 118, 196, 212
hierarchical theory of matter, 99–102, 105, 107, 119
Higgs boson, 278
Higgs field, 278
high energy physics, 5, 87, 116, 207
Hipparchus, 232
Hobbes, Thomas, 43–44
Homberg, W., 115
Hooke, Robert, 1, 3–4, 6, 8–9, 18, 28, 32, 51, 110, 112, 124, 134, 203, 242–243, 249, 251, 272, 275
Hudde, Jan, 30
Hughes, John, 252
Hume, David, 253

Humores minerales (*Mineral Humours*), 186
Huygens, Christiaan, 1, 6, 26–27, 39, 43, 133, 166, 242–243
hydrostatics, 126–127, 141, 143
hyperbola, 144
hypersonic flight, 150
hypothesis, 1, 3, 6, 8, 10–12, 87, 96, 99–101, 107, 121, 134, 139, 142, 154, 156, 161, 165–166, 178, 220, 225, 236, 251, 257, 259, 261, 274, 279

I

ideal gas, 126, 139, 150
impact theory of resistance, 148–149
impressed force, 74
impulse, 108, 127, 170
impulsive force, 77
indeterminacy, 257
Index Chemicus, 192, 214
induction, 6, 9, 14, 92, 253
inelastic collisions, 65, 70, 77–79, 84
inertia, 66, 77, 145, 157
infinite series, 20, 22, 37
infinitesimals, 24
initial conditions, 84
instantaneous velocity quotient, 37
integration, 18, 20, 23, 61, 161
intermolecular distance, 97, 114
intermolecular forces, 97, 138
internal forces, 78, 107, 185
interparticulate forces, 73, 93, 95, 102, 124–125, 195, 200, 261
interpolation, 222
interval of fits, 103, 106
Irenicum, 225
iron, 54, 56, 58, 60, 62–63, 65, 71, 99, 101, 122, 132, 138–139, 172, 174, 177, 180, 191–193, 198, 201, 203, 207, 209

J

James, Peter, 233
Jerome, St, 221–222
Jones, Inigo, 236
Jones, William, 27
Joule, James Prescott, 80
Jung, Carl Gustav, 185

K

Kant, Immanuel, 258, 267
Keill, James, 195

Keill, John, 23, 27–28, 45, 195, 252
Kelvin, William Thomson, Lord, 67, 73, 80–81, 84–86, 259
Kepler's inverse square law of light attenuation, 57
Kepler's laws of planetary motion, 140, 164, 166, 277
Kepler's second law, 166
Kepler's third law, 164
Kepler, Johannes, 15, 57
Keynes, John Maynard, 182, 244, 279
Kierkegaard, Søren, 259
kinetic energy, 70, 76, 78, 137
kinetic theory of gases, 137–138
kinetic theory of heat, 70
King of Prussia, 29
King-Hele, D. C., 144
Klaproth, M. H., 210
Kollerstrom, Nicholas, 275
Kuhn, Thomas, 5
Kuhnian paradigm shift, 24, 272

L

l'Hôpital, Marquis de, 36
Lévi-Strauss, Claude, 238
Lacroix, S. F., 256
Lagrange, Joseph Louis, 36, 41–42, 47–48, 250, 255–256, 258
Lagrangian, 136, 256
Lagrangian dynamics, 257
Landen, John, 46
Laplace demon, 257
Laplace, Pierre Simon, 36, 41–42, 47, 84–85, 170, 250, 253, 256–258, 264
Lavoisier, A. L., 115–117, 119, 185, 203, 208
law of refraction, 8
laws of thermodynamics, 82
lead, 56, 59, 62–64, 74, 101, 122, 132, 167–168, 172–175, 177, 191, 193, 198–199, 205, 207
Leibniz, Gottfried Wilhelm, 1–3, 8, 11–12, 14, 17–32, 35–36, 42–43, 45–46, 48–51, 70, 73, 75–79, 83–84, 144, 152, 166, 249–250, 252, 255, 272–273
Levenson, Thomas, 245
limits, 36–37, 66, 255–256
Lippmann photograph, 102
liquid crystalline structure, 103
loadstone, 71, 139

Locke, John, 215, 221, 242, 244
Lodge, Oliver, 264
longitude, 72
Lowndes, William, 245
Lucas, Anthony, 8
Lucretius, 89, 266
lunar theory, 36, 38, 255, 275
Lynden-Bell, Donald, 48

M

Macedonius, 222
machines, 14, 70, 73, 84
Maclaurin, Colin, 42, 46–47, 252, 255–256
Macquer, P. J., 117
Maddock, Joshua, 10
magnetic attraction, 94, 180
magnetic field, 188, 258
magnetic force, 93, 173, 180
magnetic ordering, 139
magnetic pendulum, 71
magnetic repulsion, 172, 180
magnetic shielding, 138
magnetism, 52, 55, 71, 88, 93–96, 99, 101, 107, 139, 192, 253, 258–259
Magnus effect, 159–160
Maier, Count Michael, 209
Maiman, Theodore, 33
Mallet, David, 252
Manuel, Frank E., 9, 230–231, 271
many-body problem, 71, 79
Mariotte, Edmé, 8
mass, 68, 71, 82, 97, 99, 108–109, 132, 163, 165, 167, 211, 256, 264, 267, 273
mass model of dispersion, 108–109
mass transfer, 59
mass, point-centres of, 98
Mathesos Universalis Specimin, 22
matter, theory of, 2
Maupertuis, P. L. de, 41–42, 49, 253–254, 256
Maxwell's equations, 82, 259
Maxwell, James Clerk, 81, 88, 258–259, 263
Mayer, J. R., 80
Mayer, Tobias, 275
Mayow, John, 203
mechanistic philosophy, 108, 119
Mede, Joseph, 238, 240
Mercator, Nicholas, 30, 34

mercury (alchemical principle), 184–185, 191–193, 199, 206, 214
mercury (quicksilver), 53–54, 59, 65, 107, 110–111, 114–115, 122–124, 127, 135, 147, 167, 172–174, 191–193, 196, 199, 205, 207, 209, 266
metaphysics, 8, 11, 73, 92, 184, 190, 213, 215, 218, 243, 256–257, 259, 263–264, 268, 270, 273, 279
method of moments, 37, 43, 143
Michell, John, 46, 116
Michelson-Morley experiment, 278
microscopes, 103, 105
microscopes, resolution of, 105–106
microstructure of matter, 95, 98, 102
mines, 190, 209, 212
minima naturalia, 89
minimum deviation, 265
Minkowski space-time, 264
Minkowski, Hermann, 267
Mint, Royal, 4, 25, 152, 191, 209, 224, 230, 234, 243–245
moment of inertia, 163–164
momentum, 65–66, 68, 70–71, 74, 76–77, 83, 108–109, 130, 148, 256
momentum transfer, 65
Moon, 92, 227, 231
Moon test, 274–275
Moon, density of, 275
Moon, mass of, 275
More, Henry, 89, 219, 238, 264
Morrison, Tessa, 237
multiplication (alchemical), 188, 209, 214
multiplication of categories, 107
multivariate analysis, 48
muons, 91
music of the crystalline spheres, 217

N

Néményi, P. F., 166
Napoleon, 239, 258
Nauenberg, Michael, 275–276
Navier-Stokes equations, 162
Neale, Thomas, 244
Neoplatonism, 107, 219
neutrinos, 91
Newman, William R., 185–186, 188, 194, 279

Newton and the Great World System, 1, 278
Newton and Modern Physics, 243, 275
Newton's law of cooling, 57, 63–67, 70, 80, 246
Newton's law of fluid flow, 156
Newton's law of fluid friction, 146–148
Newton's law of fluid resistance, 160
Newton's law of restitution, 70
Newton's law of viscosity, 67, 141, 161
Newton's laws of motion, 14, 82, 169, 202
Newton's metal, 191
Newton's rings, 103–104, 106, 265
Newton's second law of motion, 255
Newton's sine-square law of resistance, 148–150
Newton's third law of motion, 274
Newton's universal law of gravitation, 26, 140, 159, 166, 169, 257, 260, 275
Newton's water bucket experiment, 159
Newton, Humphrey, 215
Newtonian dynamics, 2, 68
Newtonian fluids, 141, 161
Newtonian gravity, 2, 45, 254, 268
Newtonian mechanics, 8, 42, 255, 257
Newtonian physics, 169, 244, 250, 256
nitre, 54, 99, 133, 187, 201, 203, 205
nitro-aerial particles, 203
non-standard analysis, 44
nonlinear oscillators, 47
North, Francis, 10
Nostradamus, 238
nuclear fusion, 190
nucleus, 100, 104–105, 138, 180–181, 211
nutshell theory of matter, 98, 116–118

O

oblateness of the Earth, 141, 254
Observations on the Prophecy of Daniel and the Apocalypse of St. John, 2, 230, 232, 235
occult qualities, 11, 13, 171
Ockham's razor, 273
Ockham, William of, 218, 264
Ockhamism, 170, 220
Of Colours, 194
Of Natures Obvious Laws & Processes in Vegetation, 186–187, 192, 224
oil of vitriol, 54, 198, 200–201

Oldenburg, Henry, 20–22, 28, 35, 183, 214, 261
Opticks, 2–3, 9, 44–45, 56, 58, 85, 90–91, 93–96, 100–109, 111, 113–114, 118–119, 122–124, 128, 130, 139, 141, 176, 181–182, 184, 192–195, 202, 208, 227, 252, 261
Oresme, Nicholas, 15
oxidation, 202–203

P

Péligot, 210
Pappus, 45
Paracelsus, 184
Pardies, Ignaz, 6
Paradoxical Questions concerning the Morals and Actions of Athanasius, 221
Parmenides, 12
partial derivatives, 32
partial differential equation, 67
Pascal, Blaise, 30
Pask, Colin, 144
peacock feathers, 103
Pearson, Karl, 230
Penrose, Roger, 49
periodic time, 163–164
periodicity of light, 105
perpetual motion, 69, 71–73, 76, 79, 83
perturbation theory, 276
perturbed elliptical orbits, 37
philosopher's stone, 209, 211
photons, 108–109, 160, 277
Picard, Jean, 115
Pipes of Pan, 217
Pitcairne, Archibald, 23, 51, 64, 68, 205–206, 252
pitchblende, 210
Planck, Max, 69, 81
planetary orbits, 76
planets, motions not resisted, 167
Plateau, Joseph, 256
Platonic solids, 217
Platonism, 108, 219, 222
Playfair, John, 84
plenum, 252
Pochon, Marcel Leon, 212
Poincaré, Henri, 47
point-atoms, 116
point-centres of force, 95, 116
point-like particles, 99, 107

polarization, 160, 181
polonium, 210
Pope, Alexander, 252
Popish plot, 223
Popper, Karl, 6, 247
porisms, 217
potential energy, 76, 97, 108, 136–137, 276
Prandtl, Ludwig, 158
Praxis, 184–185, 209, 266
precession of the equinoxes, 231–232, 276
Priestley, Joseph, 88, 98, 116–117, 258
prime theorem on quadrature, 22
Principia, 2–3, 5, 9, 11–12, 15, 22, 24–26, 30–33, 35–47, 53, 57, 59–70, 75, 78, 84–85, 87, 89–90, 92–93, 95–97, 99, 101, 111–113, 116, 125, 128, 131, 135, 137, 139, 141, 146, 156, 158–159, 161, 167–171, 182, 186, 195, 199, 204, 208, 217, 225, 238, 251, 253–255, 261, 272, 274, 276
Principia for the Common Reader, 48
principle of invariance, 59
principle of least action, 109, 257
principle of least time, 109
principle of reciprocity, 150
principle of relativity, 259
prisca sapientia, 83, 216–217
prisms, 8, 195
problem solving, 38, 143
prophecy, 119, 218, 220, 224, 228–229, 234, 238, 240–241, 246, 279
proton, 104, 266
prytaneum, 227
psychology, 241
Ptolemy, 217, 228
Pythagoras, 217, 226

Q

quadratures, 18, 20–21, 23–24, 34, 37
Quaestiones quaedam philosophicae, 71, 89–90, 142, 159, 171, 241
Quakers, 117
quantitative analysis, 185, 197, 208
quantum field, 263, 269
quantum field theory, 263, 269
quantum mechanics, 4–5, 68, 136, 260, 263, 268–269, 278
quantum theory, 4, 69, 105, 268, 277–278
Quarles, Francis, 15
quasicrystals, 217

quaternions, 264
Queen of Prussia, 26
quintessence, 193, 228

R

Rée, Jonathan, 234
radiant heat, 57, 65
radiation pressure, 57
radiation reaction, 55
radioactivity, 71, 210–212
radium, 210–212
radius of curvature, 37
rainbow, 265
Rankine, W. J. McQuorn, 80, 86
Rayleigh, third Lord, 160
reaction propulsion, 152
recursive method, 6, 208, 224, 275, 277
redintegration, 187, 194
reflecting miscroscope, 105
reflecting telescope, 7, 34, 191, 203
reflection, 53–54, 96, 103–104, 106–109
refraction, 54, 95–96, 107–109, 114, 159, 265
refraction force per unit mass, 114
refractive index, 194
regulus of iron, 175, 192
relative humidity, 132
relative motion, 126, 150
relative velocity, 161
Remarks upon the Observations made upon a Chronological Index of Sir Isaac Newton... publish'd at Paris, 72
renormalization, 158
renormalization group, 158, 170
replacement reaction, 199
resistance, 60–61, 66, 70, 85, 139–140, 267
resistance in continuous media, 151, 156
resistance in fluids, 3, 70, 140–148, 157, 161, 167, 277
resistance in the heavens, 13, 167–168
resistance in water, 141–142, 147, 156–157, 167
resolution and composition, 44
rest mass, 109
Reynolds number, 143, 147
Robins, Benjamin, 160
Rohl, D. M., 233
Routh, E. J., 153
Rowning, John, 116

Royal Society, 7, 17, 22, 27–31, 33–34, 45, 58, 72, 88, 140, 182, 214, 240, 251–252, 254
Ruffner, James A., 64
Rumford, Benjamin Thompson, Count, 51
Rules for Interpreting the Words and Language in Scripture, 238
Rutherford's theory of the atom, 100
Rutherford, Ernest, 210–211

S

Sacrobosco, Johannes de, 15
sal alkali, 197, 200, 203, 205
sal ammoniac, 99, 174, 192, 197, 199–200
salt (alchemical principle), 184–185
salt of tartar, 133, 198–201, 206–207
saltpetre, 187, 192
salts, 124, 133, 172–173, 180, 188, 200, 203–204, 208
Sarton, George, 237
Saturn, 135
Saunders, Richard, 215
Scala graduum caloris, 65
scaling law of Newton, 97
Schooten, Frans van, 215
Schrödinger, Erwin, 104
scientific journals, 7, 24
Scott, W. L., 118
sea salt, 187
second differentiation, 31
second law of thermodynamics, 65, 69, 73, 79–83, 85–86
self-fulfilling prophecy, 225
self-similarity, 92, 170, 216, 273
Sendivogius, Michael, 184, 186, 192
'sGravesande, Wilhelm, 72, 115
Shaw, W. A., 243
shear stress, 161
Sherburne, Sir Edward, 34–35
Sheynin, O. B., 230
ship design, 150
silver, 53, 101, 122, 173–174, 191, 207
silver coinage, 209, 245
Simms, D. L., 279
Simpson, Thomas, 46
skin-friction drag, 161
Slare, Frederick, 202
Sloane, Sir Hans, 28
Smith, Adam, 234
Smith, George, 226

Smith, George E., 157–158, 276
Smith, Robert, 251
Snell, Willibrord, 34
Snobelen, Stephen D., 240
Soddy, Frederick, 210
Solar System, 69, 76, 79, 84–85, 98, 140, 159, 254, 275
solar wind, 150
solid of least resistance, 25, 38, 143, 150
Solomon's Temple, 229, 280
space, 19–20, 44, 48, 65–66, 68, 79, 82, 85, 90–91, 98, 101–102, 118, 130–131, 134–135, 137, 142–143, 167, 177–178, 195, 212, 219, 256, 264, 266–268, 273
space-time, 264
special relativity, 4, 264
specific heat capacity, 68
spectra, dynamical origin of, 54
spectroscopy, 54
spectrum, 118–119, 195, 246
speculum, 191
sphere, resistance on, 155–156
spheres of activity within matter, 95–96, 178
spheres of attraction and repulsion, 181
spherical symmetry, 165
Spinoza, Baruch, 82
spirals, 144–145, 165
spirit of nitre, 187, 198, 201–202, 206
spirit of salt, 174, 198–199, 205
spirit of soot, 199
spirit of urine, 122, 199
spirit of vitriol, 122, 175, 197, 203
spirit of wine, 122, 132, 167, 172, 198, 201
Stahl, G. E., 117
standard analysis, 44
star regulus of antimony, 191–193
Starkey, George, 184, 214
stars, 135
stars, nuclear processes within, 190
states of matter, 121–122, 169, 175, 196
Stirling engine, 81
Stirling, James, 46
Stirling, Robert, 81
stoichiometry, 185, 208
Stoics, 83
Stokes' law of viscosity, 142, 147
Stonehenge, 236
streamline flow, 153–154
strong force, 211

strong interaction, 211, 258
Stukeley, William, 190, 236
sublimation, 196, 200, 204
sulphur, 54, 117, 122, 133, 199, 201–203, 205–206, 266
sulphur (alchemical principle), 184–185, 191
sulphureous bodies, 54, 204
sulphureous spirit, 202–204
Sun, 54, 57, 92, 101, 166, 227, 231, 236, 242, 256–257
Sun, constitution of, 56, 76
Sun, heat of, 57, 65
supernovae, 190
surface tension, 106, 113, 171, 178
symmetry, 136, 263, 265–268, 273
synthesis, 1, 44, 49
synthetic, 35–36, 38–39, 43–44, 251, 258, 269
system of the world, 42, 263

T

tangents, 18, 20–21, 23, 25
Taylor, Brook, 42, 46, 68, 88, 254–255
temperature, 59–66, 68–70, 73, 82, 86, 121, 131–133, 137
temperature exchange, 65
temperature scale, 59–60, 64
terminal velocity, 142, 156
Thackray, Arnold, 116–117, 278
The Net, 98, 193
The Chronology of Ancient Kingdoms Amended, 229, 237
The Elements of Mechanicks, 126
Theologiae gentilis origines philosophiae, 225
theology, 2, 85, 92, 108, 119, 216, 218–220, 223–225, 227–228, 234, 240–241, 251, 256, 268, 270, 279
thermodynamics, 2, 51, 59, 68–70, 73, 80, 82–83, 86, 126, 279
thermometer, 2, 58, 63, 65
thermometer, oil-in-glass, 59–62, 64–65
thermometer, slab, 62–63
thin film effects, 102–103
third law of thermodynamics, 69
Thomson, Dr. James, 85–86
Thomson, James (poet), 252
three-body problem, 257

time, 19–20, 36–37, 39, 43–44, 48, 53, 60–63, 66, 70, 81–82, 84, 155–156, 223, 230, 256, 264, 266–268, 273
time, irreversibility of, 80–83, 263
tin, 174–175, 191, 193, 207, 209
To Resolve Problems by Motion, 18, 20
torque, 162
Torricelli, Evangelisto, 152
Torricellian experiment, 110
Torricellian tube, 111, 115
Tractatus de methodis fluxionum et serierum (De methodis), 18, 21–22, 37
Tractatus de quadratura curvarum, 24, 27
Traherne, Thomas, 15
transmutation, 184–186, 209–212, 214
Trojan war, 232–233
Truesdell, Clifford, 36, 41, 80, 125, 153, 276
Turgot, Anne-Robert-Jacques, 234
Turnbull, H. W., 102

U

ultimate particles, 78, 89–91, 102, 106, 118, 211
uranium, 210, 212
Uranus, 210

V

4-vector quantities, 109
vacuum, 57–58, 64–65, 69, 90, 98, 100, 110, 112, 128, 131, 139, 252
vacuum (quantum mechanical), 278
van der Waals force, 97, 181
van Helmont, J. B., 184
Varenius, Bernhard, 186
Varignon, Pierre, 40–42
Vavilov, S. I., 99, 106
vector addition, 74
vector quantity, 77, 127
vegetation, 52, 186, 188–189
velocity, 61, 66, 68, 142–149, 155–157, 161, 163, 165, 277
velocity gradient, 161
velocity model of dispersion, 108, 195
velocity of light, 81, 181, 258
velocity of sound, 131, 141
vena contracta, 153
Villalpanda, Juan Battista, 237
virial theorem, 137, 276
viscosity, 75, 141, 145, 157

vis inertiae, 13, 73–74, 92, 141, 146, 171, 202
vis viva, 42, 70, 77–78
Vitruvius, 237
void, 90
Voltaire, 239, 252–253
vortex motion, 75, 160, 163, 277
vortex, forced or rotational, 163
vortex, free or irrotational, 163, 165
vortices, early observations on, 159
vulgar chemistry, 186, 188, 190

W

Walker, D. M. C., 144
Wallis, John, 22, 24, 26, 30
Walton, E. T. S., 211
Waring, Edward, 46
water, 54–55, 59, 68, 100–102, 106, 114, 122, 129, 172–175, 177–180, 187, 198–201, 204–206
water (as element), 90, 184–185
water bucket experiment, 164
water flowing out of a tank, 151, 163
water, capillary action of, 96, 110, 112, 171
water, condensation of vapour into, 130
water, condensation to ice, 123
water, density of, 55, 101, 107, 124
water, evaporation by reduced pressure, 131
water, incompressibility of, 128
water, porosity of, 100, 107
water, pressure in, 127
water, rarefaction to vapour, 123, 129
water, surface tension, 178–179
water, viscosity of, 75
water, vortices in, 159
water-filled prism, 8
Watson, Richard, 117

wave equations, 258, 264
wave motion, 141
wave theory, 277
wave theory of light, 57, 117
wave-particle duality, 268, 277
wavelength, 104, 106, 108
waves, 44, 53, 109, 143
weak force, 211
weak interaction, 109, 258, 278
Webb, John, 236
Weierstrass, K., 255
weight, 13, 110–112, 114, 127, 131, 133–135, 142, 176, 206
Wepster, S. A., 275
Westfall, Richard S., 27, 31, 35, 197, 220, 229, 234, 271, 275–276
Whewell, William, 49, 85–86
Whiston, William, 43, 45, 240, 251
Whiteside, D. T., 35, 40, 148–149, 275
Williams, Kim, 237
wind tunnels, 150, 152
witchcraft, 213, 215
Wolf, Christian, 31
world in decay, 83
worldline, 264
Wren, Sir Christopher, 39, 275

Y

Yehuda, Abraham Shalom, 240
Young's modulus, 145
Young, Thomas, 106, 117–118

Z

Zeno, 12
Zeno's paradoxes, 268
zero force principle, 273
zeroth law of thermodynamics, 69
zinc, 191

www.ingramcontent.com/pod-product-compliance
Lightning Source LLC
Chambersburg PA
CBHW060454300426
44113CB00016B/2588